H
33
E3 ✓
v.2

DATE DUE

JAN 5 '81		
OCT 1 '84		
MAY 4 '87		

EXPLORING FACT AND VALUE

EXPLORING FACT AND VALUE

SCIENCE, IDEOLOGY, AND VALUE
VOLUME II

Abraham Edel

Transaction Books
New Brunswick (U.S.A.) and London (U.K.)

Copyright © 1980 by Transaction, Inc.
New Brunswick, New Jersey 08903

All rights reserved under International and Pan-American Copyright Conventions. No part of this book may be reproduced or transmitted in any form or by any means, electronic or mechanical, including photocopy, recording, or any information storage and retrieval system, without prior permission in writing from the publisher. All inquiries should be addressed to Transaction Books, Rutgers—The State University, New Brunswick, New Jersey 08903.

Library of Congress Catalog Number: 78-62886
ISBN: 0-87855-229-4
Printed in the United States of America

Library of Congress Cataloging in Publication Data

Edel, Abraham, 1908-
    Exploring fact and value.
    (Science, ideology, and value; v. 2)

    Includes bibliographical references and index.

    1. Values. 2. Fact (Philosophy) 3. Science and ethics. 4. Technology and ethics. 5. Social sciences and ethics. I. Title.
H33.E3 vol. 2 [BD232]   300'.ls [121'.8]   78-62886
ISBN 0-87855-229-4

For Matthew and Deborah

CONTENTS

Preface ix
Acknowledgments xi
Introduction: The Fact-Value Problem xvii

Part I: Fact and Value in Theory

1. The Relation of Fact and Value: A Reassessment 3
2. Patterns of the Use of Science in Ethics 17
3. The Place of Empirical Knowledge in Ethics 43
4. Metaphors, Analogies, Models, and All That, in Ethical Theory 57
5. Some Psychological Presuppositions of the Concept of Virtue: A Case Study in the Relation of Science and Ethics. 77
6. Contenders for Value Theory: A Behavioral and an Evolutionary Claimant 91
7. Scientific Research and Moral Judgment: A Philosophical Perspective 123
8. Six Requirements in Search of A Theory: A Study in the Relation of Law and Morality to Social Change 157
9. Toward an Analytic Method for Dealing with Moral Change 173

Part II: Fact and Value in Practice

10. Prometheus on Trial: Technology and Morality 197
11. The Scientific Enterprise and Social Conscience 219
12. The Social Responsibility of Scientists and Engineers 239
13. Knowledge and Responsibility in the Professions 253
14. The Scientist and his Findings: Some Problems in Scientific Responsibility 269
15. Preferential Consideration and Justice 289
16. Notes on Terrorism 309
17. Approaches to Environmental Ethics 325

Epilogue: The Fact-Value Dichotomy as a Chapter in Social-Intellectual History 339
Index 365

PREFACE

To reread one's own papers written over four decades is an interesting and instructive experience. Unnoticed unities, relations, lines of development, almost pop out in the process. Earlier papers take on fresh appearance when looked at in terms of their intellectual offspring, and later papers acquire a character of culmination. Lessons about methods of inquiry begin to emerge as one wonders what the papers would have been like had later ideas been present at the outset. In one of the fashionable dialects, this would, I take it, be seen as a process of self-transcendence—the self reflecting on itself as object, from a more exalted meta-standpoint. In plainer language, it is simply learning from intellectual experience, which does not differ from learning generally. Such lessons are not private, but can be shared as hypotheses for sharpening thought. I am grateful to Professor Irving Horowitz who proposed a collection of this sort for making this learning possible. I have benefited for many years from his incisive insight in social analysis of ideas.

The papers selected for inclusion in these two volumes are organized around two central problems that have played a vital role in the study of human affairs in the twentieth century: methods of analyzing concepts, and the relation of fact and value. (They cannot of course be kept wholly apart since the role of values in ideas—most glaring in ideological thought—is itself an important issue). Conflicts over how to analyze concepts have been widespread, and have reflected the diversity of schools in both philosophy and social science. Their effects on the work of social scientists have been immediate and profound. The issue over the separation of fact and value, with its attempt to establish a neutral social science aloof from social commitments, has reverberated through social theory as well as social practice. (I shall argue that it has always been a program, in fact an unachievable one, not an unavoidable metaphysics.) The first volume includes case studies in the analysis of concepts from different fields of the inquiry into human ways, as well as theoretical explorations of different phases of the problems of analysis. The second volume both addresses the fact-value dichotomy from different angles, and includes papers on special application in practical judgment.

I should like to thank those who have given permission for the republication of a number of the papers previously published or originally designed for special conferences. Individual acknowledgment is to be found on the acknowledgments page. My indebtedness to friends and colleagues whose writings and discussion played an enormous part at various times in the original preparation of these papers and in their criticism is of course too general and too widespread to allow of more than this wholesale expression of gratitude.

The two volumes are dedicated to my son Matthew and my daughter Deborah. Both are practitioners in the fields under consideration, the one contributing in his professional work as economist to combining the methods of economics, anthropology and history, the other in her professional work in special education particularly concerned with the kinds of values involved in furthering individual self-development.

In collecting papers for Volume II it was comparatively easy to find a unifying theme because so much of my work in ethical theory has centered around the relation of ethics and the sciences of man. Accordingly, the papers (in Part I) go from general studies of the fact-value problem to specific roles of science in ethics, then to attempts at reconstructing our whole view of fact-value relations—indeed our whole way of doing moral philosophy at present. Part II consists of applied papers in the sense that they deal with attitudes, policies or practical issues in technology, science, the professions, or social life. Their analysis shows the complex relations of fact and value and how to avoid the oversimplified and the ideological by a consciousness of method. The concluding paper attempts to see the significance of the fact-value problem as a whole, as grounded in the intellectual and social history of the last two centuries.

A.E.

ACKNOWLEDGMENTS

"The Relation of Fact and Value: A Reassessment" was originally published in *Experience, Existence, and the Good*, Essays in Honor of Paul Weiss, ed. Irwin C. Lieb (Carbondale, Ill.: Southern Illinois University Press, 1961), pp. 215-29. Reprinted by permission of the publisher. Copyright © 1961 by Southern Illinois University Press.

"Patterns of the Use of Science in Ethics" was originally published in *Boston Studies in the Philosophy of Science,* Proceedings of the Boston Colloquium for the Philosophy of Science 1966/1968, eds. Robert S. Cohen and Marx W. Wartofsky (Dordrecht, Holland: D. Reidel Publishing Co.), IV, pp. 350-78. Reprinted by permission of the publishers and of Professor Robert S. Cohen. In revising this paper, the writer is generally indebted for critical suggestions to Professors John Ladd and Ruth Putnam. The latter's comments on the treatment of G.E. Moore were especially helpful.

"The Place of Empirical Knowledge in Ethics" was originally written for a symposium on ethical theory in *Etyka*, in which it appeared in Polish translation, 11 (1973): 173-89. The English version is here printed by permission of the journal.

"'Metaphors, Analogies, Models, and All That, in Ethical Theory" was originally published in *Philosophy, Science, and Method*, Essays in Honor of Ernest Nagel, eds. Sidney Morgenbesser, Patrick Suppes and Morton White (New York: St. Martin's Press, 1969), pp. 364-81. Reprinted by permission of the publishers.

"Some Psychological Presuppositions of the Concept of Virtue: A Case Study in the Relation of Science and Ethics" was originally published in *Human Values and the Mind of Man*, Proceedings of the Fourth Conference on Value Inquiry, State University of New York, College at Geneseo, eds. Ervin Laszlo and James B. Wilbur (New York: Gordon and Breach, Science Publishers, Inc., 1971), pp. 63-76. Reprinted by permission of the publishers.

In "Contenders for Value Theory: A Behavioral and an Evolutionary Claimant," the part dealing with Pepper's work was originally published under the title of "Science and Value: Some Reflections on Pepper's 'The Sources of Value' " in *The Review of Metaphysics*, XIV, 1 (September 1960): 134-58. Reprinted by permission of the editor. The part dealing with Pugh's work was originally published as a review of his *The Biological Origin of Human Values*. Published by permission of Transaction, Inc. from *Society* 15, 2. Copyright © 1978 by Transaction, Inc.

"Scientific Research and Moral Judgment: A Philosophical Perspective" was prepared and pre-published as a lead paper for one of the sessions of a conference on The Acquisition and Development of Values, Perspectives on Research, organized by the National Institute of Child Health and Human Development, May 15-17, 1968, Washington, D.C.

"Six Requirements in Search of a Theory: A Study in the Relation of Law and Morality to Social Change" was originally published in *Philosophy and Civil Law, Proceedings of the American Catholic Philosophical Association* ed. George F. McLean, XLIX (1975): 150-63. Reprinted by permission of the editor.

"Toward an Analytic Method for Dealing with Moral Change" was originally published in *The Journal of Value Inquiry* XII, 2 (Spring 1978): 81-99. Reprinted by permission of the editor.

The greater part of "Prometheus on Trial: Technology and Morality," dealing with the three stages in the relation of technology and morality, was originally published under the title of "Technology and Morality" in *Praxis* I–2 (1974): 183-96. Reprinted by permission of the editor.

"The Scientific Enterprise and Social Conscience" was originally published in *Philosophical Exchange*, The Annual Proceedings of the Center for Philosophical Exchange, ed. Howard E. Kiefer (SUNY College at Brockport, N.Y., Summer 1970), pp. 39-57. Reprinted by permission of the editor.

"The Social Responsibility of Scientists and Engineers" was presented as a paper under the title of "A Philosopher's View of the Social Responsibility of Scientists and Engineers" at the First Franklin Conference on "Should We Limit Science and Technology?" held at the Franklin Institute, Philadelphia, Pa., October 15, 1974. It was published in the *Journal of the Franklin Institute* 300, 2 (August 1975): 113-24. Reprinted by permission of the Franklin Institute.

"Knowledge and Responsibility in the Professions" was presented at a conference on The Uses of Knowledge in Personal Life and Professional

Practice, University of Illinois at Urbana-Champaign, September 18-20, 1974.

"The Scientist and His Findings: Some Problems in Scientific Responsibility" was presented at the American Psychological Association's 82nd Annual Convention, New Orleans, August 30, 1974, in Symposium, "The Contemporary Heredity-Environment Behavior Controversy." "The Scientist and His Findings" first appeared in *Genetic Destiny,* 1976, and is published here with the consent of AMS Press, Inc. (eds., Ethel Tobach and Harold Proshansky).

"Preferential Consideration and Justice" was originally a paper presented at a Conference on Compensatory Justice, Athens, Georgia, February 13-15, 1975. Reprinted from *Social Justice and Preferential Treatment: Women and Racial Minorities in Education and Business* by permission of the University of Georgia Press (eds., William T. Blackstone and Robert D. Heslep). © 1977 by The University of Georgia Press.

"Notes on Terrorism" is a revision of comments on Kai Nielsen's "Violence and Terrorism: Its Uses and Abuses" in a session on Terrorism, at a conference on Morality and International Violence, held at Kean College of New Jersey, Union, New Jersey, April 22-24, 1974.

"Approaches to Environmental Ethics" is a revision of a paper presented under the title of "Rights and Duties in Regards to the Environment, or, How Not To Do Environmental Ethics" to the American Society of Value Inquiry at its meetings in conjunction with the meetings of the Pacific Division of the American Philosophical Association, San Francisco, March 23, 1978.

EXPLORING FACT AND VALUE

Introduction:
The Fact-Value Problem

If a questionnaire had been circulated in 1960 to determine the central philosophical problem of moral theory during the greater part of the twentieth century, the likely winner would have been the fact-value problem. It might have won out also as the central philosophical issue in the social sciences. It seemed clear to state—at least there were innumerable formulations—but difficult to explicate. It drew a sharp line between facts and values, or existence and value, or the desired and the desirable, or a state of affairs and its worth, or the *is* and the *ought*, or the descriptive and the normative, or what happens to be the case and what is good, or means and ends, or the real and the ideal, or science and ethics. Then it affirmed that it was impossible to go from the one to the other or derive the one from the other, particularly to go from the existence of a state of affairs to its goodness or its obligatoriness.

There had been argument about separating fact and value in the nineteenth century as well, and the beginnings of an independent general theory of value that took strong philosophical root in the United States in the first quarter of the twentieth century. But the fact-value problem took a distinctive turn by the middle third of the twentieth century: it became a highly *technical* philosophical question. (But of course this transition to the technical characterized philosophy generally.) Hence although both philosophy and social science proclaimed and argued for the separation of fact and value, the technical analyses of the issue were chiefly elaborated in philosophy. Many schools took part in this. In Anglo-American moral philosophy, the earlier work of G.E. Moore, *Principia Ethica* (1903), played a foundational role. Later positivist philosophy standardized the position on the fact-value gap, particularly for the social sciences, and dominated the scene in the 1930s and 1940s. From Europe, powerful social theorists, like Max Weber, added their influence, though in a complex way (and similarly Hans Kelsen in legal theory) to controversies about the relations of ends and means and about the inability of science to do more than aid in grappling with means, as well as about the *ought* and the *is*. The net impact of the new form of the fact-value problem was to render absolute the separation of value from fact. A basic concept like intrinsic value or good (in a moral sense) could not even be understood in terms of human processes of desire, pleasure, aspiration, or bio-psychological concepts of need. Indeed no scientific studies of human attitudes and inclinations would be admitted to throw any light on ethical values. Similarly, fact—and that included science—was isolated from value judgments and so was responsible to nothing other than its method of inquiry. The idea of a value-free physical science was carried into the idea of a value-free psychological and social science. This had been one strain in the nineteenth-century effort to build the social sciences; it became a dominant twentieth-century chorus as the social sciences won their place in the academic establishment.

The philosophical argumentation itself has many faces. The sharp separation of fact and value is argued as a matter of logic, of linguistics, of metaphysics, of epistemology and philosophy of science, of ethical theory. Some philosophers attempt to establish it briefly, with a single resounding crash argument; others broaden their case into a whole theory of the universe and its structure. Controversy arises at every step. Every argument meets criticism, every avenue is explored, new directions of inquiry are opened up, alternative positions consolidated and clarified, and the terms of the problem refined. In large measure, however, separatist moves became standardized and the phrases sloganized as the controversy progressed, the dichotomy lost its power and the whole question rather suddenly disappeared from the public view. Moral philosophers by the late 1960s and 1970s switched their analytic energies from meta-ethics or analysis of moral discourse to normative or substantive problems such as bio-medical ethics, environmental ethics, the ethics of technology, or in general problems of social policy. Scientists too, faced with problems of atomic weapons, pollution, transplants and recombinant genetics, might argue for handling their own social responsiblities rather than having governmental controls, but could scarcely plead absence of scientific responsiblility in these new situations. Values were too clearly relevant to these vital areas to be ignored.

Of course one way of greeting the passing of the fact-value problem would be to breathe a sigh of relief and forget about it as soon as possible.

But this is not very helpful in the long run. Whatever gave rise to the problem may in the future give rise to similar problems in some other garb or to fresh problems. There is a special sadness in watching a later generation make the old mistakes because it failed to glean insights from history. In addition, the situation at present is aggravated by the slow communication between philosophy and the social sciences. The positivist philosophers in their jibes at metaphysics used to say that the metaphysics of an age tended to enshrine the physics of a previous age which had already been abandoned by the physicists. So too there is a danger that the philosophical doctrines, where they have been technically enclosed, may have undergone change and may still be used in their older form in the social sciences. This is a reason why one has to work through the fact-value problem of the twentieth-century discussions even though they may now be in abeyance.

Philosophical experience suggests that the difficulties in such large problems often stem from two initial factors—a complexity hidden under apparent conceptual simplicity, and the uncertain character of the question itself. Each of these needs some consideration.

Fact and *value* are by now familiar ideas and we take them for granted. We may not realize that they are highly abstract categories with a past history of generalization or abstraction. *Fact* is at least as broad as empirical *truth*; indeed to say that something is a fact is scarcely different from saying that a proposition describing it is true. Happenings take on a kind of eternal status of facthood; by having happened, something acquires a factuality of which it can never be robbed. There are philosophical disputes as to whether there are negative facts, general facts, possible facts, and even value facts (as there is the fact that some things are valued).

The concept of value came on the philosophical scene to unite the various provinces of morality, aesthetics, religion, economics, and any other fields of human positively or negatively charged attitude. In the late eighteenth century Benthamite utilitarianism tried to unify all human action under the pursuit of pleasure and the avoidance of pain. In the nineteenth century, particularly after the Darwinian revolution, value theory sought a pattern with two contrasting aims: one was to understand man's spiritual life as a proliferation of his animal drives and impulses in the evolutionary process; the other was to set off his spiritual character in a unified way as transcending his animal nature. Thus the theory of value from the outset was entangled in all the heritage of the conflicts about the relation of body and spirit. Some of this conflict is paralleled in the views of the scope of science, especially in the relation of the physical and the social sciences. Sometimes the boundary of science is drawn with the material; the social sciences and history are the province of spirit—of *Geisteswissenschaft*, not

Naturwissenschaft. Sometimes the line is shifted: the social studies move over to science, but history remains as *global* and *ideographic*. (Psychology holds on to both in its different branches or different schools.) Eventually, in logical positivism, all lines are removed within the sciences and history: by being cognitive they are scientific in aspiration, and only the value fields remain beyond the pale as noncognitive and emotive.

As to the uncertain character of the question itself, we have already seen that it can have both logical and ontological aspects, as well as many other faces. But the matter is more far-reaching. It can also be a practical question as well as a theoretical one, and there are different kinds of practical questions. It could, of course, be a purely intellectual matter, like the simple demand to establish the logical consistency or inconsistency of a given system. The difficulty with seeing the fact-value problem in this way, we shall find, is that it does not—in spite of all the technical work on it—really get clarified adequately, it never quite loses its chameleon character, and if we look beyond its technical character we see it constantly associated with the most serious practical social-intellectual issues of the use of knowledge in the life of the times. It does have large intellectual components which have to be worked through as part of the structure of ethical theory. But it has practical components—social, moral, ideological—that can be understood only by seeing the theory in its intellectual-social background and historical development. Certainly one of the questions directly involved is the practical character of morality and the role of knowledge, particularly scientific knowledge, within it, as well as the corresponding question of the moral responsibility of scientific pursuits. But if this is so, then the dichotomy of fact and value is not simply a sharp separation of categories, but the schema for a possible culture and morality. This raises important and interesting questions for sociology, history, and philosophical inquiry. If there is any plausibility to this as an hypothesis in our intellectual history, why was the sharpest conceivable separation of value and fact projected philosophically in the early part of our century? Did it retain a constant historical meaning throughout, or did it change its character as the various intellectual arguments shifted? Was there any response to the vast changes that were taking place in the social and historical milieu of the middle part of our century? Did the fact-value dichotomy have the same social import at the time it became obsolescent as it did in the earlier part of the century?

It must be stressed that we are not faced with a choice of practical meaning versus intellectual meaning. To understand the fact-value problem in the twentieth century we have to look for its full intellectual scope and its full social ramifications. Only then can we make sense of its heritage in present problems of theory and practice.

The papers of Part I, Fact and Value in Theory, fall into three groups that have a kind of topical development.

The first group looks at the fact-value problem as a whole. The first paper, "The Relation of Fact and Value: A Reassessment" studies the forms that the problem has taken and the types of relationship that are found. The second paper, "Patterns of the Use of Science in Ethics" heads in an opposite direction: it wants to break down the question of the relation of fact (science) and value (ethics). Indeed, I had originally thought of writing that paper under the title of "twenty-seven uses of science in ethics" to demonstrate that there was no general issue of "*the* relation." The third paper, "The Place of Empirical Knowledge in Ethics" completes the general group: it expounds the view that the sharp dichotomy expressed a program for ethics, that this program had been tried and failed, and that a new program of the intimate relation of science and ethics should now be tried.

The second group shows in different ways how intimately science does enter into the work of ethics, both in ethical theory and moral judgment. One paper (ch. 4) examines how models, metaphors and analogies operate in ethics. A second (ch. 5) studies the vicissitudes of the concept of virtue in ethical theory and shows how far it has depended on the scientific status of the concept of "state of character" and personality theory in psychology. A third (ch. 6) considers some different ways ethical theory has sought assistance or foundations from scientific theory. A fourth (ch. 7) examines in detail how scientific results enter into refining the criteria that help in actual moral judgment.

The third group turns away from critique toward forging a different model in ethical theory, particularly under the impact of social change. The first paper (ch. 8) spells out six requirements of the revised model with respect to morality and law. The second paper (ch. 9) works toward an analytic method for dealing with moral change.

The papers of Part II, Fact and Value in Practice, serve both a negative and a positive role. Negatively they show that the adequate treatment of their topics does not require the sharp separation of fact and value. Positively they suggest that the fruitful lines of conceptual demarcation flow along quite different channels set by the problems at hand. They deal with various aspects of the fact-value problem, with problems of the responsibility of science and of knowledge in its social applications, with a number of moral issues that have tended to take ideological form as isolated principles ("reverse discrimination," "terrorism," "rights of the unborn")— all of which are directed to testing the new model of ethical inquiry in important social issues. Of these papers one (ch. 10) studies the changing relations of technology and morality since World War II. Four papers (chapters 11-14) deal with the social responsiblities of scientists, engineers and professionals. The central paper among these offers a basic theory of scientific responsibility, rejecting the view that it can be settled by a general principle and deciding that the responsibilities of science stem from the state of science and the state of civilization at the historical period. One of these papers analyzes the controversy over the Jensen thesis in educational psychology. Further papers on "Preferential Consideration and Justice" (ch. 15) and "Terrorism" (ch. 16) and "Approaches to Environmental Ethics" (ch. 17) are particularly devoted to showing how the complexities of the subject matter can begin to be unravelled by the kind of alternative model offered.

Part I

Fact and Value in Theory

1. The Relation of Fact and Value: A Reassessment

The problem of the relation of fact and value has the scope and the traditional complacency which invite a fresh speculative encounter. It has taken protean forms and to track them down one by one is to go into particularities that seem to encompass the whole range of philosophy. And yet, despite the diversity of forms, there is the conviction that there is a single problem. This suggests that the differences may lie in the cast of characters and the details of each scene, while the unity is in some theme that is carried through. But can there be such community in action on a stage that is sometimes theology, sometimes metaphysics, sometimes logic, but is equally likely to be science or ordinary life, a newly-furbished axiology or shop-worn grammar?

I

Let us sample some of the protean forms in which the problem appears. Its ancient theological and its modern idealist formulations bulk large in the philosophical corpus. Here Value is securely ensconced, whereas Fact is the source of difficulty. How can the fact of evil so manifest in human life be reconciled with the goodness of the divine omnipotent creator? How can the finite with its incompleteness and discrepancy find a place in the unquestioned perfection and completeness of the totality? The materialist tradition does not escape the problem, but it is the other shoe that pinches. When the physical materialists narrow down Fact to physical movements, Value has an uneasy home in the subjective, and with the rest of the human spirit it is reduced to mere appearance. The evolutionary materialists, however, turn it into a straight scientific problem: they identify Fact with material processes and Value with human striving and goal-seeking; the problem of Relation is the causal problem of the emergence of these human processes in the course of biological and social evolution. The modern axiologists multiply realms, so that Fact would indicate the scientific sensory order or at most the metaphysical order of necessity, whereas Value would loom above them as the ideal ought-to-be; the problem of the Relation is to decipher the nature of the lure that value has for portions of fact, and the foci of its magnetism. Of late, logicians have taken a hand at the formulation: they would replace Fact by fact-terms and Value by value-terms; the problem of Relation is thus revealed as the hopeless task of deducing conclusions that contain value-terms from premises that contain only fact-terms. In a grammatical cast, Fact wears the robe of the indicative; Value, of the imperative; and the problem of Relation is the puzzle of the leap from indicative premises to imperative conclusions. Sometimes the formulation is more generally methodological: whether there is any mode of verification in terms of scientific fact by which value assertions can be put to the test. Among contemporary naturalistic philosophers, many prefer to focus on material content: Fact and Value are then located as two sets of phenomena by reference to which one may pose the contrast of describing and prescribing; the Relation problem is that of beliefs and attitudes, or of the cognitive and the affective. In ordinary life, Fact is human behavior and Value is men's ideals, and the Relation problem is the sad plight of ideals in the welter of forces in human life. With the growth of the prominence of science, Fact tends to become what the scientist *as* scientist asserts about the world; Value takes its place as recommendations of social policy, and a frequent formulation of the Relation problem is whether science is inherently "value free," limited at most to studying value phenomena but incapable of making value judgments.

An analytic approach to the terms "fact," "value," "relation" tends to reinforce the impression of diversity. "Fact" sometimes refers to existence or nature and its qualities or characteristics. Sometimes it is extended to embrace the total network of truth, from which it passes readily to encompassing the whole of reality. When it has been used in so broad a sense, the fact-value problem is almost converted into "Where in the world are we to locate value?" Value itself has been equally extended. It can cover any of the elements in traditional ethics—"good," "ideal," "ought"—and with the rise of the concept of generic value it can cover "interest," "desire," or any of a host of indications of being pro or con—as rudimentary as feeling attracted and as sophisticated as applying an articulated standard. To turn to the third term in our problem, "relation" too is something of a catchall. Relations may be logical or empirical; with respect to content they may relate sense elements, complexes, ideational elements; or again, they may be physical or psychological, social or historical, and so forth. One might even look for a relation that is evaluative: for any given sense of "fact" and "value," we might ask not merely how they *are* related but how they *ought to be* related in the light of assumed standards, or in order to achieve agreed-upon goals.

It might seem that a general problem of fact-value relation presupposed a general sense for each of the three terms. This would certainly be helpful, and there are fascinating problems in the philosophical story about the rise of such conceptions—whether we look to the growth of the concept of knowledge, or the category of generic value, or the logical theory of relations.[1] But even if no general concepts were successfully established there might still be some common lessons about processes and patterns in the hosts of particular fact-value relations. The multiplicity of problems therefore cannot relieve us of the burden of deciding whether there is a general fact-value problem, although it may make us sceptical of a ready solution.

Is it necessary to track down every particular formulation and follow it through in its own area? We cannot answer this in advance. It would be worth it if we could dispose of each as we went along. But if each led us on to another turn, then we would indeed be "getting the run around." In any case, we can first see how far we would get in studying the problem as a general one. Why should we not carry out our search as the student of comparative literature pursues a recurrent theme in the stories of different peoples and ages, or as the folklorist penetrates the core of a scattered myth? Or again, as the psychoanalyst traces a basic anxiety through a multitude of diverse symptoms? Lest this suggest some derogation of the truth status of the philosophical problem, let me hasten to add a purely scientific parallel. Why not do it as, say, Hans Selye fashioned his concept of "stress" by becoming interested in a general syndrome of being sick and looking for regulative bodily mechanisms brought into play in a general way in spite of the variety of diseases.[2] We need not prejudge how far the general problem, if there should turn out to be one, is a metaphysical or scientific one, or for that matter, a historical or cultural one. We can simply use as the domain of phenomena the many particular fact-value formulations and study the way constituent elements line up, and the pattern of action that then ensues.

II

The kinds of entities that take on facthood and valuehood are too diverse to give us much encouragement in our search. They are realms or abstract entities, types of personal activity, bits of behavior (physical, psychological), etc. There does not seem to be any type that always appears on one side of the dichotomy. Take abstract entities, like actualities and possibilities: there are actualized values as well as actual facts, and factual possibilities as well as value possibilities. Or take a human phenomenon like desire: in general value theory, with the contrast of the purely physical and the selective organic tendency, desire is on the side of value; in many of the recurrent Kantian is-ought contrasts, desire is rejected as a value indicator and takes its place among the facts to be assessed. ("It is desired, but is it desirable?" is today among the most favored ways of expressing the fact-value contrast.)

On the other hand, it is suggestive to look at the way and the conditions under which an entity shifts from side to side in different formulations. Desire is in the value column when it betokens an inclination toward the alteration of what is; it is in the fact column when it is located as an item in the situation that is being canvassed from the point of view of possible action. The determining factor seems to lie in the *office* that the item is serving—whether it is structured as part of the raw materials of the situation or functions as a guide in decision-action processes. This can be seen in such complex transformations as that of *essence* from the value column, where it had a firm grip in the ancient and medieval theological philosophies, to the fact column in modern axiologies that insist on the autonomy of value with respect to all metaphysical-structural elements. As long as every form of existence was seen as striving to express its essence, the essence had a basic guiding role. The theory of evolution destroyed this status of essence; it became simply the pattern that happened to be hammered out, and so the nature of things ceased to be regarded as self-justifying.

This suggestion is strengthened by the way in which distinctions of the fact-value type spring up in areas where we might least expect them, whenever a field is structured with an eye on decision-action processes. Within the domain of ethical concepts we might have expected that every item would have a secure value status. But we find instead that the dichotomy appears, no matter how restricted the territory. Examples of such polarizations are *means* and *ends*, sharply divorced as factual instrument and value object; *prizing* and *appraising*. Where the former by comparison is demoted to the fact of liking; even the *valued* and the *valuable* where (as in the case of the desired and the desirable) to claim that something is valued is seen as a factual assertion lacking "normative force." A recent writer goes so far as to carry this division into the very concept of obligation itself—the central shrine of contemporary ethical theory. Nowell-Smith argues that even if, in the intuitionist style, we discovered a unique property of obligatoriness, there would still remain the gap between the property and the decision; we could still ask why we ought to do that which we found to possess this property and toward which we felt a special moral emotion of obligation.[3] There is no end to this game of gaps and leaps. An extreme form of this factualization of value is found in the existentialist's picture of the lone individual for whom all the structure of heaven and earth and all his own past valuations have been reduced to a factual status, and he is left to make the free creative leap to value in the act of decision.

We see the characters and we see their role. Let us now ask about the plots that philosophical ethics has woven for them. What themes can we locate? It may be disconcerting to a philosopher who has propounded a reasoned argument about the relation of fact and value to learn that he is being studied as a specimen of a fact-value pattern; but it need not be objectionable, since determination of truth or adequacy of philosophical view is not being bypassed but simply postponed in the interests of finding out more clearly what the issues are about.

In such a survey it becomes quite clear that the central theme is the fundamental tension of "realms" and the central issue is whether they are to be assimilated, kept strictly apart, or else interrelated in some complex configuration. Purely descriptively, we may recognize three types— *assimilative, separatist,* and (for want of a better term) *empirically integrative*. Within each there may be different degrees of strength or intensity. A view that connects its concepts analytically or separates them analytically is more intense than one that is content with material equivalence or material contrariety.

Some theological and idealist philosophical theories on the assimilative side reach a high intensity. Value must somehow be present throughout— because God's omnipotence, omniscience, and goodness leave no room for anything less than perfection in his creation. Or the Absolute in all its completeness must somehow merge reality and value within it. A whole armory of intellectual weapons is forged to achieve this end. Sometimes a short cut is sought whereby it can be proved at one blow that reality cannot be "value free." For example, Dewitt Parker maintains that "every concrete experience is itself a value, positive or negative, because every experience is essentially a wish, or system of wishes, in process of satisfaction or frustration." All that remains is to reduce the world itself to "fountains of sensa within the concrete experience of some center or centers, or else centers of concrete experience in their own right" and value is indelibly ingrained in all that is.[4] This is a short cut parallel to Berkeley's view in the theory of knowledge. An elaborate formulation, from the value side, of the familiar idealist thesis that we posit our world is to be found in Münsterberg's *The Eternal Values*. For example, "We demand that there be a world; that means that our experience be more than just the passing experience, that it assert itself in its identity in new experiences. Here is the one original deed which gives eternal meaning to our reality, and without which our life would be an empty dream, a chaos, a nothing. We will that our experience is a world." From which he rapidly concludes that the world of values is the only real world: "The system of values must then be recognized as soon as we ask what has been really posited by this act of world assertion."[5] It is not surprising that the world is found to be well-endowed with all sorts of structural values from identity to growth, development to absoluteness.

With similar intensity, Kantian as well as subsequent positivist and some analytic theories proclaim the necessary distinctness of fact and value. With Kant it is the phenomenal world to which science is addressed and the noumenal world to which morality points. With G.E. Moore, not to embrace separatism is to commit the naturalistic fallacy. Other slogans are: "You cannot derive the *ought* from the *is*"; "Belief is one thing, Attitudes another"; "Science *describes*, Ethics prescribes." And so on, in a familiar vein.

Suppose we avoid arguing the question and focus instead on the type of consciousness which exhibits an assimilative or a separatist pattern. A useful device is to imagine oneself holding the position and seeing what the world is like. For the assimilative, think of St. Francis of Assisi seeing the blessedness of God in everything about him. Or recall Gandhi's insistence that what we love is the existent as such, and not some quality or condition of it. Or Albert Schweitzer jumping from the will to life directly to respect for life in any and every form. We may even go back to Job, not as he was in his doubt, but in his resignation. Perhaps all of these put together will give us a glimpse of a phenomenal field in which there is sheer gratitude for the sheer sense of being. For the separatist pattern, think of a Kant badgering a mother who loves her child to find out whether she does it from a sense of duty or from natural affection. Or the free man in Russell's early essay, "A Free Man's Worship," brandishing his fist at matter rolling on its relentless way. Or, in more modest terms, of the scientist insisting that when he says atomic warfare is evil, he does it purely as an individual, not as a scientist. In all these, we get the sense of an extricated self that waits until the situation is completely mapped and then reacts, wills, feels, commands—all of his own sweet arbitrary impulse or in his own indeterminate fashion.

There are, of course, assimilative and separatist patterns of lesser intensity, which can be mapped both in their theoretical and their phenomenological shadings. But perhaps for diagnosis the extremes will suffice. Let us turn instead to the empirically integrative pattern. It is integrative because it is looking for whatever significant relations can be found in the phenomena; and it demands some kind of initial indices for the tentative differentiations that pose the problems so that it can explore relations empirically. It keeps the spotlight on the whole field. Without any special stake in assimilation or separation, it can accept what unity it finds and what plurality it finds. Phenomenologically, this empirically integrative consciousness seems to be the most common sort among men engaged in moral deliberation. For it involves some relative separation in the concepts of fact and value embodied in the distinction of situation faced and implicit standard for successful facing. Yet it involves some sense of the dependence of the standards on what the world is like and what men are like.

These three types of general fact-value patterns and many divergent forms of each have been found as existent types of consciousness among many cultures and subcultures in the history of the globe, even during the time that we know about it. No statistical study of their incidence is here proposed; nor is any argument here intended to rest on which is more common in ordinary life or in rare and sensitive spirits, or which is found in practice and which in contemplation, or which is an extreme and which occupies the stolid role of mean. Whether they arise under determinate causal conditions (psychological, cultural and historical) is a question well worth investigating insofar as history can furnish information. Whether one or another represents a superficial or surface consciousness, indicative of special problems and exhibiting a profound misunderstanding of what is going on within men in their world, is the further and culminating question—and that is a problem of evaluation.

III

We have then our recurrent themes and we want to find an explanation for them. It is not often that the problem of the relation of fact and value is formulated in such scientific terms. And yet such a formulation is probably the most far-reaching when we deal with specific interpretations of the concepts of "fact" and "value." If "fact" be taken to refer to physical description of the state of the globe at any given time, and "value" to purposive phenomena, then the question of relation is in part the historical-causal problem of the appearance of purposive phenomena in a material world, in part the functional problem of the way in which material situations service purposes, and in part the causal-functional problem of the way in which they interact and develop in their interaction. Again, if "value" be identified with selective components in a morality, we have similar inquiries about the conditions of origin and functions of ideals, phenomena of conscience, etc., in a context of goal-seeking beings, and the conditions (biological, psychological, historical) under which these phenomena take on specific qualities. Similarly, one might fruitfully inquire under what conditions of existence of social forces and cultural (including moral) patterns, certain lines of reflective justification assumed the form we call ethical theory, and how they are now supported, and to what needs they give expression. Such inquiries can be formulated for any specific interpretation of value and fact concepts provided there is sufficient empirical identification to make the questions of causality and function meaningful.

Can significant questions of this sort be raised about the recurrent general themes we have delineated—the assimilative, separatist, and empirically integrative patterns of fact-value relation? Where a pattern is identified in individual consciousness, an explanatory approach may turn to psychological findings applied in a given cultural setting. Where the pattern is identified as a cultural or subcultural phenomenon, and formulated as a question in the history and diversification of a particular culture, then explanatory approaches turn to sociohistorical conflicts and pressures generating tensions that find such intellectual expression. And again, where patterns are considered as constellations in intellectual history, explanatory approaches tend to explore the careers of intellectual constructions and look among them for concomitance and dependence. Most desirable, of course, would be a unified schema connecting these various approaches.

Hypotheses of a causal type are available in some of these areas. There are, for example, psychological studies of the way in which people separate wishes (as fact) and duties (as value) in their consciousness, and psychological explanations of such phenomena as the rigorous isolation of the affective life in particular individuals, or again, of a passively acquiescent or a particularly defiant personality. There are cultural explanations for some antitheses: for example, the sharp dichotomy of means as facts and ends as values is ascribed by some to the separation of production from consumption in industrial economies; or again, the aloofness of value from fact or ideal from reality is taken to reflect the social isolation of religion from practical-life decisions in modern western culture. There are sociological explanations of assimilative trends in terms of dominant group efforts at inducing resignation. Such causal hypotheses, of course, cannot be dealt with in this present limited context.

I should like, however, to suggest a hypothesis stemming from the domain of intellectual history. The hypothesis is that the theory of fact-value relations is in some sense a function of the theory of the self, so that the degree and quality of assimilation or separation reflects an implicit picture of the nature of the self.

Some evidence for such a hypothesis would come from specific analysis of the historical panorama of philosophical theories. Let us scan some sample highlights. In the dominant ancient philosophies of Plato and Aristotle, insofar as there is a distinct concept of the self it appears in the greater reality or uniqueness assigned to the rational-intellectual component of man. But this component is taken to provide direct insight into the structure of reality; and so it is scarcely surprising that this structure is assigned the status of both reality and source of value. In the medieval theological picture the self is identified primarily with the soul; its commitment to God by genesis and nature provides an assimilation of structural fact and value. In early modern Cartesian dualism there is a rending of the material and the spiritual, with the individual man as a meeting place of the two. The unified picture of the individual becomes, thereafter, the locus of the mediation of material existence and spiritual value. Whatever the self is pictured as parallels the pattern of fact-value relation. Where the self falls sharply into sense impressions and feelings, fact and value fall sharply apart. Where the self is a transcendental ego, value is autonomously sovereign. Where the self takes over the cosmos and sees itself as positing reality, value and reality are intertwined; where it is seen as a fragment of the Absolute, the distinction of value and fact becomes a shortsightedness from a finite position. Where the self is treated as a product in the operations of nature, value is a quality of existential interactions. Where the self is an atom pursuing its own tendencies, reality is indifferent fact that happens to help or thwart. Where the self hovers alone on the growing edge of being, value lies in the creative increment of decision. And so on.

Such a thesis has an initial plausibility. For the self is in one sense an existent, in another intimately related to values and often even defined by its path of striving. Again, the self embodies criteria of value judgment; it is in the very center of consciousness and judgment. It is the point in the cosmos where existence becomes conscious of itself. But such properties do not make the hypothesis tautologous. For the self is much wider than value. It includes processes of learning and thinking and perceiving as well as affective and volitional components.

It is possible that some corroboration for such a thesis could also be sought in other bodies of materials. For example, conceptions of the self and patterns of fact-value relations can be sifted out of anthropological reports, and insofar as they prove to be at least initially independent variables, their correlation may be sought. Or again, in the history of theories in psychology, there might be enough separateness in the pictures of human nature and the pictures of value to find clues about concomitance.

Suppose the concomitance were established, could we then give any priority to the theory of the self, in the sense of regarding it as a *presupposition* of a theory of value-fact relations? This is a complicated problem in the analysis of ethical theory. I think it can be shown that every ethical theory has as part of its structure some existential perspective—a view of the world and man's nature and conditions which acts as a kind of stage setting for the ethical theory. The conceptual framework of the theory is dependent, although not in a simple one-one relation, on the stage setting.[6] If this thesis about ethical theories is correct, then since the theory of the self plays a prominent part in existential perspectives, the concomitance of value-fact relations and self theories can properly be construed as a kind of dependence of the former on the latter.

IV

Suppose our hypothesis could be established. How would it help us in the primary task of answering the theoretical problem, of establishing the truth or adequacy of one account over another? This calls for more explicit tracing of the outlines of our methodology. We set out to solve a theoretical problem, and instead we mapped the kinds of consciousness that corresponded to different theoretical proposals and speculated about the causes and conditions of such differences, and the relations of different theoretical issues. Is this as beside the point as it might seem if a physicist facing the problem of the relation of matter and energy had gone to the history of physical theories instead of to contemporary mathematical physics and experimentation? I shall not stop to argue that history is illuminating and might have played a part in clarifying the issues; it might instead have shown simply that physics was concerned with explanatory theories for wide ranges of observed materials and that earlier theoretical formulations covered narrower materials and had narrower concepts. But this is what everybody knows—for physics. Now is the situation in ethical theory parallel? Perhaps it is more like law or religion, where a comparison of systems and beliefs over a wide scale does help restructure the kinds of theoretical questions one asks at the beginning. Instead of asking, for example, what are the properties of God, one compares the variety of properties assigned, seeks to correlate them with the problems of men in different contexts, and ends up with an understanding of the varieties of needs and feelings that find expression in religious institutions and experience. This does not, of course, answer the original question, but it clarifies it both for the theist who will then face more clearly the issue of an independent basis for answering his question or for showing how needs and feelings may be taken to have a cognitive status and for the secularist who will reinterpret religious phenomena themselves as expressions of human insecurity. Similarly, in the case of law, the original question about the eternal rightness of a set of laws gives way to an understanding of the aims and purposes of men in the regulation of their lives by certain social instruments; the question of the adequacy of a given set is not thereby answered, but it becomes asked in a different way. So it may be with the fact-value problem and other theoretical issues in ethics. If this is so, it carries a most important methodological lesson, somewhat parallel to that in contemporary metamathematics when it found that controlled discourse *about* mathematical systems could help answer questions concerning what could be done *within* those systems. Ethics and metaphysics do not have the rigor of such a discipline, and the nearest thing I can see to it is the *historical and comparative study* of problems and their formulations. So far from regarding the history of philosophy then as external, it may very well serve as a fundamental tool for advancing the solution of philosophical problems by clarifying their formulation and proposing reconstructions of issues. And what holds for history of philosophy holds for any contextual probing of philosophical ideas, causal and functional. Here no-trespass signs are often blinders.

Very simply, then, if the relation of fact-value patterns to pictures of the self is correct, a large part of the evaluation of an asserted fact-value relation consists in determining the truth of the underlying psychological account of the self. In principle, this is a scientific type of inquiry. Whether in the light of its complexity, it can be carried out by the sciences of the day is a separate question. The study of the self is dispersed, but advancing on many fronts. There are developmental studies of the child, the way in which he registers his experience and the role played by the way in which he masters elementary coping with his environment in motility and perception, the way he interacts with parental figures and the kinds of basic layers of the self that result. There are studies of the orientation devices that the cultural milieu furnishes—in part through language, in part through structured experience—to provide a basic framework for self-awareness. There are the revealing insights of what happens when mechanisms of development and orientation miss fire or go astray, in the vast array of psychiatric materials. There are large-scale studies of social institutional valuations, whether in the Hegelian language of objective mind or the sociological language of customs and traditions. And, of course, there is the growth of inquiry into such strands as will or choice or selection, on the one hand, and—although in a troubled state—emotion and feeling on the other. This does not yet add up to a unified theory of the self. But there is nothing else that is a fraction as promising. The philosopher's business is not to tie himself to an answer in terms of the present state of affairs, but to project the directions in which an answer may be sought. He can dream of the kind of unified theory that has filled in the gaps, that does justice to physical-biological bases, to psychological-developmental forces and situations, to cultural and social variables, to historical changes, and try to suggest what are the unavoidable components and what areas allow of alternatives in development.

Part of such a synthetic task is a reckoning with alternative contemporary theories of the self. On the methodological lesson here presented we may expect to find, what I believe we do find, that the conceptions of the self which eschew this arduous scientific inquiry are inadequate; they are one-shot affairs that embody some older scientific results or fragment thereof, or else are enchanted with some marginal phenomenon of human psychology which they have discovered and which instead of handing over for careful scientific scrutiny they deck with regal philosophical robes.

V

Several objections may be raised against such a conclusion. I should like to indicate and comment upon them. "You are assuming," it may be said, "that the scientific picture of the self where you find the key to fact-value relations is itself a neutral scientific result. But is this really the case? Think of the history of conceptions of human nature, or of will. Are you not simply offering a new path for the introduction of valuations, in the authoritative disguise of scientific results?" To answer this we must distinguish the ways in which valuations may be relevant. If they enter through the back door, they are simply turning science into ideology. For example, Nietzsche turns the very phenomenon of willing into a kind of self-command and casts the mantle of power expression over all decision. If we take this as scientific truth, we may be slipping into a power ideology. But we need not. We can see it as a value cast; we can also explore how far it is a correct description for the volitional phenomena of certain personality types, look for its causes in such cases, and so on. In such an inquiry, the progressive sloughing off of aspects that were taken to be correct description but turn out to be partial selective emphasis is a normal part of the progress of any science. On the other hand, it is very likely that the scientific picture of the development of the self will show the possibility of developmental differences and alternatives, and to that extent leave open gaps for policy decision. These raise problems of ethical evaluation and justification in which there are usually specific questions of truth of assumptions and recognition of shared purposes. For example, where an extreme assimilative pattern is recognized as a policy decision, we have to ask—since this pattern usually embodies a quality of resignation—whether its resignation assumes a despair of human initiative in the world in which we live. Similarly, if an extreme separatism simply covers libertarian demand for the personal right to pass judgment on any state of affairs, we have a quite specific ethics of individualism to reckon with, which amounts usually to passing judgment on a basic trend of modern civilization.

"You are assuming," a second objection will run, "that the scientific question of the self is neutral to metaphysical assumptions. But may it not be the case that the whole fact-value problem is basically one of choosing a metaphysical perspective?[7] What you are doing then, is asking us to accept the metaphysics of science in the guise of a scientific solution of the fact-value problem." This raises a tangled nest of questions, extending far beyond the range of the problem we have been considering. We need not be driven to the simplistic view that science is without any world-view conception, nor to the equally extreme simplistic view that its selection of its basic categories is an arbitrary presupposition. It may very well be that its procedure and progress provide ways of strengthening some as against other categorial sets; that is, to look at the same question from the point of view of metaphysics, the various sciences by trying out metaphysical alternatives in their categorial sets furnish ways of long-range testing of adequacy for those very sets.[8] There is, however, another and perfectly real issue involved with respect to the limits of a scientific account of the self. Suppose the progress of such a scientific inquiry showed the self to be a center of indeterminacy and creativity such that some of its works could only be studied scientifically *after* they were done, not predicted antecedently. Would this call for a metaphysical *supplementation* in terms of, say, a free-will conception? Not necessarily. It may very well turn out that the fullest recognition of the nature of human creativity can come from the subtlest scientific inquiry into its conditions and occasions, its attendant qualities and its instruments.

"In any case," it may be objected, "you have not escaped the acceptance in the background of some type of fact-value relationship. Perhaps it escapes notice because you have an empirically integrative consciousness. Are you really doing anything more than giving expression to it? Please note that you have not hesitated to distinguish scientific conclusions and policy recommendations." I suppose it is true that the outcome of this inquiry is to recommend an empirically integrative approach to fact-value relations. But it is not as a background assumption. That is, the recommendation rests on the results of analysis together with the (probable) results of the theory of the self insofar as we can anticipate the direction of its present development. That some conceptual dichotomy is involved in the way a decision-action situation is structured was suggested by our very first clue above. The whole issue is about the character and quality and mode of operation of these distinctions. And there is nothing in our conclusion that implies that science is without its policy directions, nor that policy recommendations are without their scientific assumptions. (As Dewey so often makes clear in his ethical writings, distinctions of office are quite different from distinctions of entities. The analytic way of saying it would be that distinctions on the metalevel may be analogous to distinctions on the object level without being the same thing.)

There is no gainsaying the complexity of the problems before us. If the hypothesis that a major key to the problem of fact-value relations lies in the scientific theory of the self has any merit, it can be shown best as the growth of such knowledge makes substantial inroads into the problems and formulations, and limits increasingly the indeterminacy that the philosophical tradition in this area has permitted.

Notes

1. For a consideration of some facets of the generic-value problem, see Y. H. Krikorian and A. Edel, eds., *Contemporary Philosophic Problems* (New York: Macmillan Co., 1959), pp. 459-66. For some analysis of the conditions that a generic-value concept has to satisfy, see Abraham Edel "Concept of Values in Contemporary Philosophical Value Theory," *Philosophy of Science* XX (1953): 198-207.
2. Hans Selye, *The Stress of Life* (New York: McGraw-Hill Book Co., 1956).
3. P. H. Nowell-Smith, *Ethics* (London: Penguin Books, 1954), pp 39-43.
4. Dewitt H. Parker, "Value and Existence," *Ethics* XLVIII (1938): 479, 486.
5. Hugo Münsterberg, *The Eternal Values* (Boston and New York: Houghton Mifflin Co., 1909), pp. 75-76, 78.
6. This view is explored in considerable detail in Abraham Edel, *Science and the Structure of Ethics* (Chicago: University of Chicago Press, 1961).
7. For an illuminating survey of the fact-value dichotomy as expressive of a basic set of metaphysical and epistemological assumptions, see Iredell Jenkins, "What Is a Normative Science?" *The Journal of Philosophy* XLV (1948): 309-32. Professor Jenkins diagnoses the whole issue as the metaphysical bifurcation that took place when reality was impoverished by restricting it to the so-called primary qualities. In his analysis of the different forms that the bifurcation takes he seems at one point almost to come to a halt on the question of the relation of perception and feeling as experiential elements, but passes on to insist on the unified metaphysical character of the issue.
8. For a fuller discussion of this problem, see Abraham Edel, "Interpretation and the Selection of Categories" in *Meaning and Interpretation* (Berkeley and Los Angeles: University of California Press, 1950), pp. 57-95.

2. Patterns of the Use of Science in Ethics

It is commonly recognized that the question of the relation of science to ethics has been a troubled one, especially in the twentieth century. There is an interesting history to the question. Scientists have occasionally attempted to legislate for ethics. They thought to determine morality from their biological results; for example, their conclusions about human evolution. Or else, they offered lists of instincts or needs out of which they expected a conception of the good to be constructed. They framed concepts of mental health to serve as a basis for morals, or attempted to apply to moral inquiry the latest model from the latest science. Philosophers reacted with emphasis, often with impatience. They attempted to establish and maintain a concept of moral autonomy which would free them from an incipiently authoritarian science as it had freed them from an authoritarian religion. Sometimes they fell instead into the clutches of an authoritarian linguistics, and they confused autonomy with isolation. Many philosophers felt that the whole question had really been settled when G. E. Moore worked out the idea of the *naturalistic fallacy* in his *Principia Ethica* at the beginning of the century, and separated the moral sphere from the natural sphere; in one form or another, it has won assent in the analytic world. In a more popular vein, and among perhaps most scientists, it has seemed enough to say "Science can give us only means, not ends."

My aim in this paper is not to give a historical survey, but rather to attempt an adequate formulation of the question. For in spite of—perhaps because of—the heat it has generated, formulations are loose and presuppositions are insufficiently analyzed. I find it significant that as late as 1959, so careful an analyst as R. B. Brandt, in his *Ethical Theory* should entitle a chapter "Can Science Solve all Ethical Problems?"[1] My leading idea—and it is no doubt an obvious one, or at least I hope it is—is that there is no single wholesale relation or wholesale absence of relation between science and ethics. There are numerous specific types of relations of specific results of specific sciences to specific parts of morality or ethical theory.

The formulation I am suggesting begins with a recognition of morality as an on-going process in human life—with its own phenomena, concepts, methods, problems. There is no question of starting from science, and establishing a scientific morality from scratch on the basis of scientific data. But on the other hand, there is no question of isolating morality from scientific impact. There is no part of morality—whether concepts, rules, processes, or methods—which may not be refined, improved, or altered by the application of the knowledge arising in one or a number of the sciences. Similarly, there is no part of ethical theorizing—again, whether concepts, principles, processes or methods—which may not be amenable to refinement in the light of scientific knowledge.

It is very likely that you will agree on this. And so we have reached a consensus, my argument is over—if I had one in the first place—and we can all go home. I go on for a very simple reason: I am suspicious of a too ready consensus. There has been so much argument on the question, so deep an embedding of the theory of the naturalistic fallacy in the last half century, so complete a disregard of the results of the growing science of man in the actual doing of philosophical ethics, that I find it difficult to believe that the controversies are over. I suspect rather that they will flare up again the moment we begin to apply the formulation suggested. The main aim of my paper is thus to test the consensus by beginning the kind of research program for which the formulation calls, the location of the precise points of relevance or points of entry of scientific materials into ethics.

There are two preliminary points to be noted about our procedure. We focus on the relation of science to ethics. Science and ethics are seen as two *fields* of human inquiry. We have no right to translate this directly into the relation of fact and value. The latter is a categorial reinterpretation of our problem. It is a proposed answer, and I might add, in many respects a vague one, and a dubious one. We shall reflect on it much later, but please do not start out by equating the problem of the relation of science and ethics with that of the relation of fact and value.

My second preliminary point is that we must choose whether to fix the science side or the ethics side in beginning the study of relations. We could, of course, fix both; for example, five sciences and three parts of ethics, and so study fifteen relations. Because our primary focus is on ethics, and because we expect the sciences of man to shift their borders, I choose to start by fixing the ethics side and leaving the science side open. And on the ethics side I want to distinguish three parts. The first is moral judgment, usually called substantive morals or normative ethics. The second is the part of ethical theory that commonly goes into the texts under the heading "the nature and tasks of morality," and is usually disposed of dogmatically in introductory chapters; it is the preliminary consideration of, as it were, the shape of the field. The third consists of theoretical ethics or metaethics, analysis of ethical concepts and their relations, justification procedures, and so forth. No doubt the field of ethics could be cut in many different ways, but this cutting will do well enough for our inquiry. We have then to look into the relations of science, that is, the field of the sciences of man as they exist today, to these three parts of ethics. I should like, however, to add both a prefatory section and a concluding one. The prefatory one is an unorthodox interpretation of Moore and the naturalistic fallacy in the light of our problems. The conclusion will reflect on the logical character of these issues such as the relation of science and ethics, of means and ends, of fact and value, and how far they are analytic issues or how far they are themselves scientific issues.

Moore and the Naturalistic Fallacy

This is not an $n+1$th attempt to reckon with the naturalistic fallacy, but to suggest a mode of reckoning. The naturalistic fallacy has been so prominent as a theoretical barrier to relating science and ethics that it cannot be by-passed. I have too much respect for Moore to treat him as he has been treated by so many who purportedly agreed with him, that is, to use him to support their own views by saying that what they were saying was what he really meant. Look at *Principia Ethica* as a whole, then, not just at Chapter I.[2]

Moore was rejecting a scientific basis for intrinsic value, as he rejected a metaphysical basis or a religious basis or a historical basis. His most outright statement is to be found in his review of Brentano's *The Origin of the Knowledge of Right and Wrong* when he says: "The great merit of this view over all except Sidgwick's is its recogniton that all truths of the form 'This is good in itself' are logically independent of any truth about what exists. No ethical proposition of this form is such that, if a certain thing exists, it is true, whereas, if that thing does not exist, it is false. All such ethical truths are true, *whatever the nature of the world may be*."[3] In *Principia Ethica* he propounds the view that good is a simple nonnatural quality, that the way to discover whether anything is intrinsically good is to isolate it from all extrinsic relations, contemplate it as a world by itself and simply *see* whether it has the property of intrinsic value. Ample and subtle illustration of his procedure is found throughout the book and especially in Chapter VI on the Ideal, in which he reckons with the greater or higher values. The term "naturalistic fallacy" is coined to condemn any view, like Utilitarianism, which offers an equivalence definition of good in terms of some descriptive property. (That good is nonnatural means in this context, in terms of procedure, that the whole of which it is to be asserted or denied has *first* to be completely described as existing by itself before we look for the further property, good.) It is generally recognized that Moore's most successful argument for the naturalistic fallacy is the open-question argument: it is always possible to ask, if you define good in terms of some specific content (e.g., pleasure, desire, what God wills) whether that content is itself good, and this question always makes sense. (Later writers applied it also to ought; e.g., if ought means what my community demands of me, I can always ask: Ought I to do what my community demands of me?)

A word on this argument's significance. Moore meant it to support the simplicity of good. But look at it this way: The open-question argument thus is the guarantee of the permanent possiblity of evaluation for any described content. Moore's view of good as a simple nonnatural quality guarantees the permanent possibility of evaluation. But so might other views about good. For example, we might even legislate in our theory of definition the right to change definitions if they did not explain facts, and facts here include the whole array of immediate moral reactions as well as existent principles. In fact, we might simply stipulate the permanent possibility of evaluation through a substantial concept of individual intellectual liberty to be worked into any satisfactory ethical theory. Thus what the open-question argument gives us can be achieved in many ways. It is not uniquely tied to Moore's rejection of science in dealing with intrinsic value.

Now it is important to note that Moore did not reject science directly in his theory of right and duty. In fact he incorporates its relevance in his very definitions of the terms. In Chapter V, on Ethics in Relation to Conduct, he insists that questions about right or duty are questions about the effects of our actions in making the whole world better in one way than in another. Thus while science is perfectly at liberty to come in and give us knowledge of the effects, Moore has set his requirements so high—the effect upon the totality of the world for all alternative courses of action—that he is not unreasonably pessimistic about acquiring the knowledge needed, and so he was led to acquiesce in an extreme conformism.

In sum, Moore's ethical theory represents the extreme rejection of science from ethics. Science is irrelevant in dealing with intrinsic value and helpless in dealing with duty. Moore's ethics thus gives us a very accurate and insightful picture of what an ethics without science may be like: helpless conformity to tradition in matters of duty, intrinsic value reduced to isolated reactions in terms of attractive quality. So far from constituting a proof of the irrelevance of science to ethics, it seems rather to be an effective demonstration of the plight of ethics without science, and so to call for the rejection of its own conceptual structure for ethics because of the very results that ensue from its use in ethics.

Take one important example of such a theoretical result. It is rarely noticed that Moore's construal of intrinsic value as a quality apprehended when the object is held before one's mind as a total world by itself, made the judgment of comparative value impossible and almost meaningless. How could we judge whether beauty has a higher or lower intrinsic value than friendship? If I envisage beauty as a world alone, I apprehend directly its intrinsic value; similarly for friendship. But how compare them? If I put the two worlds together, I have the contradication of one world which is two. If I put the objects together in one world, I get a single value reading, say, the value of a beautiful friendship, or beauty and friendship side by side. The only way out is to have the degree of value apprehended directly in each alone, and the readings compared, and Moore does speak of seeing a high or low value, as when he says that knowledge has a low intrinsic value though it adds greatly to the value of wholes in which it is combined with other things.[4] But if so, it is not simply intrinsic value that is a simple quality, but value in such-and-such a degree. And if degrees of goodness can be distinguished on a scale of n, we have n simple qualities. In the light of all that has been done on ordering of preferences by scientific attempts in different fields, Moore's concepts for ethical theory have simply not had the fruitfulness that warrants acceptance. When he said that good was as simple as yellow, it is a pity that philosophers did not go to the sciences that study color, and find out how complex yellow really is, what inner orderings are possible in it, how far it is either perceptible or meaningful without some context, today perhaps even what kind of op art is possible with yellow alone! If lessons emerged of the ways in which yellow depended on context, they might have been directly applied to good.

How strictly do the inadequacies of Moore's structure follow from its neglect of science? The difficulties about comparative value do seem to come directly from the isolation procedure that is a consequence of the way he construes intrinsic value in isolation from the relations that science would study. The element of conformity in duty may perhaps be altered by tinkering with Moore's definitions: for example, if we changed his definition of duty to refer to the effects within a limited part of the world over a limited stretch of time, or if we introduced qualifications specifying that duty be determined on the best available knowledge rather than ideally complete knowledge, perhaps the helplessness might be reduced. It may be suggested as a possiblity that Moore's own belief about the lack of utility of scientific knowledge is itself a lesson of science, that he is using science itself to yield a kind of conservatism when he says that attempts to change rules are never warranted. If so, then the helplessness is a consequence of the specific scientific results reported, and might be altered if those results proved unreliable.

One further argument is worth noting because this style of query is likely to reappear at many points in our account. How, it will be said, do we justify a rejection of Moore's structure because it yields isolated attraction for intrinsic value and conformity for duty? Are we not simply assuming that these consequences are undesirable or bad? What if someone actually preferred a life of isolated responses about the good and strict tradition for duty?

Now it follows from what was said above concerning the permanent possibility of evaluation, that anything can be made the subject of evaluation. Hence we would expect that any orientation imposed upon us by accepting Moore's structure could be subject to evaluation. Such possibility of evaluation is a necessary consequence of our analysis rather than an objection to it. The crux is not then in the fact of such evaluation but in how it would be carried out. Could it be done satisfactorily by Moore's procedures, for example, judging a rule of conformity by a higher order rule of conformity to conformity (as conservatism includes among its rules that one be conservative!)? Or would a more satisfactory evaluation of conformity require as an indispensable part a psychological study of the conformist attitude, and a historical study of the workings out of various forms of traditionalism? Of course any further values embedded in the idea of "satisfactory" evaluation should be rendered explicit. The type of possible circularity here, of a nonvicious kind, is familiar from the parallel procedure concerning assumptions in scientific progress: they should be rendered explicit, and may in the context of a fresh investigation be examined for their worth, the fresh context admitting of other assumptions. All that is surrendered in such procedure is the guarantee of absoluteness.

I am not aware that Moore has been looked at in this way in twentieth-century ethics—as the supreme model of the plight of an ethical theory that rendered science either irrelevant or bound to helplessness. Therefore I have spoken of it as an unorthodox interpretation. But it seems to me so obvious a one that once we see the effects of the conceptual structure in ethics, so far from saying that Moore has proved science to have no role in ethics, we must look realistically for what role it has and what further role it can be given.

Science and Moral Judgment

By "moral judgment" I mean here the kinds of assertions that are made in substantive morality or normative ethics. The usual view is that science tells us about means to ends that we aim at, but cannot help us in the selection of ends. This is a difficult, often confusing formulation. For one thing, means and ends constitute only one categorial way in which the factors in moral judgement can be construed. The history of ethics, as well as ordinary use, shows us also ideals and their institutional embodiments, intrinsic experiences and the conditions of their occurrence, systematic wholes such as an organized whole mode of life and its constituents. When one attempts to stretch means and ends over these varied types, one often gets strange results. Examples from these alternative formulations also show a second important point—that it is often very difficult to separate out means and ends. Try to sort them out in, say, an ideal of liberty, or in a systematic notion of the good life. The paradoxical point is that it requires a scientific study of means and ends to learn how far they are separable! But more of this later. Now let us adopt the formulation of the issue in the usual slogan, and see what science gives us when it gives us means, then ask how far it can muscle into the consideration of ends.

In the original and limited sense, a means to an end is a step which taken produces or helps produce the goal at which we are aiming. To discern the correct means thus involves knowledge of causes. We expect the science to give us better and more reliable as well as more extensive knowledge here. If we wish to curb inflation, what but economic science will tell us whether to regulate interest rates and on what principles? If we want to know whether certain types of threats against the people of North Vietnam will produce intended effects, surely the South-East Asian scholars who have taken an anthropological view of the character-structure of the people can make sounder predictions than one might have by introspecting on how we would respond. In short, it is obvious that science can help furnish better grounded beliefs about *instrumentalities for limited ends*.

In addition to helping us find effective means, science can help us trace the *collateral and subsequent consequences* of the use of means, beyond the fact that the end is achieved. The threats may secure the desired effect on the people of Vietnam but making them may have undesirable effects on the people of Europe. Nor can we stop with the effects of using the means. Even the effects of achieving the end have to be considered. Suppose someone offers taking LSD as a means to a certain kind of psychic experience. Who but the scientists of the mind and the body can tell us what fully goes on in that experience (including collateral happenings and their interpretation), and what the effects of the experience will continue to be?

If brief, when it is said that science gives us means, the package surely includes consequences as well, and consequences of achieving the end as well as consequences of pursuing the end by that means. We ought to add here unintended as well as intended consequences. In many cases, only an advanced science can discover these; for example, the unintended genetic effects of nuclear experiments carried on as instrumental to the end of, say, national security.

Whether one will insist on assimilating the concept of consequences to that of means does not really concern us here. It could be done because the notion of means and ends has considerable ambiguity. Sometimes means has the more limited meaning of prior causal steps; here consequences are not prior and so not means. But sometimes an evaluative use of means emerges more sharply, and means becomes assimilated to extrinsic or instrumental value; hence consequences are part of the costs of the end, paid subsequently on an installment plan. If the view that science deals with means allows us, as it obviously must, to have science help discover the events which are costs, then the role of science is extended. For to investigate costs permits also the investigation of gains, that is, of further ends which are consequences. But, it must be noted, science is here still precluded from making the decision whether a given consequence is a cost or a gain; it discovers simply whether it occurs and leaves to others the labelling.

To have extended the scope of science up to this point has already, it should be noted, allowed it to deal with ends in the sense of exploring the consequences of *holding to or pursuing an end, or achieving* an end. Suppose the question is whether to hold to absolute victory and unconditional surrender as a goal in Vietnam; a team of scientists would have the task of working out the consequent increased destruction, prolongation of war, political and economic and moral or character effects, and so forth. The purist would add that an independent value-judgment is required to affirm the destruction of a whole people to be an evil. An alternative move would be to point out that victory is both end and means rolled into one; an end in the sense of furnishing a sense of triumph or achievement (the value of winning), and a means to securing an abiding peace and the safeguarding of presumed national interests. (The scientist is given a green light on the means aspect.) We have here an important recognition that many ends are in fact instrumental as well, or else they are sub-ends in a hierarchical end scheme. But if so, they have a quality of partial means and are in that respect fit subjects for scientific concern. Only pure and ultimate ends are beyond the pale. In a tight hierarchical system, science would be unemployed in ethics only at the top.

But even the end-elements on the way up present interesting scientific questions. What, for example, is this sense of triumph which is the end-value in victory? Why do we want to win in a war, rather than stop the war, instrumental gains aside? Is there an end-value in sheer coming out on top? If so, does it lie in a sense of achievement, or in a humiliation of the other fellow, or in a reassurance of our own worth, or what? Note what is happening in such questioning. There is the assumption, or perhaps better seen as a major scientific discovery of modern psychology, that our holding of ends has a definite role within us, that it plays some organizing role in our internal economy, and that its quality in our consciousness is in part tied up with, and varies as, the role it plays. Dewey goes so far as to take the end-in-view to be itself a means for organizing and focusing our energies to secure their expression; the end is thus evaluated as means. But apart from this extreme formulation there is involved a basic recognition that the holding of ends by individuals—and similarly of social ends by groups such as nations—always takes place in the context of what Dewey calls a problem-situtation of interests, needs, pressures, demands, claims, in which the achievement of the end would function as a mode of resolution of conflicts and difficulties and release of impediments. I state this in general terms so that it will admit of both psychological and social interpretation. Such a mode of analysis reflects a scientific discovery in the psychological and social sciences about the condition of the occurrence of ends in particular individual and social life-histories. It need not be taken as derogation of the idea of intrinsic value. How can it for it has been declared neutral to that? Science can therefore be of use in ethics in helping to determine which of the proposed ends, if held to, would by its pursuit help solve what problems.

A further use lies just beyond this one. If science studies the functioning within of goals held by the individual or group, it is also in a position to understand how strongly they are held, whether they are peripheral and conditional, or central and integral to self or group identity. And so it can contribute to the old Socratic quest expressed in the injunction "Know thyself." In Sartre's example of the young man who wanted his advice on going to join de Gaulle vs. staying under the Nazi occupation to take care of his mother, Sartre asked whether the man's coming to him for advice did not already indicate that he wanted a particular answer.[5] It is of course quite possible that resistance to the Nazis was the man's guiding value. But what if it was to get away from his mother? A recent press item says that Venezuelan guerrillas use psychiatrists to check on high school students who wish to join them; presumably part of the checkup is to see what they really want, as well as what they can bear.

Certainly we have come to expect from depth psychology understanding which will enable us to construct a scale from whole-hearted conscious desire at one end, to unconscious desires at the other. And similarly, we may expect from the social sciences, in their study of the functioning of group goals, and in their analysis of ideologies and utopias, some help in distinguishing an authentic national goal from one that serves some special, often partial, purpose. If knowing oneself is a central effort in the search for the good, it makes all the difference whether a man has a clarified stand on what he stands for, or just a feeling of internal pressure in a given direction. Some aims are given up when we know what we were looking for in them. Others are more effectively pursued. And some goods, however good in general, may not be good for particular persons. It is no easy task to find out what goods a given self might not be able to take.

Let us see what uses we have so far found. Science may help us in furnishing the means to an isolated end, the consequences of the means, the unintended consequences of the use of the means, the consequences of achieving the end, foreseen and unintended, the consequences of holding to an end and pursuing it, the role of the end within the person or group holding to it, the extent to which pursuit of the end will resolve problems to which the holding is instrumental, the depth and quality (or genuineness) of the holding itself.

This is a formidable array of uses, but we have only begun. So far we have spoken chiefly of a limited end; in fact men hold many ends at once, and serious moral issues arise from their conflict. Can science help here? In one very obvious sense it is indispensable. Opposition of ends is not a sheer logical problem like contradiction. Ends are incompatible because they cannot be achieved together, not because they cannot be desired together. For large-scale complex ends, only a scientific inquiry can determine compatibility or incompatibility. Is the goal of individual success in a profit economy compatible with the goal of cohesive social sympathy? Assuming a clarification of both goals, it is a question of psychology whether the institutional requirements of the former have an educative effect which militates against the latter. (It is often said that in our society the schools educate for one, actual life for the other.) Or again, is the ideal of liberty compatible with that of equality, or that of security? A full study in political science is required to see whether we have real incompatibilities or only conflicts of specific forms which these ideals have taken at particular historical periods.

In this area of apparent or real conflict of ends, knowledge of means plays a great part in the structuring of a moral problem, that is, whether one recognizes the presence of a problem, whether it is formulated one way or another, whether it breaks up into distinct and separate problems, and so on. Knowledge of a means may shut out concern with a moral issue by making it inapplicable. This is obvious in many nonscientific contexts. John Ladd gives a striking illustration of this when he reports his inability to make his Navaho informant face up to the moral question whether he should divide his water supply with a man he meets far out in the desert, when there is not enough to see both men through. A too ready "yes" turned out to rest on knowledge of desert plants that could always be used to alleviate desperate thirst![6] Science and its application shelves many an issue by undercutting it and making it as irrelevant as whether to shout a warning "fire" in a crowded theatre in a society where all theatres are fireproof.

Perhaps more morally significant are the effects of scientific knowlege in *restructuring* a moral problem so that it is seen as a quite different one. Economic knowledge in the Great Depression at last did this to the sense of individual responsibility for unemployment. A man stopped asking guiltily what was unworthy about himself that he could not provide for his family; the full picture of the economy and its functioning restructured the problem as the obligation of the society to maintain a reasonably high level of employment. Or again, what is felt as a single problem may be unravelled as a consequence of applied science into several problems; as the applied knowledge in techniques of birth control helps separate in the domain of sexual morality the issues of having children, sexual enjoyment and companionship, social family organization. (It also alters the pattern of sanctions, but this is a distinct matter from structuring of the moral problem. We shall not have time to look at this interesting problem, as where the introduction of knowledge of germs into a society that always thought of illness as a punishment for immorality may force alteration in the sanction system.)

It is only fair to note that science may create fresh moral problems as well as shelve some, by the same procedure of increased control. A clear case is the moral problem likely to emerge when genetics offers means of determining to some degree the character of subsequent human stock. What features are to be encouraged as desirable?

In light of the inroads that science is able to make in the concern with ends, what remains from which it is excluded? Is there any problem of ends that still remains beyond its reach. Not, of course, the mere existence of the phenomenon of a man having a given end, but the deliberation about and the justification of ends?

There is, of course, the judgment of intrinsic value that Moore insisted on, the recognition that something is good in itself, when all else is excluded or detached from the object judged. Many a different ethical theory might say that it welcomed all the information on how science helps moral judgment and would welcome still more. But it would regard all this as (collectively) ancillary labor. The last word about ultimate values would remain beyond scientific reach. Intrinsic value, however, is an elusive quarry which moral philosophers have pursued without much success. Moore, it will be recalled, protected it with the logical gap laid down by the naturalistic fallacy, which had the effect of placing the intrinsic value (or in later modes, the ought) somehow beyond. Subsequent analytic ethics has worried a great deal about what the gap consists in and what constitutes the unique element on the far side; apart from Moore's intuitive grasping or vision, it has been interpreted as a morally charged definition of "good," an emotive expression, an act of will or commitment, an acceptance of guidance for practice, a practical act of commending, and so on.

To be so elusive, intrinsic value (or its obligation counterpart) has to be either extremely simple or extremely complex. Neither course escapes the reach of science.

If it is some kind of simple, then its cash-value in moral judgment, no matter what its epistemic or ontologic status, turns out to be the criterion of *attractiveness* in evaluating an end taken by itself. This fits precisely the picture that Moore gives (or that phenomenological accounts have offered), including even the inability to give a justification other than seeing. But in that case, nothing hinders scientific study from considering the conditions of attractiveness for given classes of individuals, the nature and quality of attractiveness, projective elements in attraction, the role of learning in attraction, and so on; nor from experimenting on variations and determinants and the dependence of specific qualities on specific configurations. Bentham's attempt to establish intensity as a dimension of evaluating pleasures,[7] or Ralph Barton Perry's treatment of intensity as a degree of arousal of the organism,[8] were philosophical attempts heading in the same direction.

If, on the other hand, intrinsic value is a complex concept, then it is perhaps best at this late stage in the history of ethics to confess that it admits of no easy decomposition or unpacking, that it is a highly theory-laden concept, serving as a coping-stone in diverse theoretical structures, and that it gets its meaning from the theory or model associated with it. This alternative in dealing with the concept seems to me much the more probable one, in the light of the history of ethics. It is because we have confused a highly theoretical term with a simple observational term that we have raised such difficult problems. This confusion is more significant in the long run than our inability to find pure intrinsic values of an ultimate Moorean sort.

On this analysis of the issue, there is again no simple answer to the question of the reach of science in dealing with intrinsic value. The path of investigation follows the path of analysis of the theories associated with the concept in its diverse uses.

Scientific Study of the Nature and Tasks of Morality

Every ethical theory includes somewhere or other some conception of the nature and tasks of morality. Our concern here is particularly with the tasks; the very concept of nature would take us too far afield. The conception of tasks is usually derived from common observation of the role that morality plays in human life. Although it does not at first sight appear to be a normative conception, it often involves or specifies criteria for the satisfactory functioning of moral regulation or moral discourse. For example, it is regarded a truism in contemporary analytical ethics that the purpose of moral discourse is to guide conduct, and so no analysis of moral discourse will be satisfactory which does not show how conduct is being guided. I shall not here embark on a study of the variety of roles for which morality has been cast in the long history of ethical theorizing. My problem is rather the part that science may play in such questions. But let us remember that the two characteristics noted about statements of the tasks of morality are present even before science comes on the scene. The first is that to assign a task to morality is in some sense to find out what morality is up to; it is a statement about morality which is asserted as true or correct. Secondly, it becomes the basis for criticizing how a particular morality is being carried out, or again, how adequate a particular ethical theory may be in its analysis of the morality. When science attempts to study the functioning of morality, and to develop consequent criteria for evaluating morality, it is only extending a job that common observation and reflection have begun. Not only can it hope to do it more intensively, but it can, we shall see, even help make clearer what have been the normative and nonnormative elements in the process.

The scientific study of morality—describing its phenomena, taking it to pieces, seeing how the parts work, what configurations they form—is thus a perfectly proper job of the biopsychological and sociocultural sciences. Perhaps anthropology has borne the brunt of the job. Considerable work has been done on it though not perhaps enough as yet. Its results, to repeat, are scientific propositions about morality, moral configurations, relations or moral codes and moralities to other aspects of human life. They are not moral propositions. And yet when the knowledge is accumulated—even when the outlines of the kinds of knowledge possible become distinct—this growing understanding of what enters into a morality, how it is organized, how it functions, is seen to be a *sine qua non* for formulating criteria for a regulative critique of moralities. Of course values enter into the process of formulation, but it is not as if the values came first fully developed and the knowledge simply found the means. The values themselves are systematically clarified and articulated in the process. The comparison with medicine is helpful. The initial values for medicine may be the desire to continue living, to avoid pain, and to continue functioning in some general sense. The knowledge furnished by the sciences of the body elaborates a comprehensive ideal of health, with varieties of prescriptions and hosts of indices. So too we develop the ideal of a well-functioning morality and articulate criteria for evaluating moralities in part and as wholes. In fact such a task is forced on us in the modern world by the variety of moralities in different peoples, the conflict of moral principles, and the pressure for choice and decision. Nor is our ordinary language without expressions that in some sense suggest that we are evaluating morality in part or whole. We sometimes speak of our sexual morality as being a failure, of such-and-such a morality as being one-sided or too rigid, and so on. What criteria concerning the nature and function of morality are we employing and what values are embedded in the criteria? Perhaps common-sense ones, such as that most people would find it impossible to live up to the morality, or that desirable institutions are militating against it, or that its concepts are unclear and confusing, or in some sense unrealistic, etc. Careful analysis may help sharpen some of these criteria and embedded values; growth of psychological and sociocultural and sociohistorical knowledge clarifies or stabilizes others. But even more basically, the whole outlook that is ready to think of morality as having psychological and social functions is a product of the growth of psychological and sociohistorical knowledge. Once again, it need not necessarily be in conflict with other judgments of intrinsic worth in morality, though these have to be established on their own. It can be seen as the elaboration of additional dimensions for evaluating moralities and moral elements, for example the dimension of social utility in the light of social functioning for evaluating psychological processes, and psychological utility in the light of psychological functioning for evaluating social processes and institutions.

Many moral elements turn out to have this girding or supporting relation to social forms and institutions. It is clear in virtues; for example, the great stress on punctuality in an industrial society that requires precision, honesty in a scientific enterprise or credit institutions, hospitality in primitive societies that have tenuous external communications. Where the particular social utility is not required, we may find the virtue ignored, or even its opposite praised, as the cunning of the wily Odysseus is praised even by the gods in Homeric morality. Sociohistorical studies of a whole moral pattern, such as Puritan ethics, in familiar writings such as those of Weber and Tawney,[9] have made us conscious of the way in which the dominant shape of a morality—its particular set of rules, selective cluster of virtues, and so on—may be geared to the demands in economic and social problems of a given day. Sometimes this knowledge gets built into a critical concept of "ideological" as a way of distinguishing the less authentic, more projective, from the authentic and more insightful and realistic. Sometimes the social role of the morality is made integral to its criteria of self-judgment and we may evaluate a morality in part by how it answers vital social issues.

Comparably, a psychological concept of maturity becomes increasingly useful with the growth of the psychology of personality. Since morality involves internal control of impulse, it follows that psychological study of the growth of impulse-control, its modes and patterns in relation to general psychic functioning is not without relevance to the understanding of moral processes. Of course, again, there are values embedded in the notion of effective psychic functioning, but these may be readily agreed on in terms of indices of ability to function without overriding anxiety, or breakdown, and so forth. When the discovery of a developmental pattern and these broad values come together, it becomes possible to think in terms of successive pitfalls which a personality encounters in the course of its growth, and of successful outcomes in typical maturation. Specific criteria of maturity may thus be fashioned, fusing the knowledge of steps and dangers with the broad goals; examples of such criteria are the progress from unconscious to conscious, or from fear to security, or from dependence to self-determination, and so forth. The criteria are neither wholly physically compelled, nor sheerly evaluative (as, say, a desire for the novel as against the familiar which is sometimes slipped in on the assumption that the desire for the familiar represents a kind of regressive clinging!); they represent rather an articulated scheme comparable to the articulated ideal of health referred to above.

The combination of psychological and social analysis, coupled with further scientific knowledge, occasionally makes it possible to undercut a whole value system and not to take it at its own estimation. The clearest illustration is the so-called Nazi ethics. Its rejection comes not from arbitrary disagreement with ultimate values of adventuresomeness and impulsive strength, but from taking it to pieces, seeing the falsity of many of its assumptions, seeing the distorted character of many of its psychological elements, and so on. For example, the well-known studies of the authoritarian personality, in spite of serious methodological issues and the influence of value elements in their shaping, show the kind of impact scientific knowledge may have on norms for evaluating moralities. Interestingly, Sartre's "Portrait of an Anti-Semite" [10] grasps the same central point in an intuitive phenomenological way, that anti-Semitism served not as just a particular set of beliefs or attitudes, but as a focal characterization of a personality governed by fear and self-hatred.

In general there is no reason why the scientific study of morality should not in the long run quite revolutionize our concept of morality, just as has happened to politics and law and religion and education. It does this by making us sensitive to the kind of jobs that the morality performs in a culture in relation to the whole of the life of the culture. It is not pertinent to the argument that this be a reductive functionalism which says morality "is really just this and this alone." That is a quite distinct issue, just as in the comparable developments of the scientific study of religion the existence of God is a quite separate question from the scientific determination of the projective elements that enter into traditional images of God, or the social analyses of particular uses of historical religions. So too in morality. The jobs turn out to be psychological: morality involves developing specific patterns of impulse control charged with specific types of emotion. Or the jobs turn out to be social: morality undertakes the burden of holding back aggression in the society through the use of specific internalized sanctions; or it transmits through educational mechanisms specific patterns of desired character to a subsequent generation; or it guides the society to typical patterns of goals and activities; or it furnishes principles and modes of individual decision in certain classes of cases, and so on. So far from reducing the ethical, such scientific investigation of morality may widen the scope of normative ethics. For it brings into consciousness connections hitherto unconsidered, so that if they prove subject to control they now constitute something that can be changed, and the new question is whether they should be changed. This is obvious in ordinary matters; for example, if we discover that some of our moral rules are supported by subtle threats of punishment, we may want instead to shift to the promise of rewards (or demonstrations where virtue is its own reward). But it enters increasingly into larger theoretical issues. Compare the older treatment of conscience with that found in recent psychological writings on ethics. Conscience used to be a faculty of determining with epistemic certainty what was right and what was wrong, expressing a ruling element in our nature. The chief problem was to hear its voice distinctly in the clamor of the passions. With modern inquiries into its origin and development and realization that it is constituted by a patterning of specific emotional development, questions are raised whether there may not be different types of conscience—e.g., rigid authoritarian and democratic humanistic, each with its own psychological etiology. The implication is that with sufficient understanding men may build one rather than another type in subsequent generations—hence a broadened scope of decision. And what holds in the domain of conscience holds equally in the kinds of prescriptions a culture builds into its morality—inflexible guilt-laden, or flexible utilitarian.

In all such cases, a scientific study, going beyond common reflection, is able to see what is capable of alteration and thus is able to pinpoint where normative decision is required, even if only to maintain the same path that has previously been trodden. In this way, the discovery of causal conditions may open the way to need for moral judgment where none was seen or even feasible before. This may take place even with regard to basic assertions of the functions of morality, for it becomes possible *thereafter* to ask whether fresh functions should be taken on or some older ones abandoned. The logical character of such a process which starts with discovery of existent functions, elaborates in terms of them criteria for the assessment of the morality, and ends with a call for normative evaluation of the very bases from which the criteria were developed,[11] is obviously an extremely complex one. It does injustice to it to dismiss it as mere psychology or sociology, just because it begins with psychology or sociology; or to dismiss it as arbitrary value judgment just because it ends with a normative assessment. In between lie processes of criteria-formation which as we saw in the case of the criteria of maturity, carefully fuse broad values with responsible knowledge. There are also logical processes of concept-formation, into which the criteria are built. There are empirical elements entering into the establishment of criteria of effectiveness for the functions accepted. And finally, there are normative processes of the broadest sort, joining major values with growing knowlege if criticism is directed on some initial task of morality itself—as when, for example, some functions assigned to morality are shifted to medicine; or other functions assigned to law are shifted to morality. Ethical theory has suffered much in the past from the failure to grasp the great complexity of its own processes and from an over-dogmatism in reducing them to simple value-commitments as contrasted with empirical knowledge.

Uses of Science in Metaethics or Theoretical Ethics

There remains the question of theoretical ethics or metaethics—analysis of ethical concepts, their relations, justification procedures, and so forth. I do not propose to consider the whole image that metaethics has had of itself as neutral to values and pure of empirical elements. My concern is with the latter only since I am tracking down the use of science in this area.

Let us take the kind of items just listed, beginning with analysis of ethical concepts. Consider, for example, analyses of the good. Most concepts of it tie it to some human goal, property or process, either by identification or some looser logical bond, whether it be happiness, pleasure, desire, purposive striving, or guiding and commending. It requires little insight to realize that the analysis of the concept tends to follow the structure of the materials to which it is related. A clear instance can be seen in Bentham: he assumes that all goods are comparable because he assumes that all pleasures are homogeneous. But how are such concepts as pleasure (or purposive activity, etc.) themselves to be understood? Either in terms of an older introspective psychology, or in terms of a more developed psychology which interrelates physiological, introspective, depth accounts and functional studies. (By functional, I mean such studies as, for example, assign to pain the function of signaling illness). There is no understanding of the concepts without any psychology. You do not escape probing for underlying psychology by simply following linguistic usage. If you distinguish meanings of "good" as good-of-its-kind, and good-for, and specify the logical properties of each, so far so good; but to get a deeper understanding you need to study men's activities in fashioning or determining kinds and in using instruments. The same holds for guiding and commending.

For an example of the relation of concepts, take the familiar controversies over the relation of means and ends. If intrinsic good is identified with ultimate end, the theory of the good will take shape from the study of ultimate ends and the kinds of ultimacy to be found in human thought and action. Thus Aristotle has ends as the separate objects of wish, with means the object of choice, and ultimacy belongs to the end in the series that occupies the last point and has no instrumentality whatsoever. Dewey extends the domain of choice to cover ends and sees a thorough fusion of means and ends. Fundamentally, it is because his basic psychological picture of man is different. He thinks of human motivation in terms of an integrated reconstruction of a situation in which conflicting tendencies have arisen, and he refuses to separate the cognitive, volitional, and emotive into distinct entities. Whatever the complexities of doctrine involved, the theory of ultimate ends will reflect the theory of psychological processes. Ultimate ends for Dewey could only mean those large and permeating goals which gain their strength from the multitude of aims they support and organize and by which they are continually supported. An ultimate end could not be a sheer desire or isolated will-act.

Nor is the outcome any different if we consider the problem of method. Every general methodological claim has its underlying scientific picture.

Bentham's claim for an inductive methodology in ethics rests on the view that there is a clearly specified end—pleasure—sought in all human actions. Therefore moral judgments report reliable means for achievement of this end. Emotive theory rules out deductive and inductive method in ethics because in its theory of the emotions the relation of belief and attitude is taken to be contingent. Emotions are treated as attitudes that may be affected by beliefs but not in a dependable way; ordinary life examples are invoked as evidence, based on roughly an introspective psychology. If the psychological theory of the emotions is altered, the methodological view requires reconsideration.

Perhaps the most striking illustrations of the reach of science in theoretical ethics are those in which we are not merely forced to give different answers to the ethical questions when scientific results are changed, but in which we are forced to change the questions themselves. This is clearly seen in the history of the discussions of egoism in ethical theory. Arguments in older formulations varied from attempts to show that rational egoism yielded altruism to G.E. Moore's one-paragraph proof that egoism is logically self-contradictory![12] When advancing knowledge turned on the question of what an ego is like, how it develops, what shapes its pattern, the form of the question changed. There were experiments in task completion, for example, which showed the existing differences between people who are task-oriented and are satisfied if someone else completes the task on which they have been interrupted, and people who are not satisfied unless they complete it themselves. Or again, psychological studies differentiated personality types and their etiology, for example, people who have a passive-receptive orientation, and people who show an active consideration for others. Social studies that began with contrasts of cooperation and competition found a proliferation of concepts necessary according to modes of work, types of motives, and so forth. In general, then, the question whether man is inherently egoistic or altruistic gives way to the normative question of deciding desirable types, for example, whether to encourage individualistic self-assertive attitudes or socially cooperative ones, and so on.

Many more questions of ethical theory than we are prone to think may admit of such refinement and reconstruction. There are some signs that analytic approaches may be moving in such a direction. So at least I interpret Rawls' illuminating treatment of promising as a practice—in effect an institution—in his effort to cut through the intuitionists' claim of self-evidence for the rightness of keeping promises.[13] Why have promising? This means why have such institutional procedures, and is answered by a utilitarian reckoning of the good consequences of having such a practice. Why do I have an obligation to keep this promise? To promise is ipso facto to take on an obligation; therefore the question can only mean whether I have proper excuses for dropping it. And so on. What the analytic philosopher works through to with great effort and at the end of an epoch in which it is realized that language itself is a system of practices, was long known to the anthropologist and to philosophers with a sense of history. Nietzsche, for example, in touching promises, gets directly to the question under what historical conditions promising makes sense; it presupposes that a man can know what he is going to do in the future, therefore sufficient regularity in his life.[14] Hence it can arise as a practice only with some degree, he says, of the domestication of man! But then Nietzsche always did complain that philosophers lacked a sense of history, and no man could understand moral philosophy who had no realization that there had been many moralities.

Let us try to formulate more systematically the sort of relations between science and metaethics or theoretical ethics that we have been considering. We are obviously not dealing with so simple a relation as that of scientific premise to ethical conclusion. It is rather that of a scientific presupposition involved in the very formulation of questions and statements in ethical theory. But a general reference to scientific presuppositions will not do—we have to find the precise points of entry of scientific materials into statements of ethical theory. Perhaps the simplest way to begin would be to collect the usual nonethical terms that appear in the discourse of ethical theory—e.g. "aims," "ends," "purposes," "feelings," "desire," "choice," "want," "law," "self," etc. These terms function in diverse ways: sometimes in definitions of ethical terms, sometimes as interpretations of ethical terms or as indices for their applications, sometimes locating the phenomena or situations in which morality is applied, sometimes as constituents in the statements about human life that are offered as justification for theses about the nature and tasks of morality, and so on. There is no uniform or standard way in which they function; it varies with different ethical theories. Now most of these are terms of psychology—whether individual or social—and sometimes of biology or sociology. And while any given term may be dispensed with in a particular ethical theory, any theory that had no such terms at all would rob itself of relevance to human life—as Moore's theory in fact did. If then we more or less formalized an ethical theory, we could specify which variables had such terms as their values. These variables might then be regarded as placeholders to be filled in with terms whose explication was essentially scientific. There is no avoiding such entry of scientific materials. The question is whether it is good science or bad science, yesterday's science or today's science, the best available science or hangovers from common sense precursors of science.

That the nonethical terms have an influential role in ethical theories can be readily seen by inspection. Take, again, the theoretical consequences of the concept of pleasure in Benthamite ethical theory in the light of the particular psychological approach taken to the concept itself. If pleasure is taken by an introspective psychology to be limited to individual consciousness, the ethics is individualistic, and the idea of a common good has only an aggregative character. Again, since the good is identified with pleasure and pleasure lies in individual consciousness, and since pleasure so construed is subjective, the individual is his own final judge of his good, without appeal beyond. Since Bentham regards pleasure as homogeneous, he regards goods as summable. If, on the other hand, the psychological approach to pleasure is behavioristic, preferential acts become the units of inquiry, the theory of value measurement is revised correspondingly, the concept of group preference is an easier construct than that of group pleasure. Again, if pleasure is regarded primarily as a signal of functioning well, a more biologically oriented conception may take over, and the ethics go in for ideals of health rather than maximal accumulation. These suggestions are cursory, but I hope they indicate the way in which an ethical theory is shaped by the scientific materials that enter through the variables we have pinpointed.

The aim in such an undertaking is not merely to find the relevant scientific commitments of the ethics, but to learn how changes in the ethics take place as changes come about in the underlying assumptions. In becoming conscious in this way of the role of science in ethics we are enabled to develop a regulative critique of our ethics. As scientific results change—which in the present growth of science in general and especially the anticipated breakthrough in the sciences of man, we may almost take for granted—a developed critique would enable us to plug in the new results, as it were, and look for the changes in ethical theory that are desirable.

Perhaps this type of formulation is not essentially new. Obviously a great part of the naturalistic tradition in the history of ethics recognized that there were psychological and social commitments of ethical theory, and often began inquiry by setting them forth—witness Hobbes or Hume, Adam Smith or Bentham. And in spite of different commitments they all shared the hope that once men saw the role of the underlying psychological processes in the operations of the morality, morality would be carried on more smoothly, more consciously, and get into fewer difficulties. They thus hoped that knowledge of man's nature, revealed by the relevant sciences, would help provide a critique for moral judgment in the obvious sense that a man who is conscious of what he is doing can see how to do it better. Because they assumed a permanent structure to human life and endeavor, they seldom raised the question whether a man who became conscious of what he was doing and saw how he might do it better, might want to do something else instead! The issue, of course, is whether he is unavoidably bent on that undertaking, or whether it is changeable. I do not wish to create a gap in principle between this earlier naturalism and post-evolutionary naturalism with its greater sense of change and its greater openness to varied reactions. How permanent a structure there is in human endeavor is itself a scientific quesiton, and the character of men's reaction to the consciousness of what they discover is also a scientific matter. The idea of turning the consciousness of human processes into principles of a critique of those proceses is central to both the older and the present inquiries. But we must have a keener sense of the values that are built into this transition and their justification. It is a fitting paradox that the growth of a scientific study of ethics and a search for the scientific commitments of ethical theory should itself turn out by its results to promote a sounder conception of the *autonomy* of morals. But autonomy means here not isolation, but an actual discovery of the wide limits within which human choice is capable of determining outcomes; this in turn generates a demand for a concept of rationality in terms of which decisions may be criticized and evaluated.

Concluding Reflections

In these concluding reflections I want to push ahead beyond what I have tried to do so far. I tried to set the terms in which the multiple relations of science to ethics could be explored and to show how the program of such exploration could be initiated. I said at the outset that this paper was programmatic, but I hope I have shown that this particular program has the advantage that it can be put to work, not merely argued about, and that it renders obsolete much of the general controversy about science and ethics that has held the center of the stage. But I want to go further and suggest that the carrying out of the program will extend a breakthrough in the treatment of some of the general categories and general issues that have confined work in ethical theory. We have already noted the result in considering means and ends—the recognition that the utility of the means-ends categories, the extent to which they can be sharply separated, and their scope in ethics as compared to alternative categories, are all issues in which fruitful analysis requires a scientific study of the human processes in purposive behavior. It is the results of this study which show what theoretical use may be made of these categories. Similarly, in the paradox mentioned above concerning moral autonomy, it is the results of the scientific study of man which determine the scope of possible autonomy and the domains of significant choice. I should like to extend this type of analysis to two important theoretical problems. One is the question whether scientific method can be used in ethics. The other is the question of the scope and utility of the fact-value distinction in ethics.

The question whether scientific method can be used in moral judgment is sometimes made to seem the sum and substance of the question of science and ethics. Actually, it is a derivative question. In one sense, nothing can stop someone from trying to use scientific method in ethics. But trying is not equivalent to succeeding. Certain conditions have to be satisfied, such as a minimal stability of the phenomena, some degree of generalizability, and so forth. Of course, the best way to show the method can be used is to use it successfully. I think that a good case can be made for a Deweyan type of use, at least in many types of problems. But it is very easy to conceive of a possible discovery about man which should render the use of the method inapplicable: suppose we found that man was basically irrational whenever he approached serious questions of interpersonal relations—just as Hobbes thought divisive property interests, if they cut across, would affect thinking even about Euclid! It might then be a discovery of science that scientific method could not be used in ethics. I do not wish to be deprecating the use of scientific method in ethics by hypothetical considerations; I want simply to make it clear that this is a dependent not a primary issue in the relation of science to ethics.

Now on the question of the fact-value distinction. I stressed at the outset that this was not equivalent to but an interpretation of the distinction between fields—of science and ethics; it assumes that science gives us facts and ethics gives us values, without questioning seriously whether both fields contain both facts and values. The distinction of fact and value has, as we all know, been one of the primary handles for keeping science and ethics apart. And this has troubled many philosophers. They counter with different moves. Some will see science itself as a value enterprise, or at least as committed to some values. C.I. Lewis finds the problem of right decision pervasive and therefore that there are prescriptions in all domains of knowledge, not just in ethics. Dewey gives an instrumental view of knowledge that is equally applicable to ethics. Some recent analytic writers look to the domain of institutional facts with built-in values as an intermediate zone. But I wonder whether we really need general moves to solve our present problem. We can instead recognize that a distinction of the fact-value type is a fashioned categorical one oriented to the solution of certain problems; it may be useful in some contexts and obstructive in others. It is useful to maintaining the permanent possibility of evaluation and in preventing values from being smuggled in and imposed on us. But if used indiscriminately it creates gaps where they are not needed. Perhaps we ought to explore its functioning more carefully in a manner parallel to the manner in which we examined the means-end distinction, which bears some resemblance to it. And in ethics the notion of fact does behave suspiciously. The facts of the case prior to evaluation may include all past value judgments—everything gets factualized. On a Sartrean analysis every decision is free, so all my past value judgments now are simply facts about my past decisions. I now have to decide whether to stick to my past trends of decision. In effect, what Sartre is doing is replacing the fact-value cutting of the conceptual pie with a past-future cutting, with the present as moment of decision. Still other categorical selections are possible. Much of current analytic ethics uses the distinction of observing and acting, or the contemplative and the practical attitudes, where the earlier positivist analysis used the fact-value or the indicative-imperative, or the knowledge-volition distinction. I prefer instead the attempt to experiment with a fourfold distinction of enterprises—describing, analyzing, explaining, evaluating—while insisting that each of them can cover all the materials as well as the processes of any of the others. Hence there is no partition of field or data among them.[15] In any case, the fact-value distinction has not proved helpful in much of ethics, except to guarantee the permanent possibility of evaluation, a consequence which can probably be achieved in other ways. Both concepts—fact and value—are too stretchy and results of application not always happy. We need a fuller investigation of both to see what has gone into their stretching—why all scientific propositions whether of theory or generalization or particular observation got cast within a single rubric, and why all selective tendencies from sheer desire to complex ideal and prescription have been construed in another. It is too late here even to offer suggestions in this inquiry. For our present purposes it is enough to note that if the integrity and permanence of evaluation are safely established in other ways, we can happily experiment with other categories for ethical theory. At the very least, the fact-value dichotomy can be removed from its position as obstacle to the fruitful relation of science and ethics. In some sense our present formulation of the relations of science and ethics has been carried out without involving the fact-value categories. This should add to the conviction that the fact-value distinction is not an absolute one but is context-bound. We have not, of course, even begun the investigation into criteria for determining its proper contexts and proper limits.

Finally, what about the logical status of the inquiry into the relations of science and ethics themselves? It is certainly a philosophical inquiry, but that designation does not settle the extent of conceptual-analytic, scientific, and even ethical-normative elements that may go into the solution. Conceptual-analytical elements are clearly involved, for the meanings of "science" and "ethics" and their distinction into components are matters that involve careful analysis. Ethical-normative elements are involved in a definite but perhaps peripheral way: it is desirable to find relations of science and ethics so that knowledge of the world and its ways can act as good reasons for courses of conduct. But the crucial question is whether the world indeed has the requisite type of order. And this is preeminently a question to be settled by the results of scientific inquiry. Thus in a fundamental sense the issue of the relations of science and ethics is itself a scientific issue. And this is so, whatever the outcome. Suppose for a moment that the verdict were that an ethical theory needs no change whatsoever, no matter what the psychological theory of human striving and purpose, of human choice and emotion proved to be. This would not have the status of a metaphysical presupposition or an analysis of the logic of linguistic use. It would be the scientific hypothesis of a constancy or an invariance in ethical theory for all transformations in psychological theory. But in fact such a verdict of invariance is far from the emerging picture today. The analysis of the diverse patterns in which science and ethics are related is thus of basic importance for the reconstruction of the enterprise of ethical theorizing and for fashioning an adequate ethical theory.

Notes

1. Richard B. Brandt, *Ethical Theory: The Problems of Normative and Critical Ethics* (Englewood Cliffs, N.J.: Prentice-Hall, 1959), ch. 3.
2. G.E. Moore, *Principia Ethica* (Cambridge, Mass.: Cambridge University Press, 1903).
3. *International Journal of Ethics* (October, 1903): 116.
4. Moore's principle of organic unities denies that we can go directly from the value of the parts of a whole—each evaluated separately (that is, as a whole)—to the value of the whole with them as parts. In short, every complex has to undergo the isolation test itself.
5. Jean-Paul Sartre, *Existentialism*, trans. Bernard Frechtman (New York: Philosophical Library, 1947), pp. 28-33.
6. Fears of the ghost of any who die as a result of one's action are possibly involved in some questions about food-sharing as well as in keeping promises to the dying. See John Ladd, *The Structure of a Moral Code: A Philosophical Analysis of Ethical Discourse Applied to the Ethics of the Navaho Indians* (Cambridge, Mass.: Harvard University Press, 1957), pp. 285, 388ff.
7. Jeremy Bentham, *An Introduction to the Principles of Morals and Legislation* (Oxford: Clarendon Press, 1907), ch. IV.
8. Ralph Barton Perry, *General Theory of Value: Its Meaning and Basic Principles Construed in Terms of Interest* (New York: Longmans, Green and Co., 1926), pp. 626-33.
9. Max Weber, *The Protestant Ethic and the Spirit of Capitalism*, trans. Talcott Parsons (New York: Charles Scribner's Sons, 1958); R. H. Tawney, *Religion and the Rise of Capitalism: A Historical Study*, Holland Memorial Lectures, 1922 (New York: Harcourt, Brace & World, 1926).
10. Translated in Walter Kaufmann, *Existentialism from Dostoevsky to Sartre* (Cleveland and New York: The World Publishing Co., Meridian Books, 1956), pp.

270-87. It represents "a slightly abridged version of the first part of *Réflexions sur la question Juive*.

11. It is worth stressing again that given the permanent possibility of evaluation, there is no limit to which this questioning of the functions of morals may not go. In theory, even the whole of morality as a human institution may be put on the scales and found wanting, if all the functions suggested for it can be carried out better in other ways or through other human agencies. (Morality might then cling to a place in the criteria of "better" in such investigation—apart from the separate questions of intrinsic value.)

12. G. E. Moore, *Principia Ethica*, pp. 98-99.

13. John Rawls, "Two Concepts of Rules," *Philosophical Review* 64 (1955): 3-32.

14. Friedrich Nietzsche, *The Genealogy of Morals*, trans. Francis Golffing (Garden City, N.Y.: Doubleday Anchor Books, 1956), pp. 189-92.

15. Cf. Abraham Edel, *Method in Ethical Theory* (Indianapolis: Bobbs-Merrill, 1963).

3. The Place of Empirical Knowledge in Ethics

The question that here concerns us has been posed as: What significance has the development of empirical knowledge for normative ethics?

If we take the idea of normative ethics to be sufficiently clear without further explication—roughly judgments of appraisal and evaluation, including obligation—there still remains the sense to be given to "the development of empirical knowledge." For the idea of the empirical undergoes change in different philosophical eras as different schools sharpen, or on the other hand soften, the lines between the empirical and the nonempirical. Compare, for example, the Platonic and the Humean tradition. The Platonic allows "empirical" to embrace the sensory, memory, and generalizations embodying them, but brands them all as mere belief, in sharp contrast with reason and rational knowledge. (The flavor of this is still to be found in a disparaging use of "empirical" as connoting fragmentary, isolated, rule of thumb, as contrasted with systematically organized and theoretically grounded, a use still lingering in medical literature.) Now since Plato puts ethics at the top of his upright linear knowledge scale—grasping the Form of the Good—there is no gap between knowledge and ethics, but both are cut off from the empirical down below. On the other hand, the Humean tradition, bringing to a culmination the great attacks of seventeeth and eighteenth century empiricism on the pure rationalism of intuitive and self-evident ideas as furnishing knowledge, leads us to regard all scientific knowledge, however theoretical and systematic, as essentially empirical; it puts into the same package ordinary empirical beliefs and the kind of reasonableness we would today assign to inductive knowledge of our world. In this sense, present-day ideas of empirical knowledge think of the growth of systematic knowledge in all fields—social and historical as well as physical and biological, and the lessons of application as well as mathematical physics which constitutes the most highly systematized paradigm—as empirical knowledge. On such a conception, if the Platonic continuity between knowledge and ethics had remained undisturbed and only his dichotomy of knowledge and the empirical removed, there would have been little philosophic question about the place of empirical knowledge in ethics.

This was not to be. No sooner had Hume closed the Platonic gap by establishing the empirical character of our knowledge of all matters of fact, than he himself opened up a fresh gap between knowledge and ethics higher up on the Platonic scale. At least such is the usual interpretation of his creation of the is-ought gap and his proclamation that reason is the slave of the passions. (Strangely enough, for the putative father of the is-ought gap, he added that it not merely is, but ought to be in this slavish position.) To understand this new limitation on reason we shall have to look later to the functions that are assigned to the slave in this relationship. But whatever Hume's meaning and intentions, the subsequent history of metaethics has taken him to call a halt to empirical knowledge as basically relevant when one crosses the line into ethics.

In the present day, the heart of any inquiry into the place of empirical knowledge in ethics thus lies in the historical context of the separation of empirical knowledge ("science") from ethics. Now all the metaethical questions that so heatedly characterize contemporary ethical theorizing are the reflection of a historical change in which scientific knowledge grew to tremendous proportions, the character of human life underwent tremendous socioeconomic institutional changes, but morality remained tied in its content and organizational form to the older ways of life. Accordingly, if our question were presented as a moral and social question, it would amount to asking what reconstruction in morality is required by the historical changes in the totality of human life on the face of the globe and the transformation of the conditions of life. But the question has been posed here as a metaethical inquiry. It has therefore to be pursued as largely internal to the analysis of metaethics. I propose accordingly to argue—unfortunately only in outline because of the scope of the material—for the following theses:

1. That the contemporary question of the significance of empirical knowledge for normative ethics represents philosophically the convergence of a Kantian separation of science and ethics, a phenomenological cleavage of the content of consciousness from the context of consciousness, and a twentieth century positivism that makes an absolute division of the logical, the empirical, and the valuational.
2. That in the consequent formulation as well as in the three converging strands we have not a set of established philosophical results about the relation of the ethical and the empirical, but *an intellectual program for their separation*.

3. That this program constitutes a great twentieth century intellectual experiment and that the experiment has by now been tried and failed, both on the conceptual and empirical side.
4. That underlying this program is a set of psychological constructions and theories which have for a long time been scientifically bankrupt.
5. That the opposite program about the relation of the empirical and ethical deserves an equally intensive and extended experiment on the philosophical scene, and that it is more likely to succeed because (a) it is consonant with sounder psychological trends, and (b) it is more congruent with the practical problems of morality in the contemporary world.

All this constitutes a large task which at best in the present compass admits of little more than outline.

1. Kant's philosophy is the familiar and basic source of the view that the strictly moral contains no empirical element, that it is a logically established categorical *ought*. His linking of the moral to the noumenal and the scientific to the phenomenal expresses his acknowledged attempt to rescue morality (and through it religion) from science. His argumentative ground is that science requires determinism and morality requires freedom ("ought" implies that it was possible to have acted otherwise). Insofar as philosophical progress has reformulated the problems of determinism and freedom since that time, and insofar as his harsh antithesis has disintegrated or abated, and insofar as analytic approaches have explored the inner complexity of "I could have acted otherwise," [1] the Kantian postulate of freedom is today a tenuous basis for the absolute separation of the empirical and the ethical.

In twentieth century phenomenological approaches (e.g., N. Hartmann) and in their analytic analogue (e.g., G. E. Moore), value—that is, intrinsic value or the ought-to-be—is a distinctive quality found in the field of conscious awareness. [2] In Moore's *Principia Ethica*, the foundation-treatise of Anglo-American analytic ethics, it is a discerned nonnatural quality, supervening on but not reducible to the natural or descriptive properties of the object that is before one for evaluation. Comparably, in the phenomenological formulation, the domain of value is quite separate from the domain of the metaphysically real and the domain of the sensory existent. The ground of such doctrines is the strict attention to qualities in conscious experience and to distinctions found in the field of awareness. Science (as well as metaphysics) is concerned with secondary judgments about the context of the occurrence of qualities in this field, about conditions of occurrence or concomitments of occurrence. The basic presupposition is thus the primacy of the field of conscious awareness, whether it is formulated as "realism" or as "idealism."

Common to the Kantian, the phenomenological-analytic, and logical-positivist position, with respect to our present question, is the sharpness of separation of the empirical and the valuational. It may be a separation by Kantian postulation, by phenomenological analysis, or by linguistic analysis. But all reenforce the fact-value chasm. That is why the relation of the empirical and the valuational is not felt to be an ordinary question like the relation of molecular movement to qualitative transformations, or of wind to rainfall, or of social security to hope and despair, or of love to happiness—all of which require both conceptual analysis and empirical investigation.

2. Philosophers rarely just offer programs. They tend instead to make sharp distinctions as truths about the world, or knowledge, or man. They say, for example, that there is a divine and a human, there are bodies and minds, there are particulars and universals, there are facts and values. If they run into difficult borderlines and broad borderlands, they take these as difficulties of application, not as tests for their distinctions. Rarely do we find them saying that divine-human, body-mind, particular-universal, fact-value, are proposed categorial dichotomies whose utility for varieties of fields and purposes is to be tested. Yet these proposals are not usually sheer dogmatism. They are set in whole networks of concepts and accepted principles and problems for which it is hoped they will furnish aids in solution.

Kant does not ask whether there is a purely formal and nonempirical component in ethics. He says there is and that it has been ignored. He does not consider whether the sharp distinction of formal and material is itself tenable in ethics, whether—to suggest a conceivable alternative—the formal is simply selected material functioning in a particularly useful way. He sets forth the program of extracting the purely formal nonempirical in moral judgment. He finds it, as we well know, in the categorical imperative, and when he has discovered it he is so busy asking how it is possible that he never carries his program out too far. Subsequent critics have to debate whether the program is to yield particular moral decisions or only universal laws, whether it will discover them or only test proposed laws, whether the method can only veto proposed laws or certify them, and so on. Perhaps he thought he was a Newton who gave the laws of motion with some few applications and left it to subsequent generations to work out the analysis of the varied complex movements that are found within the world. But in such a parallel the subsequent applications would serve as tests, and might be a basis for refining or rejecting the initial discoveries. Kant, in fairness to him, did try applications to law and the theory of virtue, but not too amply, nor for that matter too successfully.

In the phenomenological-analytic group, Moore's procedure in *Principia Ethica* is most instructive. He tells us dogmatically that there is only one simple ethical idea—*good* or *intrinsic value*—that it is indefinable and that judgments of intrinsic value are self-evident. He does not say that he has a program of reducing all ethical terms to this term, and that the tasks of ethics will be clarified and more successfully accomplished if we follow his program. But his actual procedure is precisely to reduce other terms in this definitional way, yet without justifying his procedure or reckoning with alternative possibilities. Thus "good as a means" is translated into "a means to what is intrinsically good"; "good as a part" into "a part of an intrinsically good whole," "good for me" into "my having it is intrinsically good"; "ought" and "right" into "will yield the greatest amount of intrinsic value in the world"; and so on. When this mode of translation turns out not to fit the judgments made in alternative theories, Moore declares *them* to be meaningless or self-contradictory.

If Moore's program had yielded great successes in resolving traditional problems of ethical theory or in giving guidance for moral judgment, the case for its acceptance would have been advanced. But the contrary is more evident. Moore has difficulty (in fact does very little) with analyzing judgments of comparative value, he finds his theory incapable of giving any guidance to reconstructing morality, so that he recommends moral conformism, and so on. Phenomenological ethics generally, too, leaves us only with an appreciation of moral essences rather than any effective guidance to conduct. Of course the program of extracting intrinsic value separate from empirical materials need not be abandoned just because of a few failures; it can always be refined and revised and tried again. But it should be recognized as simply one among alternative programs. If theories of agriculture had rested with the hoe and the wooden plow, and actually defined their discipline with reference to these implements, we would not have been surprised when men abandoned "agriculture" in favor of methods of machine-production. If ethics is so defined and delimited metaethically that it cannot guide conduct with much success, other fields will take over its job (e.g., law, economic institutions, religious institutions, military institutions). Politics and morality have not yet learned to distinguish dogmas from programs, especially in theory. The political party in power never questions its principles; if it cannot blame the opposition or find some other scapegoat for its failures, it will lament the complex inadequacies of the human situation.

And what of the positivist trichotomy of the logical, the empirical, and the valuational? The contrast of the logical and the empirical was elaborated with technical brilliance. It appeared to solve the age-old problem of necessary material truths (e.g., geometrical and arithmetical truths holding of physical space and objects) by showing that they could be broken up into analytical formulations either definitory of ideas or logical tautologies on the one hand, and empirical generalizations on the other. And they went far toward showing how regulative principles (e.g., conservation principles, determinism principles) could be dealt with as generalizations operating in a double way, at one stage broadly heuristic, in a complementary phase as definitory. But the specific analysis of the empirical, both in terms of the verification principle and the resolution into sensory atomic facts or sense-data became overwhelmed with qualifications. And the basic distinction between theoretical statements and observation statements became more and more relativized (even apart from the contempory onslaught on the analytic-synthetic distinction itself). More and more, even in epistemology, the basic positivist distinction stood out as a program rather than as an analytic requirement, with its realizability increasingly questioned.

The distinction of empirical and valuational never got far beyond the programmatic stage. Rudimentary translations of the ethical as the expressive, the imperative, the stipulative or fiat (will-act), were never really tested. The emotive theory did elaborate some of the consequences of such a program for disagreement and reconciling conflict, which alarmed the ethical community by the place they gave to the coercive and the manipulative. And there were formal attempts to work out a logic of imperatives or formal attempts at an informal logic of prescriptives in the British schools of linguistic analysis.

As was suggested earlier, the recognition that the distinction between empirical and valuational embodied one or another kind of program does not dispose of it. Rather it turns attention to seeing how far the program can be worked out so as to aid in carrying out the tasks of ethics. If we look at the dominant trends in twentieth century metaethics as a great intellectual experiment in the separation of the empirical and the valuational—or of science and value, or even more generally as fact and value—what lessons can we draw from nearly three-quarters of a century of moral philosophy?

3. If we analyze the history of twentieth-century metaethics as the domain of evidence, the inadequacies in the separatist program stand out as numerous and clear. It is worth taking a synoptic view of the tendencies.

The program has failed to identify the value phenomena sharply enough to isolate a value domain. When the naturalists argued among themselves whether to study value as pleasure, interest, desire, purpose, object of interest, direction of striving, rationally discriminated act or object of one of these, it was clear that they were offering hypotheses concerning which line pursued would in general unify and systematize our knowledge of human appetition. It was a frankly psychological inquiry into vantage points for launching patterns of guidance in morals. The separatist program could not accept this. Assaults on naturalistic ethics in the name of the "naturalistic fallacy" reduced the value phenomenon to a ghostly unnatural quality of intrinsic value which could not long maintain itself. It was translated into an act of valuing, first an expressive act, then an emotive utterance, then a linguistic act in a variety of contexts roughly bundled together as "prescriptive." Giving up any attempt to distinguish a class of expressions by terms in the language, but clinging to the prescriptive character of linguistic use as the distinguishing mark of moral language, the program moved over to study patterns of use. Emphasis now fell on the functions served, not on a separate linguistic apparatus. But even the funcitons could not keep peace among themselves. At first there seemed to be a neat division into evaluating and prescribing, but this was really only a reflection of the old problem in ethical theory of the good and the right. Some types of prescribing or laying down obligations turned out to be really evaluating and some types of evaluating (total reckoning of what was best in a particular situation) might as well be seen as prescribing. And evaluating, of course, always extended far beyond ethics, into any preferential decision.

Even the bare equation of value with preferential decision did not produce a clean partition of value and fact. The history of decision theory in this respect simply recapitulates the difficulties of the separatist program. If one starts with a separate lineup of preferences among alternatives as the value side and probabilities of each alternative as the factual side, then the mode of combining them for decision requires some assumption of strategy (e.g., to follow the highest gain or minimize the greatest risk), and from its point of view the preference scale functions very much like a set of facts. Recent tendencies have quite reversed the procedure. They start with data of decisions themselves and attempt to work out ways of sifting out scales of desirability and expectancy or probability. In terms of guiding decision, which has long been the generic aim for identifying moral function, it is clear that beliefs help to determine the result and preferences help to determine the result. It is by no means clear that the best way of analyzing such determination or helping is by separate systematization of values and beliefs. It may be by configurations in which both are found, whose types are determined by the context of the specific problems.

Once the notion of value was broadly construed, it also became clear that the sharp distinction of value and science could not be wholly maintained. It was not merely that there were values in the pragmatic base of human purposes in terms of which the general aims of science were identified, and so guiding ideas of method justified (simplicity, fruitfulness, predictability, etc.). Values penetrated into the detailed acceptance of one rather than another hypothesis, and what had seemed to enter only in questions of application became relevant in questions of assigning probable truth. At least this controversy goes on, with the purists in science driven back to the point of redefining the aim of science as simply exhibiting relations of hypothesis and evidence without telling you what hypothesis to *accept*.

The old logical standby for getting rid of troublesome issues—that you cannot *deduce* a value statement from a factual statement—has long worn thin. For one thing, a case can be made that you can do it. Let V be a value state and F a factual statement. Then, on the assumption that any non-analytic statement is either a value statement or a factual statement, what will F *or* V be? If F *or* V is construed as a factual statement, V can be deduced from the factual statements F *or* V and *not* F. If F *or* V is construed instead as a value statement, it can be deduced from F. This is, of course, trivial, but so is the abstract general point. The critical point is rather that the dictum about deducibility requires first the sharp separation of fact terms and value terms, and this is not as easy as it seems.[3] Indeed it involves a complete revision of large parts of the dictionary, since all sorts of kinship terms and role terms and terms designating human relations have value attitudes and responsibilities already within the meaning of the terms. One is tempted to say they have been "built in" or "fused" with the descriptive content. But this would beg the separatist question. Perhaps they are natural "alloys" and the program of separation is technical and artificial and may not be capable of accomplishment except in a relative way for specific purposes. (Perhaps it is like trying to separate the form and the matter of a thing, or its universality from its particularity, or its eternity from it historicity.)

Attempts to separate value from fact in the theory of how we justify our moral judgments have not so far been very successful. Emotive theory was led to reject deductive and inductive methods for supporting ethical conclusions, and was left with the view that persuasion in ethics was simply causal influence. Analytic attempts to restore some pattern of justification between factual reasons and ethical utterances argued for a kind of informal logic; they appealed to the kinds of arguments that are felt in ordinary life to show good reasons for doing something. But the only ground for regarding them as good reasons is that they are embedded in the pattern of linguistic use—a rather factual terminus for a separatist ethical inquiry! A more rigorous attempt to systematize the logic of justification for value concepts is to be found in deontic logic and axiological systems. But these are formal constructions, and the degree of their utility for systematizing value judgments in practice is a question barely reached as yet.

In general, the relativity of the fact-value dichotomy is clear in ordinary discourse. Feelings of pro or con are taken to be value expressions till we ask how they fit into our purposes. Now they become facts to be evaluated as aids or hindrances to successful achievement of aims. But our purposes too can be seen as biographical facts when we want to appraise them as worthwhile or not. And our judgments of worth, though they seem to be standards rather than facts, yet take on a fact-like or raw material-like character. They are just the standards we use when we want to work out our obligations. Such shifts suggest that value is a complex category of relating things rather than a kind of thing or even quality itself. It is more like the "price" or "worth" of something in specific types of transactions which indeed is closer to the origins of the modern value concept in its earlier history.

In these respects the value-fact distinction is very much like the theory-observation distinction. The ideal of isolating purely theoretical statements and purely observational statements is probably a mistaken one, absolutizing a relative distinction that is serviceable in certain kinds of contexts. The main purpose in this case is systematization. So too in the ethical parallel, where the context is one of furthering systematic guidance of conduct.

In ordinary life, the systematic guidance of conduct often takes place in terms of the means-end distinction. Here means is the empirical or factual component, while end bears the value mantle. In the social sciences generally, the programmatic dichotomy has been maintained by insisting that the sciences cannot deal with ends, only with means. Here again, there is the same dogmatic assumption that because we can project a program of separating ultimate ends from means, we can draw the theoretical and practical conclusions as if the program were assured of success. On the contrary, when we try to carry it out, we see that the role of the means becomes larger and larger, and that of the ultimate end recedes. In short, the empirical is necessary throughout.

Let us pursue this point in the way suggested at the outset. Instead of disputing whether reason is the slave of passions, we ask for a job inventory of the slave's assignments! I have elsewhere suggested [4] that a realistic account of science telling us about means not ends would cover a manifold and extended set of undertakings. It would, of course, have to give us better grounded beliefs about instrumentalities for limited ends. It would have to go on to tell us about the collateral and subsequent consequences in our lives of the use of such means, beyond the fact that they achieve the ends for which they are designed; and the unintended as well as the intended consequences. There would also be questions about the consequences of achieving the end itself, the qualitative character of the end and what it meant to the man or society that pursued and achieved it, what its ramifications were in relation to the internal psychological and moral "economy" of the individual or the internal general state of the society; for example, what problems it solved, what interests it expressed, what purposes it furthered. Especially when we come to deal with unconscious motivations in the individual or latent functions in the society, knowledge about the character and relations of the end leans heavily on psychological and social science. For example, to judge that an end is serving in a projective as against realistic way, or again that it serves socially in an ideological rather than a realistically constructive capacity, is to offer a complex empirical hypothesis. Beyond all this lies the discovery of points of conflict or the analysis of opposition between different ends held by the same individual or the one society, and the long-range experience of what goal patterns are mutually supportive or productive of conflict. Nor are many of the questions suggested above resolved at any point once and for all. As conditions of life and society undergo change (often revolutionary in character), the relevance of these changes to the restructuring of moral problems and the discovery of what fresh problems have emerged require continual investigation. And finally, even the end-like character of the end allows some scope for scientific inquiry; at least it can investigate the conditions, for different individuals and societies, that make some ends *attractive*, or the likelihood of a goal or activity being held as an end.

In all this, the slave who is confined to the empirical study of relations while the master's passion gives the decisive yes or no, begins to look less and less like a slave and more and more like a brain-trust commission, which is assigned not only the complete survey of the problem but the task of suggesting alternatives for action. But if the division of labor is retained, and the master will not himself sit as democratic chairman of the commission, and insists on final independent nod, we have to ask about him (and the passions). Can he read and understand the report, or will he simply be caught by the bright phrase? What are these passions which are to be served and slaved for? In Hume's day passions covered sentiments and feelings, including human sympathy, not just Hobbesian egoism, and Hume, like Adam Smith, assumed the invariant nature of the human sentiments. If today we have a fuller sense of the cultural determinants and variety of prevalent passions, and the role of institutions in creating their patterns, shall we extend to the slave's task the philosophic critique of existent patterns in terms of ideal possible alternatives and their consequences in human life?

It is quite clear that the line between the empirical rationality and the value fiat is wearing very thin. It depends now on what the value really represents in the makeup of man. Our lesson is thus not merely that the reiteration of a program in varied contexts is no proof of itself, but that we must probe to the scientific roots and presuppositions of the program itself.

4. The psychological constructs and theories underlying the program are long bankrupt. They are little more than the old faculty psychology. The names are varied: thought, sensation, will; cognitive, sensory, conative; cognition, sensation, feeling. There is this slight variation, that will or initiation of action is sometimes within the trio, sometimes kept back to a climax in which all the others operate on it.[5]

If value is linked with the emotive, and so with the emotions, and the empirical with the sensory, we cannot tell how sharp the distinction of the empirical and the valuational will be without some account of the "faculties." If sensory is not sharply distinguished from conceptual, and if emotions involve appraisal and appraisal involves cognition, then the empirical and the valuational may quite well have firm bonds.

If value is tied to the *meaning* of the situation that is being appraised, and if meaning is a cognitive category, then value may be quite cut off from emotion; emotion may be a concomitant or a cause of the situation, or an obstacle to value.

If value is tied to will, and will is an initiation of action, something on which thought and sensation and feeling all have impact, then we want a theoretical account of the will before we can tell whether it is bound to intelligence or is essentially a blind fiat. I take it that the medieval theologians were closer to the scientific problem when they discussed whether God's reason or God's will was prior, than the positivists were when they assumed unquestioningly the pure fiat character of will-acts.

Dewey[6] pointed out that the stimulus is constantly altered by experience and has a different meaning on subsequent occurrence, that the motor is steered by the perceptual which is cognitively organized. The total picture is one of an organism-in-an-environment responding to a problem that has articulated a hesitation in the continuum of experience, readapting in a reconstruction of the whole that attempts to solve the problem. Such a psychology could never make of thought and sensation and value a set of separate entities with their own laws. For this reason, Dewey's constructs in the theory of value—especially the means-end continuum and the attack on the sharp means-end and fact-value distinctions, on the refusal to relate science and ethics—were practically meaningless to a generation in which the positivist trichotomy was unquestioned.

5. A program diametrically opposed to the separatist one we have been examining—let us call it an *integrative* program since it brings value and the empirical together—begins to look much more plausible today. Two grounds were indicated at the outset: it has a sounder psychological basis, and it would prove more congruent with the practical problems of morality in the contemporary world. It thus also answers the question of the significance of the development (that is, the growth) of empirical knowledge for

normative ethics in two ways. One is that normative ethics always operates with concepts and methods that presuppose a psychological theory, and the best available psychological theory is likely to come from advancing psychological science. (The alternative, it should be pointed out, is not no psychological presuppositions, but either past and possibly outworn presuppositions or else the uneven melange of common-sense psychology which enshrines errors as well as truths.) The second answer to the original question is that an integrative approach, in bringing the empirical and the valuational together, calls explicitly for marking the points of empirical relevance in the structure of normative ethics, and opens all areas of empirical knowledge as a source for assisting normative judgment, ruling out none antecedently on arbitrary grounds. Thus we cannot tell where knowledge pointing to reconstruction of segments of morality may come from. It can be as varied as: the rapid realization of the extent of pollution and the fact that we are dealing with a finite atmosphere, which calls for a reversal of traditional and firmly-established value-attitudes; new studies in the physiology of sex, which raise a number of fresh questions about male-female relations in the light of a changed view of female sexuality; global economic changes which render obsolete older nationalistic-patriotic values; even the history of different moralities, which may reveal constancies by showing how similar needs were institutionally serviced in a variety of social mechanisms.

Such considerations apply not only to the content of morality, but also to its very structure; for example, whether a morality is basically oriented to goals or to rules, whether it employs hard sanctions, how much it leaves the individual to decide for himself and how much it structures through social forms of decision, and so on. A world of greater material progress may accentuate individual pursuit of goals, or it may instead weaken the distant-goal orientation in favor of present value. A more crowded world at a high technical level may require much stricter moral rules in many areas to assure basic care and equality. Concern with type of sanctions will probably demand a more effective type of learning with greater motivation than the educational systems so far developed have made possible. The rise of social security and forms of medical and other care for human beings articulate a social mode of decision in many of the traditional moral problems of distributive justice; while the very existence of large organized decision-systems may itself put a high moral premium on broadening the area of individual conscience and individual liberty of choice, especially in fundamental patterns of life.

A metaethics which relates its constructions to such empirical considerations and by its very theoretical structure calls for reconstructing morality as a normal procedure of human beings in a changing world, has much to support it as compared to the trends of the past in the twentieth century that have dominantly isolated morality and frozen it.

To be plausible, however, is not enough. The same kind of broad experiment with this kind of metaethics is needed that the separatist program has had, and on the same wide scale. That is, ethical theory has to work out in a refined and technical way the modes of interrelation of empirical and valuational in all of the traditional problems of ethics. [7]

I should like, in conclusion, to underline a few metaethical considerations in the comparison of the two programs—separatist and integrative—at points where there may be misunderstanding.

First, the integrative view of the empirical and the valuational does not eliminate a distinction between a belief and a value, or between describing and evaluating. It refuses to make an absolute dichotomy with impassable chasms. There are no different domains, there are different enterprises in different contexts and for different purposes. The analysis of a belief will find some value elements in its determinants, and the analysis of a value will find embedded cognitive elements. To describe is also to select a perspective for given purposes, and to evaluate involves descriptions; yet we can describe evaluation processes and evaluate descriptive procedures. The same material content enters now into descriptive enterprises, now into evaluative processes.

The denial of gaps and chasms does not do away with differences. There are gaps all around us, and we can play a game of gaps and leaps. Hume was especially adept at that: if he is correctly taken to have fathered the is-ought gap, he also found one between the past and the future (in his "problem of induction"), as well as between particulars and generals. Similarly, that something holds for England does not prove it holds for France (whether cultural or geological); this does not mean an eternal cleavage between England and France!

Second, the relation of the empirical and the valuational does not limit the possibility of evaluation, a charge often directed against naturalistic identification of ethical ideas with natural processes. In fact, the diversity of natural processes, the pervasiveness of change, and the perennial need for reconstruction, all combine to demand the permanent possibility of evaluation. Nothing is thus exempt from an inquiry of its bases of acceptance or rejection—whether moral content, moral structure, form of life, intellectual instrument, basic dichotomy.

Finally, the relation of the empirical and the valuational does not mean the "relativization of value" in the way in which the idea of moral relativism in many of its senses has carried the connotation of arbitrary choice in morality. For one thing, as suggested earlier, the idea of arbitrary fiat rests on a particular interpretation of the will which does not bear psychological scrutiny. There is no theoretical obstacle in this integrative metaethical program to long-range human goals, constant or perennial values, and the like. But they cannot be established by fiat, or sheer introspection, or isolation of the concept of intrinsic value from an empirical reckoning. The charge that the empirical in morality means opportunism is not historically supported. Not that we have lacked opportunism—but it is found in every ethics, though disguised in some and open to critical scrutiny in others.

To go beyond such preliminary considerations as this paper has presented is to work out the detail of the integrative program. A great deal has been done in contemporary ethical theory, but it is dispersed. And to call attention to the need for an integrated treatment of the program itself has been the ground for my tackling so large a question in so brief a compass.

Notes

1. See, for example, G. E. Moore, *Ethics* (New York: Holt, 1912), ch. vi, and J. L. Austin, "Ifs and Cans," included in his *Philosophical Papers* (Oxford: Clarendon Press, 1961).

2. Nicolai Hartmann, *Ethics*, trans. Stanton Coit (New York: Macmillan, 1932); G. E. Moore, *Principia Ethica* (Cambridge: Cambridge University Press, 1903).

3. For an excellent presentation of this argument, see Robert Ackerman, "Normative Explanation," *Philosophy and Phenomenological Research* XXIV (1964): 522-29.

4. In "Patterns of the Use of Science in Ethics," especially section II. (See chapter 2, above.)

5. Dewey pointed out, in his early famous article on "The Reflex Arc Concept in Psychology," that such entities as stimulus, central process, response, were being used as discrete items in precisely the way in which the older separate faculties had operated. This held also for the divison into sensory, ideational, motor-affective. In the latter, the feeling component is tied to the motor on the basis of a theory of the nature of feeling somewhat like the James-Lange theory. The article is conveniently reprinted in Joseph Ratner's edition, *John Dewey: Philosophy, Psychology and Social Practice* (New York: Capricorn Books, 1965.)

6. Dewey, see note 5.

7. Traditional naturalist ethics, whether classical or utilitarian, does not deal much with problems of change. It is attentive rather to variety. In nineteenth and twentieth century ethics, Dewey's value theory is perhaps the most developed toward the integration of fact and value in a context of change, but it only charts a general direction and is entwined with special epistemological issues. Marxian writers have gone farthest in specific socio-historical orientation, but on the problems of ethical theory the older Marxian writers gave only suggestive illustration, whereas more recent Marxians have been specially concerned, in the light of their own social problems, with relation of the individual and the state.

4. Metaphors, Analogies, Models, and All That, in Ethical Theory

It is nothing new that a successful theory in one field finds its uses, or echoes, in another, as the seventeenth century used Euclidean models in ethics, and the eighteenth used Newtonian models in economics and ethics, and the nineteenth came to use evolutionary models. Once attention is directed on this problem, many other models come readily to mind. Think of the part that has been played in ethics by transfer to it of a legal notion of contract, a medical conception of health, a psychological conception of unavoidable pressures demanding outlet, a biological conception of an organism whose members have diverse functions to perform in the maintenance of the whole. Such models are not, however, always applied at full strength. The application of a Euclidean or a Newtonian model is the extreme case. Others, less systematically worked out, may operate only in bits, often no more than by suggestion. Even when a theory in one field colors the whole intellectual climate, its effect in other fields may be only through analogy or metaphor. When the mind geometrizes, many things everywhere are felt to "follow logically" no matter how large the gaps in strict connection. When Newton reigns, even feelings begin to "gravitate together," and when Darwin rules, every field is busily finding it has an "evolution." That is why no sharp division can be drawn between the study of full-fledged models and that of metaphors, analogies, and suggestive comparisons generally. The common problem is the influence of one field in shaping or affecting another.

The present inquiry aims to raise in some systematic fashion this whole question of models in ethical theory by asking in what way and how far different models have helped shape conceptions of the nature and tasks of ethics, influenced the detail of an ethical theory, and even furnished integrative concepts for its organizing structure. Section I considers the role of models in the prior commitments of an ethical theory—its assumptions about the world and man, about human nature and the human predicament. Sections II to IV consider in increasing strength the role that models, metaphors and analogies play in ethical theories. Section II starts with *casual* use. Section III goes on with *heuristic* uses, where the comparison serves suggestive functions in developing the ethical theory. Section IV deals with *structural* uses in which the model organizes the ethical theory and furnishes, in effect, its framework. Section V consists of some concluding reflections.

I

Every ethical theory has prior commitments in that it operates with some assumptions about the kind of world men live in, what men are like, what typical situations they find themselves in, what are unavoidable problems or conflicts in the human situation, and so forth. Now these psychological or social or historical or religious or metaphysical "foundations" of the ethical theory often already contain embedded metaphors or models that played a part in the development of the underlying field. The consequences of this model now extend to the ethical theory.

Plato's psychological commitments in his *Republic* constitute an excellent example of this process. As is well known, Plato uses for the human makeup a metaphor in which each of us is said to consist of three parts—a human or rational part, a lion which represents the noble emotions of shame and indignation as well as the ambitious desire for prestige, and a dragon which represents the appetites. This dragon combines blind aggression, sexuality, and acquisitiveness; it is incapable of controlling itself, it is arbitrary and capricious, and it demands immediate expression. It is tempting to call this the *hydra* model of appetite, after the creature that Hercules fought which grew two heads to replace every one he cut off; for though Plato does not mention this myth in that particular context, according to his treatment appetite clearly grows stronger as it feeds or is permitted expression, rather than is appeased. Now Plato's model, built into his psychology, gives a definite shape to his ethics precisely because he leans so heavily on the psychological commitments. For he ties his account of human virtues to his picture of the parts of human nature and even defines the central virtue of justice as the state in which the man in us, supported by the lion, keeps the dragon in his place. The model thus shows that the ethics will be a repressive one, and (with further assumptions about distribution of proportional strength of the factors in men) Plato concludes that the mass of men have to be held perpetually in an authoritarian mold.

By contrast, the Freudian hydraulic metaphor, conceiving of the flow of libidinal energies and the dangers of their being dammed (assuming the incompressible character of fluids), concentrates on repression as the trouble spot so that the general aim is to find satisfactory outlets. And this vital difference from the hydra metaphor remains even when the concept of the Id is developed (employing other metaphors) with properties of irrationality and demandingness quite similar to Plato's appetite. An ethics employing the Freudian psychology thus can set itself a quite different goal from the repressive structure of Plato—broadly speaking, a goal of increasing rationality through insight into oneself.

We must, of course, take care not to attribute to the model greater importance than it may in fact possess in the psychological theory itself. It is possible that a metaphor such as Plato's may come after the fact to adorn the achievement. After all, he himself gives us his dragon metaphor late in the *Republic* (in Book IX), not in Book IV where his theory of the parts of the soul is advanced. It is not easy to decide how strongly it has functioned all along. The analytic discovery of the model in the presuppositions of the ethical theory itself cannot as such then settle the essentially biographical or historical question of what came causally first in Plato's intellectual development.

What has been suggested about the influence on ethics of models in prior psychological commitments may hold equally in religious or metaphysical or biological or sociological commitments. In the western tradition God is conceived of as a father. This familial model accordingly transfers the properties of the head of the household to the deity. The consequent stress on the virtues of filial piety and obedience in the religion are not surprising. And as the ethics as a whole is organized in religious terms, the familial relations are projected through the religion upon the whole of life, and a mutual interaction ensues in all those phases. In a similar process, when the world is conceived on an organic model, with individuals as "members" analogous to bodily parts, to demand sacrifice of some part for the good of the whole is taken to be "natural." The association of an evolutionary model that underscores the struggle for existence with ethical "egoism" and of an evolutionary model that underscores mutual aid with ethical "altruism" admits of comparable analysis. In all these cases, a model originating in some particular phase of life or some particular process, is brought into a domain such as psychology or religion or biology which in turn plays a background role in an ethical theory. This prior commitment of the ethical theory serves to project the properties of the model into the moral outlook on life as a whole.

II

In considering the role of models and metaphors within ethical theory itself, *casual* use is to be distinguished from *heuristic* and from *structural*. A use is casual when it occurs on isolated occasions and is not elaborated. "All life is a battle" used in a moral context no doubt conveys the general notion that the moral virtues are to be identified with the military virtues of disciplined struggle. Yet even if the assertion is repeated and becomes standardized, it may still be no more than casual if it is regarded as self-explanatory and put to no analytic use. If, however, we find the ethical theory attempting to pinpoint the type of discipline required in war, the elements of persistence and obedience and resignation, and shaping a morality along these lines, the metaphor has clearly become heuristic. (How far it is pressed is another matter—a really persistent use of the military metaphor might inquire into different types of war and show the variety of basic virtues and forms of social organization relevant for them, and even raise paradoxes about the abolition of war as a moral ideal!) Finally, if the theory attempts a systematic picture of the battle of life, if it asks who is the enemy and who gives the orders, we may end up with structural use of the metaphor. Stoic ethics seems to approximate this. Epictetus tells us to live our lives as if we were soldiers in constant danger of being ambushed, not by what happens to us but by our reactions to what happens, so that the enemy is located within ourselves. In addition to this focus on self-struggle and self-mastery, we are told to obey the divine or rational nature operative in things and so to be resigned to the lot in which we find ourselves. In such extended use, the battle metaphor serves both as a keynote and as a structural framework for the theory.

Isolated casual comparisons are no doubt often purely literarary expressions enhancing the style, and so may serve little or no role in ethical theory. When Kant remarks, early in the *Foundations of the Metaphysics of Morals*, that a good will, though wholly lacking in power, would "sparkle like a jewel in its own right, as something that had its full worth in itself"[1] and goes on to play with the analogy, comparing usefulness to the setting of the jewel that helps in commerce or attracts the inexpert, are we to look for any further significance in the comparison? Or when R.M. Hare, arguing that the prescriptive use of ethical terms is primary and the descriptive use secondary, says that bringing up children with the latter turns them "into good intuitionists, able to cling to the rails but bad at steering around corners,"[2] do we need to explore clinging and steering to bring theoretical enlightenment in ethics?

Yet even such casual analogies, invoked and then forgotten, may in their brief use do a serious job—or sometimes help avoid doing a serious job! Kant's comparison of a good will to a jewel shining in its own right suggests that intrinsic value is obvious. It therefore saves him the trouble at that point of analyzing the emotions of the beholder who sees an act of good will. Kant is quite conscious of the problem he has postponed. He adds a footnote shortly after that he may be thought to be taking refuge in the obscure notion of "respect" for the moral law; eventually, in his *Critique of Practical Reason* he goes into an elaborate treatment of the notion, and in fact it turns out to be central to his position to maintain that it is not an ordinary psychological statement.

Hare's use of analogy on this occasion is also not without theoretical significance. It says, in effect, that life has frequent novel turns, that an attentiveness to the need for making choices is central to ethics, and that a prescriptive use of ethical terms enables us to deal more readily with such turning-points. This important recognition of a pragmatic element in deciding how to use ethical terms is passed off in the casual analogy. Hare is thus able to insist that he is merely analyzing linguistic usage.

Relatively casual models and metaphors may be invoked to convey a basic philosophical outlook in ethics. For example, N. Hartmann, in his *Ethics*, often slips into a kind of astronomical model: there is an objective heaven of values to be explored by turning our telescopes on it, and all sorts of values lie there unseen and undiscovered. This kind of model carries serious theoretical consequences, for it assumes that one who does not see a given value is either looking in the wrong direction (passing by the other side of the street, says Hartmann), or is not sensitive enough or not developed enough, or in a hopeless case perhaps "morally blind."[3] This last metaphor has often been invoked by moral philosophers to provide the reassurance that in a case of ultimate disagreement about moral values, one of the parties to the dispute is wrong, so that the "objectivity" of morals need not be abandoned. Neither in the frequency of its use nor in the importance of its theoretical consequence is this metaphor casual. It is only so considered here because it is invoked usually as if it were self-explanatory, so that it falls short of being heuristic. For all it does is to suggest that there are, in some domains of the world, inabilities to perceive what is there. Whether values constitute such a domain is not settled by invoking the model. A fully heuristic treatment would have to take the parallel more strictly and elaborate a theory of moral vision and the conditions of its partial or complete absence which could be subject to independent confirmation.

The reach of casual metaphor in ethical theory is much greater than we are inclined to suspect. The reason is the generic one that so much of ordinary language is shot through with metaphor that it would be surprising if ethical terms were wholly free and pure.[4] "Right" carries notions of straight and upright, and "ought" of debt. We feel some kind of connection between what is morally right and what is nonmorally correct, and again we think of our obligations to others as what we owe them. But only a sensitive historical-minded philosopher like Nietzsche seriously suggests using the debt metaphor for heuristic purposes when he attempts to trace the idea of obligation to the debtor-creditor relation.[5] And it takes a full anthropological perspective to see that the comparison to debt may play a structural role in a whole ethics.[6] Again, talk of "moral law" carries so much of the sense of command that major philosophical confusions have arisen from the coupling of universality and imperativeness. Ideals of "harmony" often come straight from music, "equality" is frankly arithmetic or geometric, and even apparently logical concepts of "consistency," when used in ethics, carry directly the sense of being able to stand together.

To the evidence of language that moral and ethical thinking is thoroughly imbued with metaphor and analogy must be added that of the history of thought in general, as well as a comparative anthropological survey of cultural differences. There are obviously some models in the history of western thought whose use has been more than casual—most notably that of art and craft in teleological philosophies and the machine in the rise of scientific philosophies. These have tended to obscure what is probably a great variety less systematically analyzed. Reproduction furnished a model for ancient Greek theogony and cosmology. There must have been considerably more use of growth models than usually noted, in casting ideas of the good; for surely a human history so bound up with the discovery and development of agriculture could not have overlooked this in its patterns of thought. That fire played some role can be seen, if nowhere else, in the Stoic notion of the divine fire as present in every man—the metaphysical formulation of its cosmopolitan outlook. The story of the sun and of light in the history of western philosophy is yet fully to be told. Its role as the visible model of the Good in Plato's ethics, the relation of illumination to thought (the parallelism of light and active reason) in Aristotle's theory of knowledge, the sun shining without apparent loss of power as a model of emanation in Philo and Plotinus, the unification of power, light, creativity, and goodness in the sun, and the parallel combination in divinity—all these are tantalizing items that a fuller history of thought models has yet to make richly clear. Health models—the view of morality as a kind of health—are already found in some primitive societies. And if we add to the physical models and models from various crafts and processes of work others that come from human institutions, the variety seems almost indefinitely increased.

Now while many of these comparisons are casual in the senses indicated above, some models have obviously been put to work in ethical theory for analytical purposes and some for structural purposes. Illustrations of these theoretically more serious uses will be taken in the next two parts from the history of ethical theory.

III

In the actual processes of ethical theory—the analysis of ethical concepts, forms of ethical expression, modes of organizing moral data, methods of validation and justification in ethics, articulation of criteria and systematic standards—we find frequent resort to models in a heuristic fashion. Some properties of the model are selected and elaborated to suggest a similar pattern in the ethical material. Sometimes the model is simply used, and we cannot be sure how conscious its employment is; sometimes there is a conscious attempt to work it out. Let us examine a few illustrations.

Every now and then we come upon a political-legalistic model hard at work in Kant's ethical writings. We see it in his account of conscience as practically an internalized court of law in which a man is judged or condemned within his breast for violation of the law. Kant projects this picture into his religious philosophy: God is assigned the properties of holy lawgiver, good governor, just judge, precisely the properties Kant has decided that ethics needs. Some of Kant's arguments are directly transferred from the model: for example, no man can be a judge in his own case, and therefore the judge in conscience points to an impartial outside being. Kant's use of the model either influences or reflects his basic view of morality as willed law.

The conscious search for paradigms of ethical terms in some special way of using language, so characteristic of contemporary British ordinary language analysis, is an excellent example of the heuristic treatment. Most of these attempts agree in construing ethical utterances as practical, but differ in the particular linguistic activity to which comparison is made. While one selects emotional expression through language, another shifts to persuasion, a third to commanding, a fourth to commending, a fifth to performing, a sixth to advising and choosing, and so on. Sometimes all ethical utterances have been construed along the lines of one model. But more recently there has been a growth of refinement: one starts with the ethical expression and asks which model to invoke. For example, it is asked whether "ought" in its second person use ("You ought to do it") has an imperative character, or whether in its first person use ("I ought to do it") it has a performatory-decisional character. This is equivalent to asking whether the command model (as in the "Thou shalt" of the Decalogue) and the so-be-it model of performatory utterance whose exploration J. L. Austin initiated (the "I do" of the marriage ceremony or the "I accept" of the moment of contract) are primary models for ethical theorizing. The same kind of point may be made in a nonlinguistic way. Nietzsche, for example, reads a command character into the very act of willing, so that my willing anything at all is my commanding myself. In this way he is able to see willing as a central form of power striving and thus to assimilate ethics to a power domination-submission model.

An interesting illustration of an unusual comparison to serve a specific function, in this case to work out dimensions for evaluation, is Parker's use of the musical model.[7] He not merely employs a general concept of orchestration of values, but selects features of sound such as volume, timbre and pitch alongside of the more familiar ones (such as intensity and duration) to be given an interpretation in value measurement, quality being the value analogue of timbre and height of pitch. He argues (as J.S. Mill had done for quality of pleasure) that these are not reducible to quantitative compounds, nor is height (for example) merely rank based on other measurements. No more is claimed for the analogies than utility in discovery; the features discovered, he says, could have been found in any value field.

The monetary model is a more familiar one in value measurement, and has had a wide scope, most notably in Bentham's felicific calculus. The properties he assigns to pleasure and pain—the homogeneity, quantitative additive character, possibilty of transpersonal summation—are obvious properties of money. When questions arise about difficulties in applying the specific criteria of the felicific calculus, he sometimes explicitly suggests a monetary solution. Thus he invokes such maxims—certainly useful in domains of legislative reckoning—as that increase of happiness can be assumed correlated with increase of wealth. Or else he turns pragmatic and says that unless you use something like money as a standard for pleasure you will not achieve an objective morality. The model by no means exhausts the contribution of the felicific calculus, nor Bentham's contributions to the theory of ethical measurement, but it does operate as a support.

In problems of validation and justification in ethics, the most powerful model has been the Euclidean. An ethics was to take the form of a deductive system, like Euclidean geometry, and validation of an ethical statement would consist of proving it as a theorem within the system. This model has been long under attack in ethics, in spite of its fascination, and the very diversity of attacks shows how complex are the features of the model. The fact is that Euclidean geometry tied into a bundle at least five features, which may be quite distinct: the use of logical deduction; certain types of first premises in the system, universal in form and sometims described as necessary; intuitive certainty or self-evidence of the premises; no need for specific verification of the theorems once they have been deduced in the system; and the unity of the system as a whole in that it purports to cover its field by its relatively few initial premises. Now in ethics, the use of logical deduction is attacked by voluntarists and emotivists, among others; the insistence on universal premises by those who take moral rules to be frequency statements summarizing individual particular moral reactions; the intuitive certainty by empiricists; the omission of specific verification by empiricists and inductivists and even perhaps particularist intuitionists; the unity of the system even by pluralists who accept all the rest but whose intuitionism stresses a list of independent rules and problems with only contingent relations. Now there are no doubt clusters and natural linkages among some of these diverse criticisms, but the separability of the features is not affected. The history of the Euclidean model thus shows that where it exercised a powerful attraction it meant a major construction in ethical theory. In this respect it might almost be regarded as a structural use of a model except that it was so limited to the methodological question of form alone, and seemed indifferent to the kind of content—whether it was a geometry beginning with the nature of God or Substance, with premises about natural law or about movement or about sentiments.

The lessons that emerge from these examples about the benefits and dangers in the heuristic use of models may be deferred till we have dealt with structural uses.

IV

Several of the models already mentioned or described above appear to have a structural use. They provide central or integrating concepts and the relations they elaborate remain as an organizing framework for the ethical theory. The military model in Stoicism, the legalistic model, especially with a divine lawgiver, the debt model briefly mentioned, have either been of this sort or seem capable of such a development.

For a complex model which often takes us unawares in building assumptions into ethical theory and eventually provides a fairly popular framework, consider the medical model. In questions of the body it involves a well-functioning organization with integrated aims of survival and continued functioning, fairly clear indicators of the forces that upset equilibrium, self-regulating processes to counter such intrusions, and a fairly well-articulated ideal of health to govern reflection for practice. When the model is transferred into ethics, a parallel structure of assumptions is elaborated. There is a determinate set of strivings and needs whose satisfaction constitutes the good; according to philosophical predilection it takes the form of either an outright teleology or a "teleological mechanism." It interprets the doing of evil as either ignorance or the distorted pursuit of the good; it rejects the demonic as abnormal, inclines to ideals of harmonious expression, and often takes an optimistic view of man's nature. Many of these tendencies are also found in ethical theories that do not use a complete medical model, but begin with a framework of drives or needs or impulses or desires as the matrix of ethical inquiry.

In order to present the role of structural models in greater detail I shall deal more fully with a comparison of two theories: Aristotle's use of the model of craftsmanship in ancient ethics, and Ralph Barton Perry's use of a model of economic enterprise in the early twentieth century. Aristotle frequently tells us that nature works like the artist or craftsman. Perry's theory has not, of course, the stature or influence of Aristotle's, although it did play a large part in the development of general value theory in twentieth century American philosophy. But it has a transparency which makes it an apt illustration. In fact he calls his first book on ethics *The Moral Economy*. [8]

Aristotle's craftsmanship model is that of individual production of a limited end in a particular case. The builder constructing a house or the doctor restoring health by a specific treatment, as well as the sculptor making a statue out of stone or bronze, are his favorite examples. The process is guided by an inherent plan of the form to be achieved. For Aristotle, nature works that way too, though unconsciously. (This is his familiar concept of final causation, applied in biology and physics as well as in human affairs). The model refers thus to the small-scale craftsmanship of his day, not to mass machine production of an endless stream of commodities. Perry's economy model too must be considered in the light of the economy of the time at which he was writing—the first decade of the twentieth century. We expect therefore to find the features of a free enterprise business system with each enterprise pursuing its own affairs for its own gain, cooperating with others for mutual profit and with an ideology of public service as the outcome; and the whole accompanied by a sense of endless possible progress. [9] Given this brief sketch of the models, it remains to see how permeating they are in the two theories, and what different structures are the outcome. Let us compare:

1. their conceptions of the good,
2. their modes of comparing or measuring values,
3. their accounts of what virtue consists in and what are the central virtues,
4. their general spirit and view of the nature and tasks of ethics.

1. Corresponding to the position of end in the craftsman's work, the good is taken by Aristotle to be the central ethical concept. He employs a means-end framework, hierarchically organized with means, subends and final end. The final fixed end for man is happiness, involving an expression of the ordered capacities of his nature, culminating in his rationality. Aristotle's good as natural end is quite different from the types of organiza-

tion that other models in the history of ethics have encouraged—for example, overarching endlessly approachable but unattainable ideals, or a particular quality such as pleasure to be extracted from experience in endlessly increased amount. It is the fixed goal of a craftsman who knows what he wants and is working to it in a limited number of steps; Aristotle dismissed aimlessness (having no goal) as folly, and endless pursuit of the instrumental as vain. On the other hand, Perry starts with individual interests as the units of life, each partial to itself and defining its good for itself, but uniting in some form with others for greater scope and power, and so by organization becoming an economy or community of interests. There seems to be no predetermined control of the rise of interests, for the fulfillment of any interest is taken to be good; the moral good, however, is identified with fulfillment of an organization of interests. Instead of the strict means-end hierarchy of the craftsmanship model, we have the rising scale of cooperative organization from isolated interests, through reciprocity of interests, incorporation of interests through a purpose, fraternity of interests, to a universal system of interests.[10]

2. In their modes of comparing or measuring value, Aristotle's has a constructionist character while Perry's stresses the expansion of enterprise. For Aristotle, the better is the more complete; he thinks in terms of a whole of parts which one has in mind as the good, and whatever embodies more of the elements that go to make up the whole is better than what embodies fewer. In the *Nicomachean Ethics* there is in fact a dearth of any other mode of judging the better than tying degree of goodness to extent of completeness.[11] Certainly there is nothing like the quantitative scale of the hedonist, such as the Benthamite monetary model referred to above. Perry does assert as an obvious truth that more good is better than less good; but his model is not the accountant's zest for growing profit, so much as enterprise growth that subtends public service. Hence selection and gradation of interests is made on the basis of contribution to the collective body. Prudence rises to moral purpose as there is extension and organization of more inclusive interest and egoism is stupid provinciality. "Morality is only life where life is organized and confident, the struggle for mere existence being replaced with the prospect of a progressive and limitless attainment. The good is fulfilled desire; the moral good the fulfillment of a universal economy, embracing all desires, actual and possible, and providing for them as liberally as their mutual relations permit."[12] Whereas the ultimate good for Aristotle culminates in the contemplation of the eternal, as an isolated achieved end, Perry's "progressive and limitless attainment" has to be safeguarded against such isolated absorption. His evaluation of art is significant in this respect. Since the aesthetic interest absorbs us directly and is self-sufficient, "its continuous return of good

being guaranteed, it is one of the safest investments."[13] But its isolation is a danger, and it may yield a narrow concentration, inimical to progress. The value of religion, on the other hand, consists precisely in the enlargement of the circle of life to the world in its full sweep.[14]

3. There is a clear contrast in their treatment virtue. For Aristotle, virtue is the end product of learning by practice, almost as an apprentice under the eye of the master. While no doubt a familial model is invoked at some points, much of his account of the man of practical wisdom serving as the model is suggestive of master-apprentice relations. The differentiating mark of virtue is action according to the mean. Many themes do enter into explication of the idea of the mean, but the dominant one is the idea of the just-right, an artistic or craftsmanlike fashioning of the raw materials of the human makeup to yield habits of selection that will avoid both excess and deficiency. And this is the mood that pervades the long treatment of specific virtues; in each case we have a picture of raw materials being worked up into a mean-organization. In Perry, on the other hand, the virtues are derived from the scale of cooperative organization, each economy having its characteristic principle or typical mode of action—for example, intelligence for the isolated interest, prudence for reciprocity of interest, and so on up to good-will for the universal system of interests. And he charts typical forms of errors for each level as well.[15]

4. Not only is their general spirit different, but their conception of the nature and tasks of ethics shows the differences in their basic models. The craftsman makes his product according to definite and well-established procedures and transmits his craft to his apprentices. Aristotle's ethics conceives of a fixed pattern of life with fixed goals and definite virtues transmitted from generation to generation. The task of ethics is to build these virtues to cultivate the knowledge which brings the pattern to full consciousness in those capable of such rationality. The culmination of the good is the contemplation exercised by the truly wise. For Perry, the nature of ethics is simply to offer "the most competent advice as to how to proceed with an enterprise whether large or small."[16] Its tasks are to build a more and more harmonious enterprise and to ensure that progess is not ended.

It is perhaps worth noting that Aristotle and Perry even in casual remarks seem to have their models constantly in mind. Aristotle says that if the best cannot be achieved we aim at what is closest to it, just as (he adds) a craftsman does what he can with the material at his disposal! And Perry is constantly speaking of "safe investments" and "yielding a steady return."

V

The conclusions to be drawn from this inquiry and its illustrations go along three lines. First, what kinds of effects have we found the use of models and the like to have in ethical theorizing? Second, how far is their role one that can or should be dispensed with, or have they some distinctive value? Third, insofar as they are used, what criteria can be suggested for evaluating them? Obviously, such reflections deal with heuristic and structural use, not with casual employment.

On the first question, there is an obvious sense in which models seem to be doing nothing essentially different from what would have to be done without them. An ethical theory with or without models makes factual assumptions, has some value commitments, uses a particular linguistic apparatus, and shapes itself along the lines of a given structure. In the illustrations given above, such results are in part secured through the use of models. The Platonic metaphor of the dragon carried the factual assumptions of the inability of the mass of men to control their appetites without repression; the metaphor of moral blindness hypothesized that there is a type of value-vision which may be blinded; the Aristotelian model assumed that all men pursue one ultimate end. Again, on imposing specific values in the process of shaping ethical theory, insofar as a model molds an ethical theory it contributes to the orientation that morality will have in action. A legalistic model turns almost to worship of universality, a craftsmanship model condemns aimlessness, a business model makes us uneasy about lingering in the present with intrinsic values, and a geometric model imparts a rational character to our methods. Models, too, help fashion the language of ethics. The craftsmanship model leads to an equation or near synonymy of "good" and "ends" or "intrinsic ends," monetary models and business models promote "value," and legalistic models turn ethics to a concentration on "moral law" and "obligation." Finally, that models help furnish a structure for ethical theories has been illustrated in detail.

If these jobs can be done with and without models, are models dispensable in ethical theorizing? How far are they simply imaginative diversions, how far are they replaceable as analytical tools, or do they furnish genuine theoretical components in the resultant theory? Certainly they are imaginative. To think of life as an art, or as perpetual warfare under an outside commander, or as constant self-legislation for a community of equally free spirits, or as a hopefully prospering business, or as anguished steering in an uncharted sea without even a port to aim at, is an imaginative feat. Great models in ethical theory are profound creations. Whether they are diversions is a more difficult question. They certainly are used seriously, they often express basic trends, institutions, and problems of a given age.

From the point of view of intellectual history, models in ethics, like basic models in science, prove very revealing. They are worth serious study by the historian of philosophy.

The formulation that a model is an analytical tool is itself a metaphor. This metaphor suggests that the tool is not part of the product, and even that what was brought about in one way may be brought about in others.[17] In all the contributions of models as analytical tools, and even in their larger-scale structural use, we might expect theoretical progress to take the form of stating explicitly factual assumptions imported, values accepted or shared, linguistic notions elaborated, and structural framework proposed. At that point past models could be dismissed with thanks, or retired on a pension of humanistic gratitude! Future work could go in transparent scientific terms. This is the position that sees all forms of models as at most psychological aids, ways in which a discovery happened to be made or a result achieved rather than as part of the structure developed.[18]

Such a view looks at the work as finished, the results as extricated and separably formulable. It says, as it were, that when all is over and done models will not be needed. But when is all over and done? In practical terms, certainly, the kinds of arts and crafts change, and the craftsmanship model can acquire richer meaning by attention to diversities not to speak of machine production and its conceptual effects! Similarly, even during Perry's lifetime the forms of business enterprise changed; new types of business association developed, corporations and cartels came to the fore. Though Perry largely outgrew the model when he came to his *General Theory of Value*, the proliferation of forms in business enterprise might have suggested further theoretical problems. More importantly, from a theoretical point of view, to displace the models completely requires an adequate theory of elements that are being supported by the model—of ends in Aristotle, and of interests in Perry, and their relation to desire or appetite, to reflection, to pleasure and pain, and so forth. Even Dewey, who in twentieth century ethics tried most directly to grapple with the psychology of ethics and to relate ethical concepts clearly to revised psychological conceptions, constantly found it necessary to invoke models of various sizes—from his early criticism of the reflect arc model to his significant analysis of ends (in *Human Nature and Conduct*) as functioning like targets set up to shoot at rather than goals of action; or again, his modelling of ethical evaluation (in his *Theory of Valuation*) on all sorts of social processes of grading and estimation.[19]

What I am suggesting is that models can have a theoretical function of serving as *theory-surrogates*, not in the sense of replacing a theory already achieved, but of holding the place for a theory to come. We know, for example, that a completed ethical theory requires some conceptions of the self with core values embedded in it. Different models in ethical theory will suggest that these will be transcendentally derived, biologically grounded, or culturally developed; that they will be fixed or that they will be changing. The model indicates direction of inquiry and broad character and while it waits for the theory, it furnishes a kind of theory-sketch. Or again, it functions in a more general way by being so rich in suggestion that proposed theories may go off in different directions from it, none of them as yet adequate to take over alone. In any case, however, the effort to reduce and eliminate a model is a necessary one for it is equivalent to progress in discerning the precise relations required for developing theory. The question is whether that progress can in fact take place, not just be programmatically insisted on.

As long as models are employed, there is need of criteria for evaluating their effectiveness. Some of these are an obvious: suggestive power, discernment of new relations, fruitful refinement of concepts and distinctions, comprehensive scope. These are the methodological values of any directed and systematic inquiry. More specific criteria of effectiveness come from the specific role the model is playing and the detailed consequences of its use. If it imports factual assumptions, are they correct? If it brings value-attitudes, what are their consequences? If it alters linguistic uses, what refinement is secured? And so on. Perhaps the most important general requirement one might suggest for a model in ethical theory is that it should not engage in smuggling factual and value assumptions but should declare and even flaunt them; and in its general impact, it should wake ethics from its dogmatic slumbers!

The implications of such a requirement are perhaps more far-reaching for the treatment of models than might at first seem to be the case. For the prevalent tendency has been a kind of tyranny of the model. A model takes over, extends itself and dominates theoretical construction. Even a limited analytical model often does this. For example, C. L. Stevenson began in his *Ethics and Language*[20] with the phenomenon of interpersonal disagreement and built a model for interpreting ethical terms out of this, analyzing ethical language as persuasive. He did not add an alternative model in which ethical terms would be interpreted, for example, through the situation in which men sharing some basic agreements seek to widen the area of their agreement (though he does examine patterns relevant to this as a subsidiary matter in the relation of means and ends, in chapter VIII). Consequently, the model worked out on disagreement as the basic phenomenon found it hard going when it moved to the region of personal ethical decision. Would it be possible to analyze a man's internal deliberation as trying to persuade himself? What Stevenson actually did was to supplement "disagreement in attitude" with "uncertainty in attitude" and then try to interpret such individual uncertainty in deciding as inner conflict in which one part of the personality attempts to control another.[21] While this no doubt suits some types of value decision, it clearly imposes a single character on what is probably a rich variety. To recognize the possibility that a family of models may be required for ethical terms rather than a single model would be an initial way of escaping possible tyranny.

If even a limited analytical model tends toward tyranny, how much more tyrannical is the practice of structural models? Yet, if the analysis of the structural models as theory-surrogates is appropriate, the maintenance of alternatives would seem as necessary as the consideration of alternative theories. This is of central importance both in the study of past models and in the construction of fresh ones. Let us look in conclusion at each of these points.

The basic contribution of the historical study of models in ethical theory is to the development of comparative ethics. It makes us conscious of the alternative structures that have been found, and their types or families, and so the different ways in which problems of ethical theory have been formulated. Take, for example, the major contrast between the juridical model and the inductive Newtonian model. The first is oriented to seeking laws for decision with a view to regulating behavior. The second tries to find out how people actually behave, whether in the movement of their desires or the paths of their sentiments (Hobbes, Hume, Adam Smith, and Bentham alike find their place here), with a view to expediting their behavior and minimizing entanglements. If we recognize that the different types of models develop their own concepts in terms of their own demands, we are not perturbed, for example, by finding that the "ought" in the juridical sense is not formally deducible in the Newtonian model. Hume's famous passage on the problem of transition from "is" to "ought" indicates no more than the disparity of the models. What we have to ask instead is how regulative functions are formulated in the Humean model. Such comparative study breaks through the tyranny of a single model.

Similarly, in constructing, the only feasible path—unless it prove possible to avoid all models—is their multiplication, even beyond necessity. Much might be gained by going beyond lining up familiar ones in contrast, and carrying out speculative development of unexploited analogies, metaphors, and models, if only to free ourselves by their multiplicity from any hold that unnoticed ones may have in our thinking. I am not suggesting off-the-cuff invocation of neglected metaphors, as perhaps a jeweller might say that his profession has been neglected compared to the textile field: life has to be cut to size in such a way as to catch the light, only controlled skill can produce the readiness for beauty which is caught in the glow of the gem, that the rapt beholding is the prototype of the grasp of intrinsic value. I am rather suggesting that the counterposing of models from many fields of study of man can help in ethical theory by revealing multiple relations in new lights.

Think, for example, of what might be done with an electrical model to break open the question of the nature of "intrinsic value," or what is often called "an end in itself." Picture an electrical current running through a filament in an ordinary bulb. At a certain point incandescence takes place. The glow is a fresh quality arising under determinate conditions of material and electrical properties. So too ordinary human processes "light up" as intrinsic values in the interactions of personal and social life. The value quality is a special glow, not some special or added constituent. Such a model may help counterbalance other physical models that exhibit goal-seeking, for example, as patterns of machine process. But even more, it suggests by contrast that the notion of "an end in itself" already contains a hidden model—the craftsmanship model. For unless we assumed that every act had an end, why would we think of some activity that was not being directed to something beyond it as being somehow *directed to itself?* Why not rather that it is not directed to anything? It is almost as if in the monetary model we assumed everything had to have a price to have a value, and so what was not for sale must be priced infinitely high! The notion of intrinsic good has much more packed away in it than we are likely to suppose.[22]

In general, if we recognize that the world (whether in nature or in human life) is always far richer than any single model, and also that there is constant change, then attention is shifted from the conflict of models to the correlation of the various phases that the different models have grasped. Even in the case of conflicting models, in the normative disciplines at least, there need not always be a demand for a *general* choice between them. Where they represent directly opposing forms of organization—for example, social models such as socialist and laissez-faire capitalist, or ethical models such as egoist, cooperative, altruist, etc., or the strenuous goal-directed life, the Epicurean immersion in the present, the Stoic inner-oriented stern resignation—they may be looked upon as *different possible forms of organization.* We can then ask under what conditions each in fact applies, and under what conditions it is desirable that each should apply. Attention thus shifts to the question of the criteria for a domain shifting from the scope of one model to the scope of another. There are limits, of course, to such procedures of dividing the question or correlating alternatives. But sharp choice after the exploration of such procedures and with a full sense of alternatives differs immeasurably from the tyranny of one model. It is the difference, even in philosophy, between dogmatism and insight.

Notes

1. Translated by Lewis White Beck (New York: Liberal Arts Press, 1959), p. 10.
2. *The Language of Morals* (Oxford: Clarendon Press, 1952), p. 75.
3. Cf. Nicolai Hartmann, *Ethics*, trans. Stanton Coit (New York: Macmillan Co.), I, pp. 34-44, 225-31. Hartmann doubts "whether one can operate on another for moral cataract—whether ethics as a science can" but allows of moral guidance and opening of eyes by means of one's own vision (p. 39).
4. For some consideration of the roots of moral terms, see May Edel and Abraham Edel, *Anthropology and Ethics*, rev. ed. (Cleveland: The Press of Case Western Reserve University, 1968; New Brunswick, N.J.: Transaction Books, 1970), ch. 10.
5. Friedrich Nietzsche, *The Genealogy of Morals*, trans. Francis Golffling (Garden City, N.Y.: Doubleday Anchor Book, 1956), p. 202 f.
6. See the account of Japanese ethics in Ruth Benedict, *The Chrysanthemum and the Sword* (Boston: Houghton Mifflin, 1946).
7. DeWitt H. Parker, *The Philosophy of Value* (Ann Arbor: University of Michigan Press, 1957), esp. pp. 103-05 and 189 ff.
8. Ralph Barton Perry, *The Moral Economy* (New York: Charles Scribner's Sons, 1909, 1937).
9. For a discussion of Perry's theory in its historical relations, in comparison with Dewey's, and in contrast with later twentieth century ethical theories, see Abraham Edel, "Some Trends in American Naturalistic Ethics," *Philosophy in France and the United States*, ed. Marvin Farber (Buffalo: University of Buffalo Press, 1950).
10. Perry, *The Moral Economy*, pp. 78-79.
11. In the logical works (*Topics* III) and in the *Rhetoric*, there are more varied indices, largely taken from ordinary usage in preferential selection.
12. Perry, *The Moral Economy*, p. 72.
13. Ibid., p. 192.
14. Ibid., p. 253.
15. Ibid., p. 81.
16. Ibid., p. 2.
17. Our argument here is itself a good illustration of how a model works. It sets reflection going. Why cannot a tool be used to achieve results and then have its destined place in the completed product? A rock could be used as a hammer in construction and at the last moment be fitted in as a coping-stone! It is the very idea of a tool that is there being transformed in such an objection. To call something a tool probably means that it is being used to act only on other things. But we could revise the term to refer to a mode of action instead of an object. There would then be no tools but only tool-behavior of things. Compare the view that there are no slaves but men held in servitude with the Aristotelian conception of natural slavery.

18. It is worth noting as parallel to this controversy about the eliminability of models in ethics, the general conflict on the question of models in science and philosophy in recent years. Two opposing views have taken shape. One maintains that some kind of philosopical orientation of outlook—in effect a metaphysical model—is inevitable as an initial act in interpreting the world, so that one's picture of the world is in some fashion different from what it would have been by taking an alternative stand. For example, Stephen Pepper's *World Hypotheses* (Berkeley and Los Angeles: University of California Press, 1942) outlines four basic root-metaphors and shows how each develops its own interpretation of what evidence itself consists in, so that there is no outside way of adjudicating absolutely between them. Pepper proposes instead to see the advantages of these different spectacles by applying each in turn to the various fields of philosophical inquiry. Kindred tendencies are found in Whorf's linguistic hypothesis that different languages embody basically different categories of thought, so that the range of translation of ideas is severely limited (*Language, Thought, and Reality: Selected Writings of Benjamin Lee Whorf*, ed. John B. Carroll, published jointly by Technology Press of Massachusetts Institute of Technology and John Wiley & Sons, New York, 1956). For a critical treatment of Pepper's thesis, with some comment on Whorf's relation to it, see Abraham Edel, "Interpretation and the Selection of Categories," *Meaning and Interpretation*, University of California Publications in Philosophy, XXV (Berkeley and Los Angeles: University of California Press, 1950), esp. pp. 69-72.

The opposing view, familiar in the positivist tradition, holds that explanation by the use of models in philosophy represents the prescientific trend of explaining by assimilating the unfamiliar to the familiar, and therefore may limit the development of fresh modes of thought. Ernst Topitsch, for example, continuing the positivist tradition with historical sophistication, argues that the metaphysical views of the prescientific period represent the use of biological, technomorphic, and sociomorphic thought patterns that developed in ancient philosophy, in part constituting a refinement rather than a sharp break with mythological modes of thought. See his "Society, Technology, and Philosophical Reasoning," *Philosophy of Science* XXI (1954): 275-96; also his *Vom Ursprung und Ende der Metaphysik* (Vienna : Springer-Verlag, 1958).

Thomas S. Kuhn's *The Structure of Scientific Revolutions*, International Encyclopedia of Unified Science, II, 2 (Chicago: University of Chicago Press, 1962) may be seen as a counter-claim to the positivist thesis for science itself, since he views revolutions in science as primarily the replacement of one paradigm by another, where a paradigm is simply some particular accepted scientific work in all its richness which furnishes the modes of analysis and problems of work for the scientists of its time.

19. John Dewey, "The Reflex Arc Concept in Psychology," *The Psychological Review* (July, 1896): 357-370. Reprinted in Joseph Ratner, ed., *John Dewey, Philosophy, Psychology and Social Practice* (New York: Capricorn Books, 1965). For the target conception, see his *Human Nature and Conduct: An Introduction to Social Psychology* (New York: Modern Library, 1930), Part Three, Section VI. For processes of appraisal, see his *Theory of Valuation*, International Encyclopedia of Unified Science, Vol. II, No. 4 (Chicago: University of Chicago Press, 1939), IV.

20. C. L. Stevenson, *Ethics and Language* (New Haven: Yale University Press, 1944).

21. C. L. Stevenson, *Facts and Values* (New Haven and London: Yale University Press, 1963), essays IV and XI, esp. pp. 191-203.

22. In Perry's *General Theory of Value* (New York: Longmans, Green & Co., 1926), to find something intrinsically good represents no more than the effort to maintain or perpetuate an object of interest. Apparently the theory still wants to make sure that there is no real resting place to hinder progress.

5. Some Psychological Presuppositions of the Concept of Virtue: A Case Study in the Relation of Science and Ethics

The concept of *virtue*, together with that of *good* (value) and that of *obligation* (right, duty), are the Big Three in the history of ethics. When John Laird published his *An Enquiry into Moral Notions*,[1] it seemed obvious to him to parcel the field into "aretaics," "agathopoeics," and "deontology"—that is, the study of virtue, of the good, of duty. But this was the learned historian's perspective. Actually, *virtue* had long before receded from its eminence in ancient ethics and in eighteenth century ethics, bowing out to *good* and to *obligation*.

The three concepts have no doubt systematic relations. Moral concepts have a collective job to do, and when one recedes, the others take over, or perhaps the one recedes because the others are driving it out. Now just as military strength cannot be understood without the economic and industrial base, so the metaethical relations of ethical concepts cannot be understood without seeing the background economy. This background is psychological and cultural as well as methodological and intellectual. Contemporary philosophers have paid attention chiefly to the latter pair, and the analysis of ethical concepts has been guided chiefly by epistemological considerations. The present paper wants to straighten the bent stick by bending it back in the other direction. I choose the concept of *virtue* as a case study because its theoretical eclipse is itself an interesting phenomenon, because its psychological presuppositions are closer to the surface, and because there is some tendency today in philosophical circles to revive the concept. The same kind of job could be done for *good*—in fact, Dewey did it—by looking to the underlying psychological theory of purposive action and goal-seeking, which in effect calls the metaethical tune for this concept. *Obligation* (right, duty) has been, on the whole, more resistant to such analysis because of the sheer complexity of the strands that are built into it.

A few historical notes will be helpful. The basic starting points for the analysis of *virtue* are to be found on the one hand (the moral-code side) in the veritable parade of virtues on the face of human social history; on the other hand (the theoretical or, if you like, the metaethical side) in the diversity of philosophical treatments of the concept.

The descriptive parade need not occupy us long. For we are all generally familiar with the panorama of different virtues that have been central in different societies, cultures, subcultures. We may recall: the Greek cluster of wisdom, courage, temperance, justice; the Spartan war-virtues of courage, discipline and obedience, taciturnity, dogged adherence to tradition, loyalty, as contrasted with Athenian individualism, receptivity to the novel, attitudes of freedom and equalitarianism, readiness to discuss and to exercise initiative; the medieval Christian outlook with a stress on inwardness, purity, love, obedience, patience, humility, and the familiar history of vices that issue in the deadly sins; the Calvinist-puritan ethic with its constellation of thrift, diligence, sobriety, justice, sexual purity, and the striving for success in one's calling, the liberal virtues of individualism, initiative, creativity, independence, readiness to change, prudence, rational orderliness, humanitarian equality, and so on. These are bare suggestions of a wealth of constellations in the western world alone, to which the study of other areas of the globe and the comparative study of primitive cultures can add tremendous variety. If, however, some philosophers among us would rather consult the dictionary than the spread of history and comparative culture, they can find an equally extensive list, though less organized. For example, among the a's alone one can get going with: abject, abstemious, abusive, affable, ambitious, amiable, etc.

The diversities of theoretical analysis reflect the general philosophical approaches, as they seek a unified understanding of the concept. Yet they also show clearly, when lined up, how the presupposed psychology of the soul or the self furnishes the guidelines within which the concept is developed. Thus Plato starts off, in the Socratic dialogues, by asking whether all the virtues are really forms of knowledge; he ends up with a whole theory in which the unity is found in the soul's quest for the Good, and the several virtues then reflect the different parts of the soul in their effort to carry out their functions.[2] In general, teleological philosophies look to virtues as the ordered dispositions or states of character which will enable the human being to achieve the good. Aristotle,[3] analyzing virtue into its genus and differentia in order to furnish a formal definition, locates the genus as *state of character*. The differentia he finds to be *aiming at the mean*. Accordingly, he analyzes each virtue by locating first the raw materials built into the character trait, such as feelings of fear and confidence blended in some degree in courage, or social activities of giving and taking integrated in liberality. The differentia is seen in the formula that governs the blending: if the mean is achieved we have the virtue of courage, if excess in one or the other direction we have the vices of timidity and rashness; so too for liberality as against prodigality and miserliness. Aristotle's analysis gives the appearance of a plurality of virtues, each constituting a stable relatively isolated system in its operation, fully actualized in the virtuous man and embodying a more difficult balance in the continent man.[4] In the later development, in the utilitarian philosophies, there is a similar structure for virtue. Stable character traits are formed, under the operations of association as the psychological principle according to which experience settles into patterns; in the case of virtues as qualities of a person, the governing principle is that of utility, that is, character traits whose operation yields the general well-being.

There is another tradition in ethical theory which gives a less set and less dispersed account of virtue. The psychological base is some more unified, more central process that is going on and lies at the heart of being a man. Virtue becomes narrowed into one supreme effort. Thus in ancient Stoicism[5] the spirit in its pursuit of serenity and independence wages a continual internal war not to attach itself to what is beyond its power; virtue lies solely in effective detachment. In Augustine, too,[6] there are no achieved virtues that can bask in their intrinsic worth and be desired on their own account, for such would be vices inflated with pride; the moral worth of every act lies in the character of the will embedded in it. Similarly, in Kant virtue is a moral strength of the will. The concept of virtue enters Kant's ethical scheme at a late point. A holy will has been defined as one governed completely by the moral law, and obligation has its origin in the fact that men are moved by inclinations that are in conflict with the moral.[7] The task of continuous progress toward the unattainable idea of the holy will is morally enjoined on men.[8] Hence, virtue lies in a continuous battle in which it is always at the starting point and in which not to advance is to lose ground.[9] Again, in Nietzsche, though in a quite different sense, it is will that is seeking expression, now as domination; hence virtues express nothing more than the quality of domineering strength or weakness.[10]

There is a third tendency in the philosophical presentation of virtue in which the focus is shifted from the cultivation of stable traits in the pursuit of the good or the quality of the inner struggle to the milieu of sociocultural interaction. Virtue is approached in terms of men's appreciative and critical responses to human behavior, but along different lines. Eighteenth century moralists, under the spell of the Newtonian model, looked for the underlying laws governing the responses and moved toward utilitarian theory.[11] Contemporary phenomenology takes the appreciative response to be the apprehending of values of personality, and looks for the essence in each cluster of virtues.[12] The materialist and naturalist approaches move toward a sociocultural base for understanding virtues; these are patterns of character or self stabilized and approved because of the institutions and forms of goal-striving they support. In Marxian historical materialism they are tied to the class-formations and in evaluation to the direction of social development.[13] Dewey, accepting the shifting content of virtues and the fact of unavoidable constant change, redefines the concept in methodological terms to indicate the central features required in the expression and pursuit of human interest.[14]

In spite of this long tradition, it is striking that twentieth century philosophy does very little with the concept of virtue. Sometimes it is reduced to a subclass of values: virtues are values of personality alongside of social values, action values, religious values, etc. More often, in analytic approaches, it is wholly subsumed under obligation. For example, E. F. Carritt says: "If we define virtuous dispositions as those which lead people to do impulsively and effectively what reflection would generally or often show to be obligatory we seem nearer the truth."[15]

This twentieth century eclipse of the concept of virtue can be seen to stem from both methodological and psychological considerations.

One very important cause was the positivist restriction of metaphysical entities in which such notions as self were swept away and even dispositional terms had for a time difficulty in surviving. The fear was that the language of potentiality or power or disposition would be used in explaining the phenomena, yielding obscurantist accounts. Bentham long ago had already described a disposition as a fictitious entity, attempting to express what is permanent in a man's frame of mind.[16] Because the idea of character or self or personality is constitutive within it, the notion of virtue was particularly susceptible to this attack. The extreme psychological behaviorism of the earlier part of the century had the same effect. Once we translate the language of virtue into specific sets of acts, then the ethical apparatus which deals with right acts or obligatory acts hopes to handle the materials without any additional theoretical concepts.

The empirical critique of the concept of virtue has often leaned heavily on Hartshorne and May's *Studies in Deceit*.[17] These investigators carried out tests of honesty behavioristically. They gave maze puzzles in which the subject was cheating if he kept his eyes open. They asked students to mark their own test papers and had paraffin-paper techniques for detecting where the student corrected his previous answers. They asked subjects to check books read and included false titles on the list, and so on. Because there was insufficient correlation of results in diverse tests, they concluded that deceit is not a unified trait. Deceit and honesty are rather specific functions of life situations. Deception is a method of adjustment used by a child when conflict arises in the environment. Qualities of man as an organized socially functioning self, not isolated virtues, distinguish a good man from a bad one. Transactional or situational qualities rather than intrapsychic traits fit the picture of the results.

To this shift in perspective was added the anthropological concept of cultural patterns, impinging on the relatively plastic child in the process of his growth.[18] Institutional and cultural forms and the practices and goals they embody furnish the unity of character. The outcome is, of course, men of a given sort, but both the explanatory principles for understanding that sort and the normative principles for evaluating it are sociocultural. A major practical base of the concept of virtue has always been the fact that the new generations arise and education is required to develop character. Education itself, however, may find that it does not do this job successfully by concentrating on the formation of single traits, associating praise or dispraise with them. Rather, it may find that it makes more progress if it cultivates insight, or uses role-playing to develop sensitivity in understanding others' positions in human situations. Thus there may be no distinctive role left for the concept of virtue. The materials may be better cut and sorted along the lines of other concepts.

This stage in the history of the concept of virtue constituted a low ebb. It seemed that methodological, scientific, and practical considerations combined to render the concept useless and even misleading; to say that a child did not steal because he was honest was no better than the old jibe that opium put one to sleep because it possessed the dormitive principle. Still, this was not the end of the story. But the philosophically significant point is that the factors that came to the aid of the concept were of the same type as those that had shunted it aside. To these we must now turn.

First, there was a methodological retreat on the strictness of positivism and extreme behaviorism. Dispositional terms became reestablished with methodological finesse, and generally, it became permissible to construct theoretical terms and leave open as a separate issue whether the particular ones should be given a realistic interpretation (as "atoms" and "genes" eventually were) or a purely instrumental interpretation for purposes of calculation and systematic simplification (which is all one could claim for "average man" or "gross national product"). What was purged was obscurantism. As Laird had pointed out in the case of *virtue* as early as 1935:

> There is no good reason for regarding a man's moral character as a deep and holy well in which all that really matters is a secret sediment at the bottom. If the moral character were a name for the *unknown* cause of his moral thoughts and actions, it would indeed be secret and mysterious by definition; for it would be an unknown quality. But if it is essentially a statement of our knowledge about what *would* happen as well as an account of our knowledge about what *has happened* in the moral way there is nothing more mysterious about virtue and moral character than about anything else that concerns morality. [19]

Such an account does not settle the question of the viability or utility of the virtue concept; rather it leaves the settlement open to scientific and theoretical investigation. In abstract terms it leaves us with the following situation: each of the big three ethical concepts would appear to have some type of linkage to a subset of psychological phenomena. *Good* is linked to appetites, desires, purposes, pleasures, etc., whatever the precise logical character of the linkage; the concept is viable if there is sufficient stability in this subset, whether a stability of goals or of criteria for ordering and evaluating goals. *Obligation* is linked to phenomena of "conscience"—a range of feelings and phenomenological qualities—as well as to rules and practices functioning to regulate claims within the group; and similar issues of logical analysis and of stability arise as in the case of *good*. *Virtue* is linked to ideas of character, personality, and self-pattern. The question is whether these three subsets of psychological phenomena are relatively independent or whether they constitute a systematically related field. And if the latter is the case, the further question becomes pertinent, whether the concept of virtue is in some way reducible to the other concepts (or any of the three to the other two) so that its properties and the phenomena to which it is linked are explicable by the others and their associated phenomena. The viability of the concept of virtue is thus a special function of the kind of results we get in the scientific exploration of character, personality, and self.

It is to be noted that the kind of stability and relative independence a viable concept of virtue would possess need not be that of a separately existing entity. It is quite compatible with virtue being a function of the total state of a field. Suppose it is granted, and indeed it seems very likely, that the concepts and phenomena we are dealing with constitute a systematic field. Or in general, let it be granted that any individual (thing or person, property or trait) is part of a wider milieu so that all ultimate explanation of its behavior or its special features is in principle a function of the total field. It may still be the case that, given certain constancies in the field, the behavior of one element in the field can then be understood in terms of its internal state alone. Even in the physical domain there will be differences of such a sort. The smoothness of a sheet of paper may be a function of the moisture of the field, if it is isolated, but of the weight on top of it, if it is at the bottom of a ream. On the other hand, the hardness of a metal may be usefully seen as a function of its inner constitution as long as the temperature fluctuates only within certain limits; near the surface of the sun its hardness would not be a comparably useful concept, but in practical technological contexts and in ordinary scientific contexts it may do. So too for *virtue*, if the field conditions are of the requisite sort. Let me take a few illustrations from the psychological literature which might seem to have a bearing on the answer. The aim here is not to give a conclusive answer but to restructure the way questions about virtue have been discussed in order to show how an answer may be sought. In this sense, the paper is (alas) simply programmatic.

First, then, for the concept of character. On the whole, it has been criticized as too commonsensical, too crude, to bear the burden of scientific work and philosophical precision. Lawrence Kohlberg[20] has criticized the virtues and vices found in the conventional language as turning moral character into a "bag of virtues"; he relies largely on the Hartshorne and May results and kindred studies to affirm its lack of utility. Dewey's treatment of character as a set of habits[21] has the same pluralism as the Aristotelian and the common conception, but Dewey's view of habit is more dynamic. When we recall that Dewey gives intelligence a reconstructive role in the problem-situations where habits are in conflict, and that he uses the notion of habit in a stretchy way to cover the idea of self as well as that of intelligence itself, habit would seem to get us not far beyond the idea of relatively stable congeries of dispositions whose stability depends on the dynamics of the field. Gordon Allport, it is interesting to note, regards character an an evaluative concept; he says: "Character is personality evaluated, and personality is character devaluated."[22] If this is so, then the notion of virtue can have no greater stability than that of personality.

How far will the scientific exploration of personality carry us in our search? Allport, after examining fifty definitions, settles on "Personality is the dynamic organization within the individual of those psychophysical systems that determine his unique adjustment to his environment."[23] This would not be very helpful to the concept of virtue, for several reasons. It ties the study of personality to the study of differences, rather than to the constancy of a pattern among individuals; traditionally, at least, a man does not grow his own special virtues, but is praised for having the ones that are hoped for among all men. (Would it be logically inconsistent to say that all men had the same personality, or would one then have to say that men lacked personality?) Again, the reference to psychophysical systems assumes the constancy can be exhibited in the physiological and psychological terms; the kinds of constancies appropriate to virtues might very well require cultural and sociohistorical terms as well. In a more comprehensive formulation of personality by Kluckhohn and Murray,[24] prefatory to a collection of studies on all behavioral-science levels, the conception of personality is cast more in terms of certain functions to be carried out in a human being; it is the constancy of these functions which gives unity to the concept. Illustrations are: reduction of conflict by scheduling, by social conformity, by identification; reduction of aspiration tensions, and so on. Human personality is treated as a formation embodying a compromise. The analogue for virtues in moral theory might very well be to list a set of functions in internal organization, interpersonal relations, and goal-pursuit, and see the virtues as appropriate modes of internal harmonization, of reducing interpersonal conflict, of regulating aspiration, and all of these stabilized as habits of individuals.

Along such lines, one would be tempted to look for some deeper constellation, in which personality takes its origin—for example, a depth account of the ego and ego-functions. In fact is seems as if the resurgence of the idea of personality itself drew considerable strength from the psychoanalytic theory of determinants. For the latter conveyed the promise that internal operations would be found distinctive enough to secure a patterned outcome whatever the external (sociocultural) factors; this was, in effect, a promise that psychological constancies would be established across historical-cultural lines. And even where culturally variant character traits were to be found, their significance would lie not in the differences so much as in the role they played in the internal economy. This was the way, for example, in which in the earliest Freudian formulation, attention had been focused almost exclusively on the inner career of libidinal energies. Again, the Freudian theory of succeeding oral, anal, and genital stages in child development tied specific virtue patterns to each; for example, orderliness, frugality, obstinacy, punctuality and cleanliness were regarded as anal traits. We have here a multiple thesis. In part it is causal: particular zones become at different times the focus for discharge of excitation, and the situation of toilet training becomes the mode of interpersonal relation in which the designated traits arise. In part it is diagnostic; excessive rigidity in these traits is taken to involve an original disturbance at that particular stage of development. Perhaps it is also phenomenological; no one wholly leaves behind earlier stages, so that normal exercise of these character traits involves some of the emotional quality of its origins. Contemporary work has developed such an approach with greater attention to the permeation of cultural factors. Erikson continues the developmental orientation;[25] he finds critical steps in the child's progress, mastering of which is productive of basic character-attitudes. Thus when the infant successfully lets his mother out of sight without undue anxiety or rage, we have the achievement of trust as against basic mistrust; experiences in connection with major bodily functions show us the setting in which autonomy is developed as against shame and doubt, and initiative is developed as against guilt, and so on. Cumulatively there is the development of individual identity and ego integrity. The psychological traits here involved are readily recognizable as the content of major virtues in moral usage. Erich Fromm[26] attempts directly to construct character-types through fusion of psychoanalytic and cultural categories, for example, the exploitative orientation in which a man takes by force and cunning, or the marketing orientation in which a man experiences himself as a commodity with self-esteem depending on the extent to which he can sell himself. Abram Kardiner's concept of a basic personality[27] also attempts a fusion: he points to a relatively stable constellation of traits in each person expressing largely the individual's early familial relationships and methods of rearing. These play a secondary role alongside of primary economic relationships, and are reflected most clearly on a large-scale screen in the projective nonreality systems of folklore, religion, etc. The basic personality constitutes, in effect, the psychological profile of the society.

Such approaches appeared to be stepping-stones in the attempt to develop a more integrated model for the study of personality which would take account of the social, cultural, and historical as well as the biological and innerpsychological. The shape of the outcome is not yet clear; no doubt it depends on the general form that an integrated theory of behavioral science will take. In more recent psychological study the concept of the self has reemerged as an inclusive concept, of which the ego is one part, and focus·lies on the growth and patterning of the self-system in response to nature and society. Comparative anthropological study (as in the work of Hallowell[28]) shows the norms which get built into the self; modes of self-reference, location of position, sense of the self in time, interpersonal orientation, etc. Psychological study turns to direct features of the self-system. For example, the current work of Witkin and others[29] relates perception and cognition to personality; it distinguishes the field-dependent and field-independent characteristics of individual selves and examines the conditions of self-formation which achieve weak or strong boundaries between the self and the field. Current studies of shame and guilt in this context interpret them clearly as modes of response in self-formation and self-regulation, rather than as emotions with a life of their own.[30] Studies of sensory deprivation and effects of drugs probe for chemical bases and differential psychological effect on the self. Such comprehensive studies, as well as the sociocultural work on the relation of self-form to institutions and social practices, considerably enrich the concept of self and increasingly give substance to the concept of virtue as prefigured in the more dynamic moral philosophies we noted above.

It is in terms of the outcome of such developments that the viability of the concept of virtue in moral philosophy has to be settled. There is no point in forcing it. Any purely philosophical formulation would constitute in effect a projection of possible results, and would be tested by subsequent actual results. If the basic psychological picture that emerges has room for relatively independent self-features over and above a focus on the transactional character of self-expression—given specific constancies in the field relations—then the virtue concept is worth restoring to central importance.

Let me speculate a bit on possibilities. Under what field conditions would the concept of virtue become critically central to moral philosophy? Under what conditions would it become wholly inapplicable? Under what conditions would it have limited specific application? Here we have to recall the initial emphasis on its systematic relations with other ethical concepts. Suppose the nonself portion of the moral field becomes so complex and so changing that there could be no stable rules of obligation and no stable goals to serve as guides in regulating conduct. Short of an intuitive theory to deal with immediate situations each on its own, moral philosophy might seek to find its stability in the maintenance of a certain kind of self. This is, in effect, the Stoic model; or the situation in which the only moral advice we can give is to consider all the facts, be sincere in one's judgment, and so on. It is worth noting here that Kohlberg's criticism mentioned above is operative only against very specific virtues. In fact, his own theory of moral development leans heavily on ego factors of growing insight and intellectual development, which in the old Aristotelian language would have been described as "intellectual virtues"! Suppose that, on the other hand, there are stable goals, even if only in the minimal sense of the ones necessary for survival, but the complexities and flux of human existence bear heavily on the self. This is an extreme supposition, apparent only in extreme cases. For example, Robert Lifton's recent picture of "protean man"[31] is generalized from the kind of total shifts of self-form that he found to characterize subjects in Asia who had undergone successive upheavals. Especially in Chinese subjects he found shifts under extreme stress from older ways to revolutionary selves, to counterrevolutionary selves, in each period with a total transformation. Lifton thinks this kind of protean man is the emerging self-pattern, given the conditions of the contemporary world. Now under these conditions, short of some notion of the intrinsic value of the passing self-forms, moral philosophy would have to center moral guidance on a basic goal-orientation.

Finally, what are the practical needs, resting on constancies in the field of human life under the shape it seems likely to take in the contemporary world, which might support a limited set of virtues of different sorts on different levels of generality? I suspect very much that the level of generality that will crumble is the middle range one — the virtues that have a degree of conduct referred to which is too specific for the changing situation and insufficiently broad for generalized guidance; examples would be the virtues of patriotism in its national reference, chastity in its Victorian sense, and obedience in its unreflective accepting sense. But both highly general and highly specific virtues may still have a central place. Respect for others, loyalty as a sense of commitment, sincerity in interpersonal relations, a sense of responsibility for others, collective participation in common welfare, and so on, would exemplify the broad type. Many of these would almost have the kind of generality which would give them a constitutive place in self-formation, alongside of the characteristics of ego-strength like a sense of realism and an ability to see a situation from other than the immediate perspective. On the other hand, the very specific virtues would be tied to the very specific necessities of the existent mode of life; for example, care and caution in complex technology might have the absolute character that punctuality often sought in an industrial plant; precision and clarity may have a comparable base. In these senses, virtues may continue to have a place in the common-sense way in which they have played a part in education. But instead of constituting a miscellaneous bag, they will each come with an explanatory and justifying preamble, pointing to the constancies and necessities on which they rest.

It is worth noting that in these respects the problem of virtues is quite parallel to that of generalizations in obligation and in the good. There have been attacks on the notion of rules in moral obligations in specific situations. Rules can be misleading by holding up the hope of a systematic classification of duties and deductive applications to particular cases. Thus, Dewey condemned rules (in the sense he used the concept) in favor of principles of analysis. Recently there have been attempts to reinstate rules in a constitutive sense, on the assumption that morality is carrying on an enterprise and every enterprise, like every game, has its constitutive rules which tell you how to carry it on or play it, as distinct from its rules that simply epitomize its lessons of experience about how to play it well. Generalization concerning obligations may very likely have the same outcome I have suggested for virtues—viability for the highly general (almost constitutive) and for the very specific, where strict adherence is required by the state of life. On the other hand, in the case of the good, it is the most general goals that have proved empty (e.g., the idea of happiness, or of pleasure), and the middle range and the very specific goals seem to be coming into their own.

I suggest as a concluding diagnosis of the concept of virtue, in the light of its psychological and sociocultural bases as we have considered them, that it remains a good bet for fresh philosophical analysis and systematic revival. But this assumes there will be some degree of field stability, as well as some methodological sanity which does not insist on reducing everything to momentary states or all of value theory to instantaneous preferences in momentary choices; and it posits attention to pragmatic factors or practical needs of the age as part of the determining factors. In short, the concept of virtue has be dealt with not in terms of linguistic uses alone or of common-sense psychology, but in terms of the results of the fuller scientific picture of the underlying phenomena to which the moral concepts are linked.

There is a final warning which must always be kept in mind. As knowledge of the underlying phenomena advances it is always possible that the materials underlying virtues and the materials underlying obligations (or either with values or all together) may yield a unified picture at a deeper level, and be parcelled out in a novel way; if so, new concepts may make inroads on both virtue and obligation. Something like this seems to be happening when the idea of *commitments* as both expressing self-formations and issuing in obligations straddles the field of the older divisions. Even without something like this happening, and with maintenance of the present Big Three and their linkage to the psychological phenomena explored, it is possible that some hitherto neglected component may rise to the top and take its place alongside—conceivably emotion and feelings alongside of the entrenched phenomena, though it does not seem likely in the light of present psychological theories of emotion. Or again, a deeper discovery of functional relations among the components may gear one of the concepts to the service of another—as, for example, virtues have so often been geared to basic social aims in the historical parade we surveyed above. Whatever the long-range outcome in the politics of ethical categories, we will never be able to make sense of conceptual changes unless we see them as subject to the same kind of analysis as we have here been pursuing for virtue.

Notes

1. John Laird, *An Enquiry into Moral Notions* (London: George Allen and Unwin, 1935).
2. Plato, *Republic,* esp. Book IV.
3. Aristotle, *Nicomachean Ethics*, esp. Books II–IV.
4. In Aristotle's ethical theory the plurality proves only provisional, or else a failure to achieve the unity of virtue found in the good man (the man of practical wisdom).
5. Epictetus, *The Manual*, included in *Essential Works of Stoicism*, ed. Hadas (New York: Bantam Books, 1961).
6. Augustine, *The City of God* (New York: Hafner Publishing Co., 1948), Book XIX, 25.
7. Immanuel Kant, *Foundations of the Metaphysics of Morals*, trans. Beck (New York: Liberal Arts Press, 1959), pp. 30-31.
8. Immanuel Kant, *Critique of Practical Reason*, trans. Beck (Indianapolis: Bobbs-Merrill, 1956), p. 33.
9. Immanuel Kant, *The Metaphysical Principles of Virtue*, trans. Ellington (Indianapolis: Bobbs-Merrill, 1964), p. 69.
10. Friedrich Nietzsche, *The Genealogy of Morals* (Garden City, N.Y.: Doubleday Anchor edition, 1956).
11. For example, David Hume, *An Inquiry Concerning the Principles of Morals* (New York: Liberal Arts Press, 1957).
12. Nicolai Hartmann, *Ethics* (New York: Macmillan, 1932), II.
13. Frederick Engels, *Anti-Dühring* (New York: International Publishers), ch. 9-10.
14. John Dewey, *Theory of the Moral Life* (New York: Holt, Rinehart and Winston, 1960), ch. 4.
15. E. F. Carritt, *Ethical and Political Thinking* (Oxford: Clarendon Press, 1947), p. 85.
16. Jeremy Bentham, "*An Introduction to the Principles of Morals and Legislation*" in *The Utilitarians* (Garden City, NY.: Doubleday, 1961), ch. 11.
17. H. Hartshorne and M. A. May, *Studies in Deceit* (New York: Macmillan, 1928).
18. For example: Ruth Benedict, *Patterns of Culture* (Boston: Houghton Mifflin, 1934); Margaret Mead, *Growing Up in New Guinea* (New York: William Morrow and Co., 1930).
19. John Laird, *An Enquiry into Moral Notions* (London: George Allen and Unwin, 1935), p. 17.
20. Lawrence Kohlberg, "Moral Development" in *International Encyclopedia of the Social Sciences* (New York: The Macmillan Co. and The Free Press, 1968), Vol. 10, pp. 483-94.
21. John Dewey, *Human Nature and Conduct* (New York: Modern Library, 1930).
22. Gordon W. Allport, *Personality* (New York: Holt, 1937), p.52.
23. Ibid., p. 48.
24. Clyde Kluckhohn and Henry A. Murray, eds., *Personality, In Nature, Society, and Culture* (New York: Alfred A. Knopf, 1948), chs. 1-2.
25. Erik H. Erikson, *Childhood and Society* (New York: W. W. Norton, 1950), ch. 7.

26. Erich Fromm, *Man for Himself* (New York: Rinehart and Co., 1947), esp. ch. 3.
27. Abram Kardiner, et al., *The Psychological Frontiers of Society* (New York: Columbia University Press, 1945), esp. chs. 8, 14.
28. A.I. Hallowell, *Culture and Experience* (Philadelphia: University of Pennsylvania Press, 1955), esp. ch. 4.
29. H.A. Witkin, et al., *Psychological Differentiation* (New York: John Wiley and Sons, 1962).
30. Cf. Helen B. Lewis, *Shame and Guilt in Neurosis* (New York: International Universities Press, 1970).
31. Robert J. Lifton, "Protean·Man," *Partisan Review* (Winter 1968): 13-27.

6. Contenders for Value Theory: A Behavioral and an Evolutionary Claimant

Our earlier chapters explored the relations of science and ethics largely to confront the separatist view that denied any fruitful relations. Once the right to relate them has been won, the more interesting constructive task faces us: in what ways should a theory of value make use of science?

In speaking of science on the one hand and value theory on the other, and how the second should make use of the first, there is no intention of reinstating the fact-value dichotomy in a new garb. "Science" and "ethical theory" here designate *enterprises* rather than *domains* or *entities*; both involve facts and values (purposes) in differently integrated complexes or unities of preanalytic wholes out of which facts and values are abstracted. There is no contradiction in rejecting the fact-value dichotomy as absolute while maintaining the distinction of purposes in the two enterprises of science and ethics, the one descriptive-explanatory, the other evaluative. [1]

In general, when we look for a scientific basis for ethics, we mean simply that we seek to rest ethics on knowledge and take science in its modern sense to be the paradigm of reliable knowledge. This, however, covers a number of controversial claims. One, for example, is about the "reach" of science: the ways of inquiry and the resources of the sciences extend to and are relevant in various ways for dealing with moral matters. Another involves the hypothesis that other proposed "ways of knowing" often invoked for morality—intuitive, phenomenological, interpretive—prove in the long run not to be different from the fuller complex scientific procedure but simply fragmentary or less integrated (though sometimes more intensively explored) parts of it. [2] The thesis rests on the historical view of the growth of science and its self-understanding. Our concern here is, of course, only in its face toward ethics and how ethics made use of it. For a long time science was spoken of in a collective sense with physics (and astronomy) usually in mind, consonant with the mechanistic model of the seventeenth and eighteenth century which was reductionist in spirit. Even the hedonistic ethics that was incorporated into utilitarianism could invoke a physiology like Hartley's underlying the psychology of pleasure and a theory of physical properties of physical impulses to explain the intensities of different pleasures. By the nineteenth century, and certainly by the twentieth, however, the fractioning of science had broadened the options for ethics. A project of finding a scientific base for ethical theory had to choose whether it would build on the data of the cultural sciences, or the historical sciences, or some selected combination. Often it mattered which of conflicting schools were adopted, for example whether the resources of the behaviorists or the Gestalt psychologists, whether the evolutionists or the personality theorists in cultural anthropology.

In the nineteenth century, the job of drawing the map of the sciences was taken to be the same as finding the natural joints of reality (except by those who denied reality had natural joints). One has only to think of Comte or Spencer, Marx or Mill. There were fixed categories of physical, biological, psychological, and social science to correspond to matter, life, mind, society; there were evolutionary levels of development and emergent qualities of matter, or reduction of them all to sensory phenomena. In the first third of the twentieth century the psychological and social sciences had still to fight for autonomy and use heavy metaphysical armor, for the belief in a future reduction to physics was still strong even among behaviorist psychologists. Interestingly, after the mid-century, when they were long firmly established and had no worry about the right to exist, their problems multiplied from within: conflicts of method, divisions into schools, and the very richness of advances into new fields worked havoc with all neat borders and classifications and permanent schemes. Indeed the veritable explosion of new sciences and hyphenated sciences has characterized not merely the study of man but the older physical and natural sciences as well. From the standpoint of the ethical search for a scientific basis, the experience of the last two decades in particular has shown that there is no settled system of the sciences to draw on, nor any systematic development of interdisciplinary research. The psychological and social sciences have shifting domains and a shifting character. There have been occasional approaches to partial unification, as in the so-called "behavioral sciences," and some attempt at task-oriented unification in the "policy sciences," but there have also been methodological splits amounting almost to fission. And there have been interdisciplinary studies, though too diverse in type to have a single impact: some have a common end but no common operations; some are more integrated in that the solution of a problem in one field depends on the value of a variable to be furnished by another field; and some are more fully unified by the constant cooperation of all participants fully cognizant of one another's work. In any case, there is no fixed scheme of social sciences, each with an assigned contribution.

The Basis for Ethical Theory

Let us look back now from the side of ethics to see which scientific resources have been invoked to furnish a basis for ethical theory. Clearly, whatever showed promise was tapped. It was hoped by those who regarded man as a machine that physics would furnish the plan of the human machine and the rules for keeping it in good condition; and we have seen (ch. 4) the role played by physical models in ethical theory. The Darwinian revolution in biology was followed by Herbert Spencer's later evolutionary ethics and the theories of Social Darwinism, but also by Kropotkin's opposing biologically grounded ethics of mutual aid; and of course there were endless variants of evolutionary ethical patterns suggested in the literature that followed. Freudian psychology spilled over into ethical theories as the various psychoanalytic revisions and movements developed his insights or else phases he had neglected. Anthropological comparison of preliterate cultures quite early gathered data about moralities and began to move from their variety and functions to theories of moral relativity or moral development or central moral function. Economic conceptions of value played a generative role in the development of the theory of generic value, while political science and law in their concern with the state and authority helped fashion the basic models of community and contractualism which were to govern moral theory in the greater part of the twentieth century. Historical data furnished the matrix in which conservative, liberal, and Marxian theories of history set their accounts of morality, with a vastly different potential for individual action and social policy.[3]

There is accordingly no easy way to determine in advance what will be useful. This is something to be learned or experimented with, and the whole history of the relations of science and ethics constitutes materials for such learning. It is also a field full of ideological traps as we see even nowadays from the frequent use of ethological data to provide subtle argument against equality for women or of animal territoriality to suggest the permanence of the sentiment of nationality, just as in the past lists of instincts were dragged out to support the existence of established institutions of property, family, and war. Such phenomena simply add to the importance of the general question of how a theory of ethics should make use of science.

Four ways may be distinguished in which moral philosophers and scientists have used or proposed to use scientific materials for ethical theory. One has science chiefly setting *constraints* on ethics. A second has ethics calling on science for *assistance*. A third has ethics going to science for *foundations*. A fourth has ethics seek to make its *methodology scientific*. Let us consider these in turn.

Constraints

The recognition of scientific constraints is an old matter, consonant even with the separatist outlook. It is a minimal recognition of science: a theory of ethics must not be inconsistent with the facts as science finds them to hold of this world. There is no point holding to a moral ideal which is incapable of achievement because it is self-contradictory or contradicts the facts of the world. That is why, for example, the theory of human nature has played so prominent a part in ethics. If man is inherently selfish, then morality cannot ask him to do the impossible and be unselfish, it can at best adjust motivations to have a coincidence of interests. (On the other hand, holding the ideal may be offered as part of the evidence that he cannot be wholly inherently selfish.) It is usually assumed that a man cannot be obligated to do what it is impossible for him to do, though he can be held responsible for not doing it if the impossibility is a consequence of his own previous action or neglect. There are qualifications to the general formulation, with respect to the kinds of impossibilities: for example, the maxim must not be used to breed acquiescence in an evil social order, just because it is difficult to change it without sacrifice. Again, in general, the need for the means is a constraint on the pursuit of the end. On the whole, this first type of use of science for ethics is too familiar to require extended comment.

Assistance

A program of scientific assistance for ethics *starts with the ethical materials* and develops their ideas and problems and calls in scientific assistance at well-defined points where it thinks the scientific methods or results may help out. It treads cautiously and its use tends to be piecemeal rather than wholesale. It need have no preferred investment in any one area of scientific resources, it leaves empirically open whether the assistance will be on a small scale or widely generalizable or even momentously large. For the most part, moral philosophy generates the problems for which scientific aid is invoked; but science in its impact on society and other aspects of human thought may have generated problems which then took moral shape. In the twentieth century moral philosophers quite naturally looked to lively scientific fields that might impinge on their problems—for example, economics on distributive justice, evolutionary biology on the kind of human nature that evolved, and psychoanalysis on the development of conscience in the individual. Occasionally an account combined studies: for example, in presenting character types Erich Fromm clearly put together a psychoanalytic picture of stages and a Marxian picture of social influence.[4] Thus, the "hoarding" character is anal in the Freudian sense on the inner side and capitalist-acquisitive in social behavior. Whatever the value of his typology, Fromm did help contribute an added dimension for distinguishing authenticity and maturity as criteria in moral judgment. In *Ethical Judgment: the Use of Science in Ethics* (1955), I attempted systematically to gather the contributions that the different sciences of man (biological, psychological, social and cultural, historical) had made to moral judgment on a variety of questions, of both theory and moral attitude.[5] This involved no decision as to which among conflicting schools in each was correct; ethical consequences could be drawn from the methods and techniques and (for tentative consideration) from the results of each. The general thesis was that there were many questions in moral judgment which required the scientific answers in order that the moral judgment should be more definite. (Chapter 7 in the present volume is a good example of this approach at work, and of course *Ethical Judgment* illustrates it more fully). One could in this way dream of a scheme for an ethical theory in which parameters were indicated for the different sciences to fill in their values and to make the ethical theory more complete—that is, if there were a scheme of the sciences with some stability. But this, as we have seen, has not been forthcoming.

Foundations

A search for scientific foundations for ethics *starts with some scientific* materials in one or another area of scientific inquiry. This is quite clear from the way the theory is constructed, or sometimes from the systematic way in which the foundations are first presented and the ethical terms defined by reference to the scientific ideas, usually in a somewhat reductive fashion. The chief theoretical aim is then to show that the resulting scheme can do the jobs that people ordinarily expect of their ethics. It is this kind of program that has most alarmed traditional moral philosophers and redoubled the intensity of separatism. Perhaps the most impressive construction along these lines in recent decades was Stephen Pepper's *The Sources of Value* (1958), resting essentially on E. C. Tolman's psychological analysis. It appeared fully formed while others were arguing over the naturalistic fallacy and the relations of science and ethics. It was almost as if in the midst of a debate over the feasibility of a machine someone simply went ahead and built it. The only question was: would it work? The criteria of an ethical theory working are not, of course, as simple as those of a machine performing a given (visible) task. And suppose Pepper's theory has inadequacies. Do they come from associating ethics with science as such, or with particular scientific resources and data, or is it the manner in which the material is employed? Certainly so imposing a construction merited a full and careful critique. Our study of Pepper's theory is included in this chapter both for the important questions involved about the terms on which different scientific materials relate to ethics and as an illustration of the approach of a program of scientific foundations.

Scientific Methodology

An attempt to have ethics make its methodology scientific is in one respect quite old for our century. It was propounded by John Dewey in the philosophic assimilation of the Darwinian revolution. Given continuing and possibly accelerating change in human life and human knowledge as a normal feature of which we become aware, focus moves from the shifting content to the methodological and from fixed principles to intellectual instruments constantly tested in the solution of succesive problems. With morality reoriented as part of the equipment for human survival and common well-being, the unified contribution of science to ethics is methodological. This was offered by Dewey while the social sciences were still in the making, but it fits neatly the picture of their subsequent shifting character. The contribution of the scientific content to ethics on this view is miscellaneous, not the work of one or another privileged science; it comes from all the sciences and all the technologies too, and is unified by the context of the problem to which in the particular case it is being addressed. It is the whole scientific outlook, with its method and organization of inquiry and the attitudes and habits attendant thereon, which has contributed in a unified way to moral theory.

This decisional orientation in ethics, embodying a scientific orientation, gained in strength with the development of decision theory in logic. This accentuated the methodological side, although the starting point had been the influence of the Darwinian biological outlook in philosophy. Interestingly, with the recent breakthrough in biology, and the fresh bid of biological science for a hand in molding ethical theory, while sociobiology went along the path of a scientific foundational program,[6] the methodological turn appeared in different biological form. It was given, as it were, a concrete "embodiment" arising from the fact that in evolutionary development the brain is the organ of decision. Hence ethics should be understood in terms of decision process as seen by reference to the development and processes of the brain. Now since computers are attempting to model brain processes, computer problems and processes, as well as computer needs as revealed in computer science may give us an insight into value theory. This view has been expressed by G.E. Pugh in *The Biological Origin of Human Values* (1977) which, though associated by him with the sociobiological ethical theory is basically best seen in the light just indicated.

It is important to note that there is no incompatability between the different ways of using science for ethical theory. The program of scientific assistance begins cautiously but draws no impassable limit. The program of scientific foundations, whether Pepper's or any other form, is a complex hypothesis which might conceivably turn out to be correct but it looks at present as if it would have to expand its scientific base considerably, perhaps beyond recognition, to accomplish its aims. It would probably have to move into an historical-evolutionary hypothesis integrating the contributions of a variety of scientific fields and dealing with a complex conceptual apparatus as well as an immense factual sweep. Pugh's theory, insofar as it is a foundation program, shows the need for this. It is also quite possible that an evolutionary approach, by integrating morality and all its data within its materials, would transcend the bounds of the original question which generated the problem of how ethics uses science and set new terms for the inquiry.

To suggest more intimately the analytical problems of this area, I have here included a review of Pugh's book, as well as the study of Pepper's book.

Stephen Pepper's *The Sources of Value* belongs to the scientific wing of value theory.[7] Pepper is firm throughout in his conviction that to understand value in any or all of its aspects — whether its behavioral paths or its introspective appeal, whether its theoretical concepts or its linguistic formulations, whether its individual manifestations or its social configurations — one has to see what is going on in the phenomena and their relations, in the processes within and between men in the natural world, and see it in the fullest detail that science can provide. Pepper's great strength lies in the way in which he attempts to utilize concrete findings to piece together a framework that he thinks will answer the traditional problems of value theory. He is not content merely to sing the panegyrics of scientific method. Whatever the adequacy of the framework itself, he has shown us, as few works on the contemporary scene have even attempted, the necessity for tracing what we may call the *specific value-theory potential of scientific findings*. And in doing so, he impels us to reckon with the whole problem of the terms on which value theory uses scientific materials and the role that scientific theories can be expected to play in value formulations.

I

"Since this study makes no pretense at being a definitive theory," Pepper tells us at the outset, "the process of thinking which led to each successive stage may turn out to be more helpful to other students of the subject than the actual conclusions reached. The chapters ahead may be truly considered as a report of one man's exploration and survey of this field as he penetrated deeper and deeper into its factual relationships." [8] Yet when the story is over we can find a dramatic unity within it. Pepper's hero is the purposive act. The first half of the book is devoted to exploring the character of his hero. Purposive activity is approached as docile adaptive behavior. Men are bodily organisms with a given repertoire of initial drives. Drives have a determinate structure or rhythm, culminating each in its own inherent quiescence pattern. Drives act out in numerous ways under different conditons and pressures. There are derived drives and mutations. But the basic pattern persists throughout. Pepper gives us a rich and full analysis of the structure of appetition and aversion. Basic drives with their impulse pattern and embedded needs and tensions are inventoried as well as analyzed, and there is comparison with instinctive behavior and chain reflex in other animals. There is significant stress on the gap between drive and quiescence in which human learning emerges, along with recognition of blind drive behavior, and so an insistence on not carrying the analysis of value in cognitive terms too far. There is careful treatment of the anticipatory set involved in purposive activity and the role of mediating judgments; of the goal object and why it should be carefully distinguished from the ultimate quiescence pattern inherent in each drive. There is similar careful treatment of aversion, and convincing criticism of the views that assimilate aversion to an appetition for a negative goal. There is a whole theory of "injectives" or energy mobilized to face obstructions and drive frustrations; aggression, for example, is interpreted as an innate drive, not an appetition, whose goal is always a subordinate goal within some other operative purposive activity. [9] There is a study of derived drives and mutations that take place at various points in the appetitive structure, considered in the light of the theory of learning and competing accounts of motivation. There is a recurrent attempt to reckon with gratuitous satisfactions in the effort to show that such pleasures can be seen as telescoped appetitions within the structure already developed.

All this is biology and psychology, but Pepper ties it intimately with the discussion of controversies in value theory—detailed issues of the locus of value or the precise role of cognitive judgment in value determination, and so on. Pepper's procedure makes quite clear the way value enters the picture. He starts out by indicating a broad field of extremely varied items ("anything good or bad"), eschewing dogmatic views that there must be a common property to all these and resting content with a unified intent—"the problem of how to make well-grounded decisions in human affairs."[10] Looking for some promising starting phenomenon whose scientific study would yield at least a partial organization of the field, he reviews such major candidates as affection (pleasure and pain) and conation (desire and purpose), and decides to start with purpose. After his thorough examination of purpose he is not then claiming to discover for the first time that pleasure and achievement are values, but to see the natural structures in which they arise and to which they are attached. The logic of his procedure implies an independent inventory of "values" as an initial set of preanalytical data. (In fact, he does not carry out an inventory; it might have been helpful if he had.) Because his fundamental inquiry has been correlational in the sense indicated, he is able, on completing the picture of purposive behavior, to ask "Where do the 'values' lie in these purposive structures?"[11] What he seems in effect to be asking is: how much of the traditional field of values do the mechanisms explored account for, and how far do the distinctions in these mechanisms and their laws light up the relations and criteria for assessing these values? He finds that he can account for, refine, relate, and systematize values of the conative, achievement, and affective type. Purposive activity is thus seen by Pepper to embody natural structures—he uses the term "selective systems"— which furnish "natural norms" by which values are ordered. He describes a selective system as a natural mode of selection with a "split dynamics." Unlike the wind pushing the leaves along or the sieve sifting particles according to size, purposive activity—let us stick to the initial paradigm—involves a drive whose quiescent pattern provides a norm for assessing the goal, and the goal is the basis for assessing the subordinate steps; food is good insofar as it satisfies hunger, and the work involved is justified by the food we get as a consequence. We are told formally: "A selective system is a structural process by which a unitary dynamic agency is channeled in such a way that it generates particular acts, dispositions, or objects (to be called 'trials'), and also activates a specific selective agency (to be called 'the norm') by which some of the trials are rejected and others are incorporated into the dynamic operation of the system."[12] The core of this conception, as I understand it, is simply that we can find determinate conditions and processes for understanding both why certain values appear in human lives and why in varying situations of conflict some regularly prevail over others, or appear as strong tendencies to prevail. In some respects it seems to be working along the same lines as contemporary inquiry into homeostatic systems, and might even be interpreted as referring to a special class of such systems, of the more complex and structurally differentiated sort. The natural norms would then resemble the rules that constitute the "plan" of operation of the system.

Having found purposive activity to account for a large segment of human values, Pepper in effect goes on to ask why we cannot find other selective systems to take care of the rest of the field. This is the story of the second half of the book. We follow the career of our hero, purposive activity, as he moves slowly out into the world. He finds himself pushed around by his fellow purposes in the same organism, and there emerges the value of a balancing prudence. An integrative tendency produces a personality with norms that emerge as roles (ego-ideals, levels of aspiration) and conscience, and the ideal of maximum integration. Then he is truly out in the world, and social situations face the person with the others; a norm of maximum satisfaction for all concerned adds itself to the assembled pressures. The fact of cultures contributes stresses on conformity through institutions and on an integrative tendency that is most evident in the tensions of social change. From here on the scene shifts, not to social history as we might expect, but to the whole evolutionary vista, dipping down again to biology. Survival values now come into plain view, not simply the survival efforts of the living individual but the natural selection operative in the population. In fact, the plot works up to a surprise ending—it would have the impact of a Hitchcock ending if the secret had not been given away as early as Pepper's previous *A Digest of Purposive Values*, his preliminary study published in 1947.[13] Although Pepper has listed quite a number of selective systems that generate norms, when he comes to examine the lines of "legislation" among them—in effect, their dominance and submission relations—it turns out that one of them has been pulling most if not all of the normative strings. At bottom, Pepper finds that there are only two main dynamic agencies: "the purposive drives of the individual organisms" and "the reproductive process of an interbreeding population."[14] As the story unfolded I was reminded of Freud's battle of the Titans—Eros and Thanatos, the life and death instincts. (I am referring to the style of plot, not a strict parallelism of properties.) Since Pepper does not furnish us with Greek gods or mythological heroes to represent his Titans, let me call them, as one might name chemical ingredients by initials, simply PD for purposive drives and NS for the natural selection agency. Now just as Thanatos is used to explain much that is irrational in human life, so in Pepper's account the authoritarian conscience among the personality norms and the remarkable hold in human history of "irrational religion" become intelligible as machinations of NS. For such religion has operated to maintain social solidarity, and conscience expressing it has led the individual to sacrifice himself for the group. And just as Freud speculated that Eros might itself be doing the work of Thanatos (because pleasure comes with a release of tension), so Pepper is ready to see the repertoire of PD as itself an adaptive feature in the operations of NS. In conscious social life, too, Pepper sees a logic in this struggle: under relaxed conditions PD has its day as we pursue individual happiness, but in social emergencies NS ruthlessly dominates. His normative solution for this see-saw points toward an adjustable social structure "that can be centralized in authority during emergencies, and decentralized to make way for liberty and opportunity for individual satisfaction in times of peace and safety."[15] He sees this as a social invention of a new era, aiming at maximizing happiness, "subject to the legislation of human survival values over affective values."

So much for the synopsis of the plot. Pepper's cunning of nature may well remind us of Hegel's cunning of reason. But I suspect the reader will be asking himself two different questions by this time. One is objective: what are the scientific credentials for such a view of value? The other is subjective, and in the mood of the drama critic: can he identify himself with the character whose odyssey of self-discovery we have traced? Can he recognize himself?

II

The use of science in value theory is a complex enterprise which involves many distinct problems and can be carried out in different ways. There is first the familiar nest of methodological issues—the conflict of methods on which sciences as well as philosophies sometimes fall into "schools." With respect to scientific findings, we must distinguish: taking the concepts of a specific scientific account, invoking some of its generalizations, actually putting these to work in specific value-theory issues, and facing in some systematic way the problem of relating the use of different sciences and their findings. Each of these carries certain cautions. A careful appropriation of concepts involves some consideration of their logic in the light of controversies within the scientific field itself. The generalizations are not to be slavishly followed; they can be compared with alternative or competing accounts, with the kind of evidence that comes with them, with data in the value field that might itself serve as evidence concerning them or suggest other possibilities. The use of these materials for value issues can itself take different forms—merely suggesting an analogy which is not translated into value terms, taking a value example and analyzing it in terms of the scientific findings about its constituent factors (one of Pepper's favorite procedures), using a conceptual scheme from the science to analyze value concepts, and so on, all the way to the possible extreme of making deductions by combining the scientific generalizations with value statements. (And all this can be done with different degrees of rigor.) Finally, on the relation of the materials from different sciences, there are perhaps the most serious problems of strategy. Can one use a method of "successive approximation," bringing in one science after the other, adding insight after insight? Or should one open the doors at the outset to all the relevant sciences in the very initial identification of phenomena and the mustering of alternative organizational concepts? A strategy of selection can make or break a value theory that aims at scientific affiliations.

In the light of such separation of issues, we may now take a critical look at Pepper's use of scientific materials. First as to method: Peppers's outlook is behavioristic, with major reliance on E.C. Tolman in psychology and R.B. Perry in value theory. But this emphasis on the objective in description of values does not carry with it a rejection of other methods. While the phenomenologically inclined may wince when Pepper, simply in a footnote,[16] identifies "phenomenological" and "introspective," his actual employment of phenomenological description—for example, in discussing pleasure—is ample, and his interweaving of the introspective, the behavioral, and the physiological is quite self-conscious. His point is rather that he finds introspection most useful when it is set in an area that is previously explored behaviorally and, where possible, physiologically as well. What we have here then is not a methodological bias, but a quite reasonable hypothesis about the likelihood of success in using a given ordering of methods.

With respect to the use of scientific concepts and findings, there is greater unevenness. The most intimate use is that of behaviorist psychological materials. Here Pepper appropriates concepts and revises their structure, invokes specific theoretical generalizations, and argues for some against others. For example, he adopts a multiple drive theory of motivation as against a single drive approach and as against an ideomotor theory, revises Tolman's list of drives, and offers a modified tension theory of pleasure and pain.[17] It may be that he is occasionally too free in assuming specific inherent mechanisms for discovered behavioral and phenomenal differences. Yet in spite of his fundamental reliance on a drive structure of appetition, Pepper does not go into the many controversies about the logic of the concept of drives itself. There is no discussion of potentiality notions here (although potentiality is dealt with in a wholly different context—the potential object of value), nor of anthropological claims for the penetration of cultural factors into the very heart of what appear to be inherent drive properties, nor of sharp challenges that the whole picture of drives as aiming at the release of tension represents a pathological process.[18] There is simply the assertion that the conditions of quiescence of a drive refer to ultimate determinate facts. Nor is there consideration of the possible inherent mutual discrepancy of some drives, unless the interplay of purposive drives and natural selection dynamics be so construed; on the whole, he appears to rely on natural selection to keep the repertoire in line, but does not consider that even a moderate degree of discrepancy might have serious value repercussions.

Perhaps the most far-reaching assumption in its value theory impact is that the purposive structure of appetition is projected into derived drives.[19] Now there is considerable discussion of the phenomenon of mutations in drives, and a variety of mechanisms mediating transition is suggested. Later on, too, in dealing with personality integration, the drive concept itself becomes considerably thinned out into the concept of a "perduring disposition."[20] One is left wondering, therefore, what justifies Peppers's antecedent assurance that the structure of purposive activity, which fits basic drives with determinate conditions of quiescence, will fit the vast array of particular purposes in all their social and cultural detail after mutations have taken place in which former means now function as ends, when there have been sublimations, when an integrated self operates and infuses constitutive goals, and so forth.

The scientific materials of the second half of the book are less intimately utilized than the psychological behavioristic studies. From Kurt Lewin, Pepper takes the concept of an organism's life space as a scheme of representing the conflict of aims in an individual, but no scientific results seem to be invoked. Such points as that a choice between purposes is different from a choice between means to achieve a single purpose, or the subtle question of why a man should be prudent or an organism aim at maximization, could very well be raised and Pepper's own answers be given without reference to the representational scheme. In dealing with the psychology of personality, Pepper expounds Erikson's rich and fruitful scheme of child development, which is a theory of the stages of maturation and of danger points in personality distortion. But he makes no specific use of it. The general idea of personal integration is carried out rather in a behaviorist mode, with drives and habits as "atomic elements of a personality,"[21] yet no behaviorist materials dealing specifically with integration are brought in for consideration. Neither are they much considered in the scientific materials subsequently introduced: they tend to be used less for specific findings and specific hypotheses, as were the materials of the first half, than as illustration of some general concepts that have some value theory pertinence. Thus reference to Freudian materials serves chiefly to fasten attention on conscience and the mechanism of repression. In the social sciences especially, scientific materials are scarcely at all invoked; for example, the rich materials of cultural anthropology are referred to only to establish the bare facts of cultural variety and cultural patterning. And yet specific theses in Pepper's theoretical framework clearly require psychological and sociocultural evidence—particularly the way in which he interpreted conscience and traditional religion as part of the operation of natural selection producing social solidarity and inducing sacrifice; under certain conditions, he says, "it is necessary that individual men should perform individually imprudent acts even to the sacrifice of their lives."[22] Would he have concentrated on the single thread of sacrifice and solidarity if he had gone more fully into the variety of even psychoanalytic interpretations of conscience, and into the diversity of functions that anthropology and history have shown for religions?

Perhaps the most serious problems in Pepper's use of science arise, however, with respect to his fundamental strategy of selection and relation of scientific materials. Pepper's procedure is to start with the intensive scientific study of a single purposive act and to build up through the wider constellations of interpurpose stabilization, whole-person integration, social (many-person) situation, cultural configuration, and population evolution. But what is the nature of this building up? Actually, there are at least three possible interpretations. It may be an *individual-genetic* process, a *theoretical-analytic* process, or an *evaluative* process. Confusions are bound to arise if the results of one analysis are applied to the others. At some points at least, Pepper's procedure suggests the individual-genetic: "every human society has its undoctrinated babies continuously flowing into its midst, and these keep arriving with the old original repertoire of instincts not conforming to the social pattern. To mold these into conformity, the social sanctions will always be required."[23] We are also told that the child is "liberally endowed with self-regarding impulses."[24] Such remarks make it seem that the building up process is a temporal accretion of fresh levels in individual development, an outward extension prompted by internal conflicts and external pressures. On this approach one could not assume, but would have to show that the structure found in the earlier stages would persist unchanged in the later stages, I have already commented on Pepper's assumption that the structure of purposive activity remains unchanged in passing from basic drives to derived drives.

On the other hand, Pepper's initial formulation of procedure is theoretic-analytic: a method of element-analysis is proposed, starting with the simplest relatively isolated examples and working on to the more complicated[25]—occasionally this is seen as successive approximation. But here too there is no clear consideration of the atomic character of the elements, and whether they remain identical (and in what respects) as constellations are formed. Moreover, since the approach is analytical and not temporal, it might well have been compared with constructions that work in a different direction. Take, for example, the recurrent problems Pepper has with the question of gratuitous satisfactions.[26] His conclusion is that gratuitous satisfactions (and simple riddance) are *"properly to be regarded as purposive structures in spite of simplicity*. They are purposive structures reduced and telescoped to their pivotal acts alone."[27] But what kind of telescoping is this? If it were genetic order in individual development, Pepper would have had to reckon with actual alternatives—for example, with such a scheme as Freud's in which the pleasure principle is initially on the scene and the reality principle grows up with its gaps and detours. But if this is a thesis of analytic propriety, then what is the difference analytically between gratuitous satisfaction as a telescoped purposive structure and purpose as an expanded and deferred satisfaction? Is it a thesis of the fundamental scientific role of one or the other? Then what precisely are the testable consequences? When Pepper says, "A complex appetition is simply a search for a consummatory act that unfortunately is not gratuitously given"[28] one begins to wonder what the whole issue is about. But such wonder strikes at the very way in which he has expounded and uses the analytic method.

The third possibility is that the building up is an evaluative process. The single purposes are no longer to be regarded as initial equipment nor as first approximations, but are invoked simply because in every evaluative situation we find that there are networks of existent individual purposes present, whose relations play an evaluative role. This is in fact the way it seems to turn out at the end when happiness and group survival come to the top. And in between, Pepper occasionally lets directly evaluative considerations enter; for example, in considering the relation of achievement values and consummatory affective values, he is eloquently persuasive on why we should not gulp.[29] But such evaluative comparison of different values can be carried out without a theory of development or of inherent elements and configurations. That it is rendered considerably more effective by a theory of supporting mechanisms, as Pepper shows, is another matter.

What would have emerged then, if Pepper had followed an alternative strategy in the relation of scientific materials? Suppose one had thrown the field open more widely at the outset instead of beginning solely with the psychological-behavioristic. Why attempt to lay down a definitive structure for all appetition to begin with? Why not simply recognize that a variety of phases and elements are found in the basic drive mechanisms, and leave open for comparative research the kinds of rhythms that may occur, the conditions of their occurrence, the internal costs of their maintenance, and so forth? Should tension or pleasure or even achievement be treated as unitary concepts for all personalities and all cultures? One thinks of depth psychological treatments of anxieties as different, dependent on whether they stem from the ego's relation to the superego or the ego's relation to the id; or again, the classification of pleasures as essentially distinct in terms of the intrapersonal economy. May not aesthetic pleasures and whatever mechanisms underlie them occupy a far larger area in the individual's life than the purposive stress in Pepper's account might suggest? Are whole alternative structures with a quite different patterning of tension, rising course of achievement, or consummatory pleasure speculatively possible? For example, Bateson and Mead find the Balinese having a pattern of inherently satisfying rote sequences rather than conative sequences with climaxes; there is an environment of fear but a kind of enjoyment of fear. Bateson compares it to the "acrobat's enjoyment both of the thrill and of his own virtuosity in avoiding disaster."[30] There may, of course, be claims for certain structures as natural and others as developed under certain conditions. But can we be sure without a more systematic theory of the development of the self and a wider consideration of comparative anthropological evidence that even "gulping" is not the natural terminus in the structure of a purposive act and that the apparently independent mechanisms that spread out pleasure in consummation are not cultural habits under conditions of greater well-being and security? Surely any general structure for purposive activity presupposes answers to comparative questions; even the assumption that the pattern found for basic drives holds for derived drives cannot be established from one culture alone.

We may speculate even further. Why should we start on the individual behavioral level? Suppose we had begun on the sociocultural level with value-facts as cultural facts (as Dewey tends to) and then worked out in two directions, to the historical and evolutionary vista in one way and the individual psychological and biological in another. Would we have ended up with the same general theory of value? Perhaps, too, one starting point is more fruitful for ethics and another for aesthetics; we cannot assume that value as such determines the same starting point for every "species" of it.

The conclusion implicit in these questions and speculations about alternatives is that there is a serious risk in starting with a special set of scientific (in this case behaviorist-psychological) materials, assuming we have an isolated system, and attempting from them to reach value theory conclusions. It is not, of course, a problem that philosophers themselves can solve. Scientists too lament the separatism of their fields and the partitioning of schools—witness recent integrative efforts under the name of "the behavioral sciences." The least that can be done is to let the diversity of scientific materials in to raise questions *at the very beginning*; this will mean here that conclusions about the structure of appetition will be seen to involve assumptions of sociocultural invariance, rather than be in some sense prior to the impact of the social fields. It would be best, of course, if we had an integrated theory of man to which to relate our value theory. But since the sciences have not reached that point, it would seem sounder philosophical policy to steer our value theorizing toward the point at which they are heading, rather than anchor it to some existent segment or even to a juxtaposition of separate approaches.

III

Even at the risk of some retracing of steps we have to ask what is Pepper's portrait of the self. I doubt very much whether many of us will find ourselves identifying with what I have spoken of above as the hero of his account—even when he becomes a person-integrated organism. It is commonly recognized that the theory of the self is one of the most difficult parts of a naturalistic theory of value. I do not think this is because, as antinaturalists so often insist, a naturalistic theory is allied to scientific accounts, but because the ususal naturalistic theory follows too restricted a strategy in relation to the sciences. But there is a further complication—the possibilty that value considerations enter into the theory of value itself. Pepper's work carries throughout a marked individualism. I am not referring to that evaluative individualism which is part of the liberal tradition as well as of the scientific wing of general value theory, but to a theoretical atomistic individualism that seems to be intimately bound up with the biopsychological way in which the stage is set for value theory. (Perhaps it is helpful to keep in mind a contrast with Dewey, who shares the evaluative individualism but not the theoretical atomism.) Philosophically, this tradition is familiar enough—from Hobbes through the greater stretch of British empiricism. But we must pinpoint it in the scientific materials, so that it will be seen to characterize a certain selection of scientific approaches and not assumed to be an inherent property of the liaison of value and science as such.

What are the fundamental units—the cast of characters in the drama of value processes, with which Pepper's theory operates? They are, as we have seen, not persons, nor even organisms: the organism is taken for granted, of course, in its physical unity, and persons come on the scene only in the second half of the book. The fundamental units are *well-formed purposes*—appetitions, aversions, drives blind or with articulated structure, telescoped purposes, and derived drives but with assured purposive structure. The important thing to note is that these are not presented as the causal background, though they may serve such functions too; they are the actual basic participants in the drama of value processes. Everything revolves about them. And since they have an initial within-the-individual character, value retains this fragmentary individualism. It was only late in the performance that we detected the whiplash of another character—the dynamic agency selecting for survival. But if this reverses action and makes the other characters cringe, it has not altered their nature; they simply wait for the storm to blow over.

The richness of Pepper's treatment of the purposive structure may obscure the thinness of the treatment of the theory of the self. This is not exactly Pepper's fault; as is well known, the theory of the self in contemporary psychology is in a very confused state. Pepper's treatment reflects both the behaviorist and the Freudian pattern—an account of integration of internal pre-self or sub-self forces. His atomism is reflected in a sensitivity for the "rights" of drives that lose out or impulses that are repressed. The latter are more easily disposed of; to the question of why wanting to commit suicide is not as legitimate as wanting to play golf,[31] Pepper answers by referring to the evidence of the self-frustrating character of repressed conflicts. In the case of the clash of drives—each, he says, is on its own unless blocked by another—he has the problem of explaining prudence: "Now, why does a person prudently consider all the drives competing for his favor and try to get the maximum amount of satisfaction possible in view of the total situation? Of course, this is just what the resultant act achieves in the structure of a life-space. But how is this motivated?"[32] The terms "person" and "his favor" surprise one in this context, because there has been no account of the person. The answer offered is a mechanism according to which "The resultant act is the shortest path for relieving the complex of tensions in view of the *total* situation"; as Pepper goes on to say, "The dynamics of life-space or the personal situation is thus intrinsically prudential. A man naturally tends to look after his own interests."[33] But the shifting from "person" to "organism" to "man," even with some reference to thinking and to the judgments of reality, does not really amount to a theory of a unified thinking and judging self. What we get at most is an analogy to a banking system in which there is a central store of energy and each drive has determinate credit drawing power. Now while we have come to expect analogies in value theory to serve as theory-of-self surrogates, it is all the more necessary to recognize that they are simply holding a place for a theory-to-come, and so call attention to those branches of psychological science in which there is an effort to develop this theory. Pepper does recognize the integrative tendency of the personality as a distinctive operative system, but he treats it merely as a tendency to integration; it looks like a kind of broker theory of personality analogous to the old broker theory of the state. Actually, the content of the eventual theory of the self and its development may show—what phenomenological clues tend to suggest—that the development of the self modifies *from the very beginning* the character of desires, purposes, and their conflict, so that an isolated treatment of them below the self-level may have much that is misleading.

The major question on which Pepper concentrates in this widening picture of the way in which purposes build up into constellations is the gap from one individual to another. Here there is nothing to correspond to the reassuring fact that conflicting purposes are all within one organism, and it is precisely here that the within-the-organism individualism of his initial stage setting begins to exercise its fullest force. The laws of learning operate only within a given organism; hence the difficulty in securing recognition of another's pleasures and pains on a level with one's own. Social interaction is exemplified at length in the situation of economic exchange between persons with preexistent individual demands; it is as if the paradigm for social life is wanting something for oneself that someone else has. It is interesting to note that throughout the whole book the illustrations used are of individual-oriented purposes; a man is fishing, a geologist is thirsty, a man is lost in a blizzard, Sonny wants the car tonight, and so on. There is practically nothing, whether in illustration or analysis of emotion—apart from the general reference to sacrifice prompted by conscience and irrational religion, noted above—of interpersonal relations of an outgoing or affiliative sort. There is one paragraph on mutual love [34] but only in the context of the Freudian genital stage. (There is no analysis of sympathy; one wonders whether Pepper would have given a Hobbesian account.)

Now this is an area in which there have been considerable attempts to revise the Freudian treatment of interpersonal relations in which others are dealt with as objects for the satisfaction of inner needs and the resolution of inner tension. Academic social psychology develops concepts of affiliative needs; neo-Freudian schools distinguish exploitative from productive relations; whole schools of psychoanalytic theory stress the priority of interpersonal relations in explaining human individaul reactions; phenomenological schools carry out a scrutiny of interpersonal emotions; existentialist philosophies pinpoint the "authentic" versus the spurious in I-Thou relations; anthropological studies point to the character of interpersonal relations as a function of the type of social structure and familial setup. How is it possible to neglect this major field of the nature of the interpersonal by focusing so intensely on the intraorganic?

What is more, Pepper seems to bypass the whole philosophical tradition from Hegel and Marx to T. H. Green and Bosanquet, which saw the nature of man as a social being. He does not look at anthropological evidence for the permeating character of culture in the individual makeup. When he does deal with Dewey as central in the understanding of the social aspect, he takes Dewey's chief contribution to be the recognition that the social situation serves an evaluative function. That is, he omits the whole genetic side of Dewey's outlook, Dewey's fundamental insistence that man is primarily a sociocultural being, not a biological or an individual-psychological being. He does not note that Dewey has a different initial stagesetting for value theory in which value facts are themselves sociocultural facts.

Pepper's theory thus clearly lacks any integrated account of the nature of a person. At most there are the all-too-determinate drives that babies are said to come with, plus the embedded survival mechanisms. And there is even a jockeying between these two components. On the one hand, "a repertory of purposive drives is an adaptive feature for a species in some particular life zone"[35]; on the other hand, "man must not get the idea . . . that the survival values are not as much a part of him as the affective values. The dynamics of survival are just as deeply implanted in his body and behavior as the repertory of drives that give him his capacity for happiness."[36]

An individualistic-happiness component and a totalitarian-survival component biologically embedded in the constitution of an organism are old themes in the philosophy of man. It would be a pity if they took on scientific robes by being allowed to entrench themselves in a selective stress on the materials of biological evolution and behavioristic psychology plus Freud. In Pepper's book neither the remaining psychological fields, nor the whole range of the social sciences get an adequate day in court to contribute what they can to the theory of value.

IV

How far in the domain of conceptual construction has Pepper broken through the contemporary sidetracking of naturalism in value theory? Do his concepts of *selective system*, *natural norm*, and *legislation among selective systems* establish a framework for a happy marriage of fact and value in the midst of so many contemporary philosophical pressures for divorce? And how comfortably can the *ought* or *should* settle into this family?

Pepper develops his concepts at various places; not all are systematically explicated; the central concept of selective system, he tells us frankly, came to light only when the book was half done. Perhaps as a result of this experimental development there is a certain lack of firmness in the outline of the concepts. One rushes in to do the work in another's place. Sometimes selective system seems almost to coalesce with natural norm; but according to the original definition of the latter, a natural norm consists of a dynamic agency operating through a selective system in the selection of some against other values,[37] and this would not make sense if there were no distinction between natural norm and selective system.[38] Selective system, too, is a bit unsure of itself vis-a-vis values. Most of the time—and this would seem their appropriate role—selective systems order and assign priorities among different lines of values. They also, in a causal sense, explain the existence of some values. But in at least one important context—again, in dealing with natural selection—we find Pepper arguing from the fact that natural selection resembles in its operation a selective system to the claim that what it institutes has therefore a right to be called "value." Pepper thus seems to waver between looking to the selective system to *institute* values and finding it to explain the occurrence and strength of independently identified values.[39]

Nevertheless, I think that Pepper has a good conceptual team, if he will keep the different notions more strictly apart, and establish jurisdictional lines. For the concept of values there need only be a clearer underscoring of their independent identification, as suggested earlier. On selective system, natural norm, and obligation, I should like to add a few comments.

The concept of selective system will probably have to be sharpened to do the kind of job for which Pepper attempts to use it, that is, to delimit a group of isolable determinate systems sufficient and exclusive for the value domain. As it stands, Pepper's list of seven systems, modified by the discovery that two are basic, leaves both their relations and their independence insufficiently clarified. As he uses the notion of selective system, I do not see why it would not apply to much more. For example, why could not even vision be construed as a selective system, with determinate mechanisms of rejecting stimuli, with personality components entering into selection, and with natural norms of clarity and distinctness? Nor do some of Pepper's selective systems quite seem to fit the definition.[40] Nevertheless, the fundamental significance of this concept remains: it points to the importance of identifying those typical processes within a man, and those typical transactions of man with his environment and his fellows, and those typical long-time structures that can play a role in the occurrence and support and mutual relations of values, including the kind of ordering they manifest. It will however, have to be considerably refined to distinguish the different scientific relations which it telescopes—where it is describing the causes of the occurrence of values, where the way in which natural processes are at the service of some values in one or another functional relationship, where they bind particular values or provide automatic scene-shifts so that one value replaces another, and so on.

The concept of natural norm seems to be concerned primarily with kinds of selection and evaluative criteria for selection among values. The "naturalness" of the norms comes from the existence of the determinate processes (selective systems) that support and persistently advocate them. Reduction of the drive in a purposive act, maximum of achievement and satisfaction for the personal situation, conformity to conscience in the personality structure, and institutions in the cultural pattern are illustrations of such norms. As I understand it, we could discover these in the value field by tracing lines of dominance—that is, which prove to be authoritative over which. Most of the norms Pepper lists do not need the notion of selective system for their discovery. But we really do not understand what is going on and why one stands out over another unless we trace roles and relations in the underlying selective systems. In doing this, Pepper has an apparatus that is clearly capable of avoiding the traditional dangers of sanctifying natural trends. For example, conformity to conscience is offered as a natural norm, but if we found its voice variable and discovered mechanisms of control it could even be eliminated.[41] And certainly institutions that function as natural norms through prevailing successful sanctions are historically capable of alteration. Where the traditional moralist would say, "This is a functioning institution, but is it good? Should it exist?", Pepper's way of asking such a question seems to be "This is a natural norm; but is there some other natural norm legislative over it?" For example, he offers as a modern self-realizationist slogan: "not fit yourself to nature, but fit yourself for many natures or even fit yourself to make nature fit you."[42]

Pepper's problem here in dealing with lines of authority among existent norms is very much like the problem Bishop Butler faced when he wanted, from the comparative inventory of observable constituents of the human makeup—passions, benevolence, self-love, conscience—to lay down lines of authority without outside appeal to his religious beliefs. Butler seems to me to have reached a phenomenological resolution: authority is a quality that conscience simply wears which the others do not have. Pepper's solution is to trace the behavioral lines of strength and victory and the extent of their dependability in the operations of underlying mechanisms. I do not believe that there is any question that could be asked in another value-theory framework that cannot be asked in his. His has the advantage of pointing to further lines of evidence and inquiry. Whether a still stronger position could be reached by a more systematic integration of the phenomenological and the behavioral is a separate question.

On the concept of obligation, Pepper has barely three pages, but it is sufficient to indicate the kind of interpretation he gives to it. "Ought" refers to the relation within a selective system of the trial act to the sanctioning corrective agency.[43] This is extended to the legislation of one selective system over another. There is no difficulty in this as long as there are determinate lines of this legislative authority. Pepper, as we have seen, believes that there are, ultimately emanating from the dialectic of the two basic norms of maximum satisfaction and survival, and the difference between secure conditions and emergency conditions.

There remains, however, the question of whether he has left room for the actual decision concerning what is good or obligatory as distinct from the judgment of what is likely to be found good or obligatory. We have seen that he does seem to take a normative position favorable to an adjustable social structure, but he presents this as an emerging invention explicable in terms of the structuring of human problems, and so he has not violated his system. Presumably he could not be asked to propose normative *decisions* without going into the content of the selective systems in all their specialization in the modern world. Moreover, this is a work in value theory generally, not on forging specific techniques for facing varieties of ethical decision. And yet, since the unified intent specified for the whole field was how to make well-grounded decisions in human affairs, it would have been appropriate to reckon more specifically with decision-processes. Particularly interesting would be those types of decisions that arise in the more difficult situations. For example, what should the individual do who finds himself in sharp disagreement with his fellow citizens in what is obviously an emergency situation and when his fellows are ready to use the sharpest sanctions for nonconformity? How will it help him to realize that to persist in his principled disagreement will make him the victim of the tag-end of natural selection operating through social conformity? Or is he betting on his own stand representing the integrative tendencies and the conformists the social lag? These are important issues in justification of one's stand in principled conflicts so common in the modern world, and so comprehensive an approach as Pepper offers could well work out its first-person implications.

In a thoroughgoing naturalism of Pepper's sort, the actual fact that men evaluate, plan, decide, and, as Dewey so often pointed out, the growth of large-scale agencies of decision and planning in modern life, have shown that decision itself is a phenomenon that requires greater scientific exploration. I almost think that Pepper, even within the terms he sets up, could have included human decision-modes as a type of selective system on its own; I doubt whether the conscious cognitive dealing with values (not to speak of analyses of value) can be wholly hemmed in within the framework of a purposive structure cast in bio-psychological terms. It seems to require a stagesetting for value theory that is fully cultural and historical. It is in this direction rather than the underlying biological evolutionary processes of natural selection that we may look for the rounding out of value theory. And perhaps this would more adequately take care of the recurrent claims for the "autonomy" of values that the critics of naturalism are so fond of invoking.

The tension indicated in the end of the Pepper study between the cultural and historical setting of the decision systems by which people increasingly guide their lives and the underlying biological evolutionary processes is precisely the theme that emerges from Pugh's book, *The Biological Origin of Human Values*.[44] But his theory, by grounding the human decision system as itself a biological process, hopes to narrow the gap and furnish reliable criteria of value. On the wave of evolution, biologists had dominated the early part of the century, but the anthropologists dominated the 1930s and into the 1940s, the psychologists peaked at mid-century, and an ensemble of behavioral scientists with a strong sociological contingent took over particularly when values became an enterprise attractive to the social sciences. Then ethology got into the picture, and recently, as might have been expected after the cracking of the genetic code and the upsurge of genetics, sociobiology has its innings. Except for perennial Marxisms, historical approaches remain modest. It should be added that occasionally a scientifically-minded philosopher has pleaded that scientists should put their contributions together in an integrated model instead of each promising to do the whole job alone. (Julian Huxley was about the only one to offer a scientific view of ethics in which an evolutionary perspective integrated psychological, social, and historical theory as part of the working apparatus.) But this has usually been frustrated by the conflict of schools and disciplines; and in any case the fission and fusion of social sciences has been too rapid in recent decades to provide a firm base for integration.

The significance of Pugh's book lies in the fact that it does propose an original and insightful model of an integrative sort for value theory. It offers a unity in an evolutionary framework, within which it weaves together fairly traditional themes of biological ethics. The title of the book and the praise bestowed by E.O. Wilson (on the jacket) may suggest a kinship with the current sociobiology movement, but Pugh clearly is not a part of that controversy. Of course the evolutionary focus on group survival is central, and that is likely nowadays to be given a sociobiological turn as gene survival. Pugh finds much to be located in the genes, but this seems to serve as scarcely more than a way of saying that something is built in, for his model comes from computers and decision systems, not the growth of organisms. Perhaps this should not surprise us. The whole decision approach in contemporary philosophy has had an extraordinarily unifying effect in theory as well as in practice. To think in terms of decision (as for example Dewey's treatment of ethics) brings in relevant data from different fields. And as for practice, it was after all the demands of World War II which in their urgency forced interdisciplinary cooperation and team work among the sciences.

Pugh utilizes significant properties of computer decision systems to illuminate what is going on in evolutionary processes, after an initial warning that "design" is not to be transferred in a blatantly teleological way. Now even in the simplified case of a single ultimate objective, a system cannot just make each decision by scanning all possible alternatives at every point to select the best in the light of that objective; such a process is both too costly and there is not enough time (calculated in years of operation required for each next moment's decision). Hence criteria are needed as surrogates for the ultimate end. These are the values that drive the system, and from which in the interaction with complex conditions secondary values arise. Now in the evolutionary process, in a comparable way, an animal form is developed in the group effort to survive, and in the case of man the brain grows from a nervous system. This suggests treating the brain as a computer decision system for the task of survival, with its various component specializations carrying out the functions that different computer and program components bear in decision processes. The brain is thus seen as a value-driven decision system that exercises judgment in terms of innate values that are themselves surrogates for the ultimate objective of group survival and give direction to all human development. Part I explores the human decision system in comparison with artificial decision systems, Part II examines the structure of human values, and Part III goes on to reflections on values for personal decision and social policy.

The riches of the book are to be found in the first part, in the detailed transfer of lessons from the artificial to the natural system and the ideas the model generates—whether in the long run they hold up or not. The speculations are fertile and never dogmatic. Indeed Pugh says (at the end of Part II) that the whole value concept is itself probably "only an approximation to the human motivation structure"[45] but for the present it gives us a framework for research. The practical fields from which the comparisons come are as varied as problems in scheduling production, bomber flights, and desegregation busing plans. Fortunately his approach is that of an experimenter, not a pure mathematician: he shows why both the artificial and the natural systems will go for what works satisfactorily rather than for the expensive perfection of the optimizer! He carries his model into suggestions about the development of the brain. I do not know how they will fare among the brain physiologists but his philosophical treatment will prove quite congenial to the naturalistic tradition in philosophy: for example, free will is given a meaning as a phenomenon of choice in a preexisting set of values. His initial attitude to primary values has an advantage over most run-of-the-mill accounts of human nature in terms of lists of needs and drives: it incorporates them in a wider theory and can try them out as hypotheses within its framework. Moreover, by including the values that enter into learning, into common sense and into the formation of a world model, as well as by stressing the continuities between animal and human, the theory achieves some of the same force that philosophical pragmatism brought to the naturalizing of mind. Finally, it should be noted, the status of the primary values makes them primary only within the short range—that is, the experience of lives lived—not within the long range of the evolutionary process in which they are being tested as surrogates or instruments for carrying out the job of group survival.

The theory has a double potential in application: it could be an instrument of evaluation and criticism, even of basic values, from a philosophic-historical vista; or it could incline to veneration of the innate and perennial values. Pugh does recognize that there are environments in which the value system does not operate effectively; for example, he finds transience and impermanence destructive to values. He says, "Rather than attempting to adapt human nature to an inhuman pace of change, we might do better to harness technology so that it works in support of the enduring human values." [46] (But of course it would not follow that what people usually mean by the enduring human values are what he has listed as the innate biological primary values!) He does envisage the possibility of changes in our innate values to suit social conditions with which they are incompatible, but thinks the tampering too dangerous in the light of lack of knowledge so that it is wiser to adapt the social environment. [47] The invocation of the innate or primary values as criteria and as ultimate bases of judgment is so preponderant and the reminder that these judges are themselves somewhere beyond human consciousness being judged so much of an afterthought, that Pugh apparently feels it necessary to affirm explicitly—and I think sincerely and correctly—that his theory is not a status quo theory. [48] On the whole, the theory does not realize its potential as a critical instrument. I think that there are several reasons for this, in part external, but in part reaching back into the theory itself.

First, there is the selection of content for the primary values. Pugh finds that there is a complex system embracing basic emotions and sensory-affective reactions, traditional biological drives, and social tendencies as well as selfish ones; significantly he also includes the values at the base of intellectual endeavor elicited from an examination of our learning mechanism (identified as Bayesian). He is hospitable to McDougall's conception of dynamic instincts and some of his list, he accepts Wilson's belief in altruism as innate, he takes the current picture given by the ethologists almost wholesale, but equally incorporates some of the current psychological views of man as affiliative. The result is an aggregate which at times seems indiscriminate, at times selectively arbitrary, a listing which we are told (since he is no specialist) is only illustrative but which nevertheless at times gives the appearance of aiming at comprehensive completeness.

Furthermore, the mode of analyzing values is uneven. Sometimes there is breakdown into elements (for example, four distinct value-components in the act of elimination); sometimes we have an unanalyzed unitary value (for example, desire for dominance or for approval) which literally clamors for analysis. We are not told why some things find a place while their kin in the research literature are wholly missing—for example, shame is there, guilt is missing. And the sense of arbitrariness is increased by occasional unexplained theses definitively asserted; for example, that most social motivations come in just two dimensions—joy versus sorrow and pride versus shame. [49]

Second, there is no attempt to examine any of these from the point of view of the other psychological and cultural sciences than ethology, animal psychology, and a corner of anthropology (concerning the horde). Thus there is nothing on what psychoanalytic theory or cultural anthropology have done with shame or domination and authoritarianism, and this in spite of the fact that in his own theoretical program Pugh is not reductionist but finds a place for a pluralism of scientific perspectives. He is of course concerned only with the biological origins of values—that is, finding what is biological in the values. But all these psychological and social science studies would be very important if he takes seriously the task on which he spends so much time—getting plausible derivations of "macroscopic" values on the commonsense level from the biological elements in the hope that his theory would thereby be rendered plausible. (Whether it would or not, without the more precise "bridge laws" between fields—like those the physical sciences employ when chemical phenomena are thus derived from physical laws—is the critical question for all such hopes.) Many of the derivations are simply likely stories off the cuff. Some are ingenious—for example that humor which employs incongruity appeals because it fits the pattern of the counter-instance which is a basic intellectual value in proof. But even ingenuity is not enough; Freud and Bergson had ingenious though differing explanations of the same phenomenon.

Third, one wonders why the concepts employed have not been analyzed more thoroughly. Pugh shows great analytic sophistication at various points—for example in his treatment of the language of commonsense morality [50]—yet he is content to accept the concepts of altruism, selfish and social, that have caused so much confusion in contemporary thought, without attempting to refine them.

Fourth, there is the problem of sources. A physical scientist is in one sense at the mercy of contemporary psychological and social scientists when he enters their field and looks to see what is going on—particularly if he brings the habits of the physical sciences in which the latest works are likely to be the most advanced. In the sciences of man he may instead encounter an old ideology in the latest fashionable clothing. It is quite likely that Pugh's application of his theory has suffered somewhat from a dependence on the rather intense anti-social, anti-cultural character of present-day sociobiological theory.

The net result is a picture of basic or primary values that becomes increasingly honorific as it goes along, that loses its critical force and finds room for traditional biases (most marked in the question of sex differences), that is on the whole near the frontier of a broad liberalism in outlook not because the theory points there but because the author stands there. It should, however, be added that throughout this second half of the book there are many sharp analyses of theoretical issues as well as of some questions of contemporary social morality. In the latter, since Pugh is taking the standpoint of a scientist, he selects particularly those new questions on which our cultural tradition has provided no guidance.

Perhaps the basic difficulty I am trying to pinpoint comes earlier in the theory. It may be that Pugh's admiration for the evolutionary process as a good designer of decision systems has focused on success and insufficiently on failure and imperfection. He may not be taking seriously enough the occurrence of internal conflicts and frictions, even his own warning that the evolutionary process which produced the innate value structure took a very long time so that the structure may now be out of date and more appropriate to a primitive human society than to an urban society.[51] Thomas Huxley in the nineteenth century took that approach with respect to innate aggression and saw human ethics as battling against evolution. But Julian Huxley in the mid-twentieth century called on us to recognize that ethics itself was part of the evolutionary process. And if we take this seriously, how would it be translated into Pugh's terms? It would require considerable reformulation, since for him ethics is secondary, responding and responsible to primary values. But Pugh actually allows the brain to have changed since man acquired speech, and so the design of the mechanism is accommodating itself to leave place for more learning. Now Pugh assimilates the prewiring within the body, which gives us appropriate sensory reactions, to learning—thereby extending the concept of learning. But this works both ways: learning is therefore filling in by nature's methods what was left to be done on the spot at a later time. It is, in short, part of the design that man (in the old adage) should be born half-made and left collectively to finish the job. How he finishes the job is to be included in the account of the machine, but this extends the scope of the machine. Just as the air is included in the account of the operation of the lungs (John Dewey's favorite illustration of a transactional approach), and the light in the account of the eyes at work, so the social milieu should be considered in the account of the mind at work. In this perspective a fully integrated social-biological transactional account is called for as itself the picture of evolutionary development. There are hints of such a view in Pugh's recognition that in more recent stages of evolution primary values developed within a human social environment already rich in secondary values.[52] To carry it through completely would largely change the whole scenario for the examination of values. It would not be the search for biological primary values as separately operative in common value reactions, but a much more complex investigation of the development of common values as biosocial phenomena on the basis of biological *and* sociocultural knowledge. Primary values would themselves be seen as capable of evolutionary development, in much the same way human feelings can grow more refined, or conceptions such as health can be articulated from rough feelings of pain to complex understanding of conditions of well-being. So, too, at certain points in the evolution of the human species even the ultimate aim of survival became transformed, and the goal of the conscious part of evolution became not just to live but to live well—not just life but the quality of life—even if there be a hidden source (hypothalamic?) that plays the card of life alone. It is thus misleading to raise the question of man consciously changing his primary values by devious means. Nature has already done it by social means!

We need not pursue the philosophical questions that a transactional approach gives rise to. But it clearly seems to me the direction in which Pugh's major theory should move if it is to overcome its inner hesitations.

Notes

1. For a detailed presentation of this view, see Abraham Edel, *Method in Ethical Theory* (Indianapolis: Bobbs-Merrill, 1963), esp. chs. 7, 10, 13.

2. For the comparable kinship of different methods of analyzing concepts, see Volume I, *Analyzing Concepts in Social Science* (New Brunswick: Transaction, 1979), pp. 1-41.

3. It is worth noting that conservative theories of history tended to take an anti-scientific stance in the nineteenth and early twentieth century, but nowadays frequently invoke scientific support.

4. For Erich Fromm's account of character-types, see his *Man for Himself, An Inquiry into the Psychology of Ethics* (New York: Rinehart and Co., 1947), ch. 3.

5. Abraham Edel, *Ethical Judgment: The Use of Science in Ethics* (Glencoe, Ill.: The Free Press, 1955). This starts with an analysis of the problem of ethical relativity. The sciences are then examined to see what assistance they can give to answering the questions raised in the analysis of that problem.

6. See Edward O. Wilson, *Sociobiology: The New Synthesis* (Cambridge, Mass.: Belknap Press of Harvard University Press, 1975).

7. Stephen C. Pepper, *The Sources of Value* (Berkeley and Los Angeles: University of California Press, 1958), pp. xiv, 732. I am indebted to the National Science Foundation for a grant in 1959-60 during which I did the study of Pepper's work that appears in this chapter, originally published under the title "Science and Value: Some Reflections on Pepper's 'The Sources of Value'" in *The Review of Metaphysics*, Vol. XIV, No. 1 (September 1960): pp. 134-58.

8. Ibid., p. 2.
9. Ibid., p. 161.
10. Ibid., p. 14.
11. Ibid., p. 269.
12. Ibid., pp. 667-68.

13. Pepper, Stephen, *A Digest of Purposive Values* (Berkeley and Los Angeles: University of California Press, 1947).

14. Pepper, *The Sources of Value*, p. 674.
•15. Ibid., p. 685.
16. Ibid., p. 704, note 4.
17. Ibid., (respectively) pp. 143 ff., 131 ff., 152 ff., 349 ff.

18. For example, Kurt Goldstein contrasts this with sound life in which, he says, "the result of the normal equalization process is the *formation* of a certain level of tension, namely, that which makes possible further ordered activity" (Kurt Goldstein, *The Organism*, American Book Company, 1939), pp. 195-96.

19. Pepper, *The Sourches of Value*, p. 148.
20. Ibid., p. 461.
21. Ibid., p. 481.
22. Ibid., p. 591.
23. Ibid., p. 431.
24. Ibid., p. 464.
25. Ibid., p. 41.

26. These problems take many forms. Cf. p. 53 on a presumed difference in their physiological base; pp. 196-198 on why pleasure should not be regarded as the terminal goal within appetition; p. 188 for a critique of Perry's view that satisfaction is the feeling of the harmony of goal object and drive; p. 144 on the way in which pleasure involves slowing up achievment in the consummatory phase of appetition, whereas achievement demands should speed up the terminus—which, together with introspective evidence leads Pepper to regard affective values as resting on an independent selective system; p. 693 where Pepper notes that aesthetic pleasures have not been dealt with specifically, except in the indirect way in which all affective value in the consummatory field is regarded as aesthetic.

27. Pepper, *The Sources of Value*, pp. 262-63.
28. Ibid., p. 265.
29. Ibid., pp. 352-53.

30. Gregory Bateson, Comments on Margaret Mead's paper, "The Comparative Study of Culture and the Purposive Cultivation of Democratic Values," in *Science, Philosophy and Religion*, Second Symposium (Conference on Science, Philosophy and Religion in their Relation to the Democratic Way of Life, Inc., New York,

1942), p. 96. The full account of the Bali pattern is in Gregory Bateson and Margaret Mead, *Balinese Character: A Photographic Analysis* (New York: New York Academy of Sciences, 1942), cf. pp. 47-48.

31. Pepper, *The Sources of Value*, p. 515.
32. Ibid., pp. 432-33.
33. Ibid., pp. 434-35.
34. Ibid., p. 480.
35. Ibid., p. 677.
36. Ibid., p. 680.
37. Ibid., p. 286.

38. For a context in which the coalescence causes confusion, see the discussion of natural selection. A question is formulated: "Is evolutionary natural selection a natural norm instituting values?" (p. 613). And on p. 654 we find "the natural norm of natural selection" evaluating social structures and making a selection. But in the summary of selective systems (p. 663) natural selection definitely appears as the last of the seven, and so regarded it is assigned as its norm "the continuance of an interbreeding population" (p. 667).

39. Pepper says, "It would be arbitrary to deny the term 'value' to the selective action of the life processes in natural selection while allowing it for the selective action of purposive structures" (p. 615). Later on he says, "Suppose it is admitted that survival value is properly denominated a value in that it arises from a selective system comparable in its dynamics to systems motivated by purposive drives" (p. 636). But in dealing with purposive structure as embodying selective systems he used a quite different logic: "value" referred not to the selective action of the process, but to conations, achievements, and affective experiences. Purposive structures showed that they were relevant to value theory by explaining the occurrence and helping to refine the standards for these important classes of independently identified values. This is quite different from arguing, as he does in dealing with natural selection, that because the operative system resembles purposive structures it is a selective system and therefore what it institutes are to be considered values. On the contrary, it is because men do value survival, both individually and in group decisions, that it is worth raising the question of whether natural selection is a different selective system or simply a complicated organization of interacting purposive structures in a changing environment. In this context, I should have thought that on Pepper's view natural selection was the proposed selective system, survival (of individual or group) were the values, and some features of the operation of the selective system that scientific exploration reinforced—such as adjustment or conscious adapation, or controlled remodelling of environment, with an empirical study of where each was dominant—the natural norms.

In general, in dealing with these conceptual problems, I think Pepper has paid dearly for his sharp rejection of the kinds of inquiries that the logical analysts have engaged in by regarding them as linguistic tinkering. It would have been more consonant with his scientific approach to pursue the sharpness of their methods while pressing for and criticizing their tacit scientific presuppositions.

40. The consummatory field is regarded as a selective system, but when it is characterized in relation to the definition that is emerging (p. 666) we are told that "the trials are the acts of maneuvering to find the optimum conditions for satisfaction." This makes it sound exactly like a purposive structure with increase of pleasure as its goal; when this question was first discussed in the book (pp. 354 ff.) attention was called simply to a tendency toward maximization of pleasure in the consummatory phase of other purposes. Pepper may very well be correct about two

separate physiological mechanisms; what is at issue is whether he is not straining the description of one process to fit the definition of selective system.

41. Pepper, *The Sources of Value*, pp. 504-05.
42. Ibid., p. 613.
43. Ibid., p. 367.
44. George Edgin Pugh, *The Biological Origin of Human Values* (New York: Basic Books, Inc. 1977), pp. xii, 461. The discussion of Pugh's work that appears in this chapter was originally published as a review of his book in *Society,* 15, 2 (January/February 1978): 92-95.
45. Ibid., p. 337.
46. Ibid., p. 445.
47. Ibid., p. 413.
48. Ibid., p. 419.
49. Ibid., p. 291.
50. Ibid., pp. 356-60.
51. Ibid., p. 373.
52. Ibid., p. 347.

7. Scientific Research and Moral Judgment: A Philosophical Perspective

This paper is concerned with the relation of science to moral judgment. While it will have to deal with our understanding of moral judgment and even different interpretations of the concept of morality, its aim throughout is to see the impact of scientific ideas and scientific research on the criteria which can be applied to sharpen and refine the moral judgments that we make.

The kinds of scientific study of morality that I have in mind may be illustrated from four vital currents in the last century. One is the sociological naturalization of traditional moral theory, as in the work of Durkheim and Piaget. A second is the psychoanalytic resetting of morality. A third is the expanded anthropological study of values in recent decades, cutting across institutional studies on the one hand and personality studies on the other. The fourth is the sociohistorical treatment of morality in relation to other phases of a society in its development, as seen for example in Max Weber's work on the Protestant ethic in relation to capitalism or Marx's historical hypotheses about morality as superstructural in relation to economic life. Of course there are many other paths of work—for example, biological studies of instincts or evolution or comparative studies of aggression, or social study of institutions—which throw light on problems with which we are concerned in moral theory, though not initially directed on morality. Any knowledge which adds to our understanding of the nature and development of man, his world-milieu and his predicament and problems, adds to our view of the matrix in which morality emerges, takes shape, and operates. There is, moreover, no predicting where the fruitful research on morality specifically will next emerge. It may be in the scientific study of educational techniques and learning processes, or cosmic yearnings and forms of religious experience, or breaking points in historical transformations, or linguistic careers of moral terms, or in the mathematical techniques of decision theory.

Our traditional notions of morality, too, have taken different shape. One is in terms of aspiration for the good life. Another is in terms of a system of laws which we cannot but respect and find binding. A third is cast in terms of the self and its appropriate growth and character. A fourth, more recent, is in terms of moral deliberation and choice. Later on I want to examine these as models in moral philosophy, and for purpose of convenient reference we may label them as the *goal-seeking* model, the *juridical* model, the *self-development* model, and the *decisional* model.

The underlying thesis of this paper is that the isolation of morality from the fruitful currents of scientific inquiry has given to moral judgment the character of being merely expressive or emotive, but that even the limited growth of knowledge relevant to morality in the scientific work that has been done changes the picture to some degree. It makes moral judgment more effectively *judgmental* by furnishing *criteria* that make possible more definite answers where without it there appeared to be merely bias or arbitrary selection. These criteria concern both the conditions in which morality is being applied in situations of choice and decision, and equally the establishment of priorities and other orderings among the values that are being applied. It is such criteria that enable us to say that one step in the process of working out a moral judgment is more suitable, fitting, appropriate, relevant, or correct than another. It is the clarification and systematic increase of the criteria that make the difference between the arbitrary expressive and the rationally grounded. And it is precisely at this point that scientific research can be of help.

How Science Helps Establish Criteria in Moral Judgment

It is well known that both in philosophy and in science there has been a sharp conceptual dualism of *fact* and *value*, or the *is* and the *ought*, and that this allegedly unbridgeable gap has been used to bar in wholesale fashion attempts to study the specific relations of science to ethics, as well as the relevance to ethics of achievements in the philosophy of science. In the familiar slogan, science can do no more than furnish facts and means; it is closed out of the reckoning of ends. Since I am not here concerned with any general theory of the relation of science and ethics, but with discerning points of specific impact, I shall bypass the fact-value problem with only the following comments:

- The fact-value dichotomy cannot be equated with the science-ethics distinction. Science contains values as well as facts, and ethics contain facts as well as values.

- The typical problem of moral judgment rarely poses us with going from facts to values, but from facts and some values to other values.
- The philosophy of science shows us the tentative and theory-laden character of even large-scale dichotomies, and that they require assessment in terms of the jobs they perform. The fact-value dichotomy is helpful where it prevents the smuggling of values through apparently factual concepts—for example, to declare something a "need" is to say that if it is not satisfied certain *undesirable* consequences will ensue. But the fact-value dichotomy is misused if it bars specific investigation of how our concepts of needs are built up to be serviceable in moral judgment and the precise role that scientific knowledge plays in such a process.
- As long as the intransigent dualism is maintained, it was suggested above, moral judgment appears to have an expressive or emotive character, or else a dogmatic character issuing from simple fiat or expression of arbitrary will. The question is not whether moral judgment can be like that—for it often is—but whether it has to be like that, or whether its being like that is a consequence of either lack of deeper knowledge or failure to use knowledge we already have. No utopian claim for rendering moral judgment completely determinate is here involved. There is simply the attempt to see how far criteria are rendered available for a greater determinateness.

Moral judgment involves taking the morality which we have and applying it to situations in which we find ourselves. (In this process we may, of course, be brought to reconsider and reconstruct portions of the morality.) Without entering into a fuller study of the phases of moral judgment, we can see that it involves at least what we may call *structuring the situation before us*, that is, determining the character of the problem and what moral principles or notions are relevant; *interpreting those principles or notions* so as to be able to operate with them; *establishing an effective ordering* among the values or principles that constitute our alternative choices, in terms of criteria of decision; and envisaging available means for achievement of the path or paths contemplated.

I shall not dwell on the last of these aspects, the use of science for furnishing more effective means—since it is taken for granted nowadays—except to point out that this covers a great deal of ground. For it includes not only prior causes but discovering subsequent consequences and so posing fresh problems of estimating costs and benefits. Sometimes it even involves the establishment of techniques which quite remove serious moral issues that previously offered incompatible choices. For example, the invention of insurance affected moral judgment on where burdens should lie by removing (in socializing burdens) the hard choice between either letting them lie where they fall, or else imposing them on others than the ones affected. Similarly, the development of contraceptive techniques quite undermined the previous hard choice between sexual abstinence and risking pregnancy. The relation of scientific techniques to moral solutions in both personal and social morality would well repay much more detailed study that it has hitherto received. Of course sometimes the scientific discovery complicates the moral problem and even forces conceptual changes; as, for example, the transplanting of a living heart, if it is not to be construed as a case of killing A for the sake of B, may involve redefining the death of A in terms of criteria of absence of brain electrical phenomena rather than phenomena of the activity of the heart.

Letting go, then, the question of the impact of science on means, I want to discuss its impact on the other aspects of moral judgment indicated, and in special detail the development of criteria of correctness. I shall deal with:

1. structuring the problem,
2. interpreting moral ideas,
3. ordering concepts that assign a special status, such as natural, normal and abnormal,
4. ordering concepts that furnish criteria of authenticity,
5. ordering criteria on a developmental theory, such as maturity,
6. ordering criteria in terms of which judgment is rendered, such as moral and immoral, right and wrong.

Structuring the Problem

There is no systematic theory which tells us how to describe the situation to which we are to apply our morality with its large inventory of rules and goods and virtues to select from. And yet some formulations are obviously better than others, and some give rise to more fruitful hypotheses or directions of solution. (It is like the comparable predicament in a scientific inquiry: what questions are to be asked and what hypotheses explored.) It is asked, for example, whether it is wrong for the government to support illegitimacy in welfare recipients by contributing for each child born. But this may not be the correct principle to invoke. A sociological analysis of the situation may center it as an issue of family structure in the typical recipient group, or an issue of lack of economic opportunity. Different formulations of the moral problem are a consequence, or different moral problems are regarded as appropriate. The discovery of what is appropriate or correct in the initial formulation thus proceeds from a great deal of knowledge of the context and whatever theoretical knowledge we have in that field.

Sometimes the logic of such a judgment can be reduced to finding the best means to remove a troublesome phenomenon—for example, in urban ghetto rioting the contention that "violence should be met with violence" can be judged as a less effective means than anticipating and removing grounds that precipitate the riot. What makes such a judgment possible is not merely the general adage that "prevention is better than cure," but the specific interpretation of the aggression in the violence as a result of massive frustration—in short, a specific application of the complex frustration-aggression theses in psychology. But not all cases of structuring the situation may be so readily translated. For example, when Mussolini invaded Ethiopia, he justified his conquest by saying that all the other great powers had colonies but Italy had none, therefore it was unjust. I recall some contemporary writer of that period comparing this to the story of the child who burst into tears on being shown a picture of the Christian martyrs being thrown to the lions because one scrawny lion in the corner had no martyr!

Obviously, some moral structurings of a situation are less suitable, fitting, appropriate, relevant, correct than others. When we speak of the central problems in a given situation, age, or context, we are making selections or structuring the problem in a way in which we believe issues should be tackled if the maximal good of the people involved is to be achieved. It presupposes values, of course, but linked to an analysis of what is going on, whether in sociohistorical, cultural, or psychological terms. And so the illumination that guides the structuring comes from the strength of the theoretical understanding of the situation. This is clear too in an individual's moral struggle to decide. In Sartre's well-known illustration of the young man who came to him for advice on whether to stay in Paris and take care of his mother, or go off to join the anti-Nazi resistance abroad, the startling thing is that Sartre asks why he came to him for advice rather than to his priest or some other professor.[1] Clearly, Sartre interprets this choice of advisor as indicative of a partial decision—for the young man would suspect the kind of advice he would get—and therefore as an attempt to avoid the responsibility of decision. Insofar as Sartre has discerned a prevalent psychological tendency in the human predicament, his answer makes more sense in the case than a moral weighing of mother-child affection as compared with resistance to oppression or national self-determination as an ideal—in all of which too, however, one could well consider the impact of scientific knowledge concerning the relationships and the ideals.

I have raised this issue of structuring the problem first because it is often overlooked how influential the initial structure selected is in moral judgment. It plays the same role in ethics as the way the question is formulated in science. Although considerable philosophical analysis of different kinds of problems would be required to sort out how values enter into such a judgment of correct formulation, and how psychological and social knowledge of man and the situation enters, and how both are related, there is little reason to doubt that the knowledge gained by scientific research plays a formidable part. But there may here be no uniform or general answer to the question of correctness. It is just as in the diagnosis of an illness from the symptoms—not merely the specific disease, but whether to route the inquiry along organic lines or psychological lines. The judgment of correctness rather than arbitrary preference hangs on the extent and scope and systematic character of the scientific knowledge. This first topic of structuring thus underscores the massive role of scientific understanding in furnishing specific criteria of fitness, correctness, etc. We shall find that this lesson recurs in every phase of moral judgment we examine.

On Interpreting Moral Ideas

I have not in mind, in this phase of moral judgment, the interpretation of our central ethical concepts of right and ought, of good and bad, of virtue and vice. To see how scientific knowledge helps our understanding of these is a more complex task on which some suggestions may emerge later. I am thinking here of the material terms that enter into moral rules and specific virtues, and the criteria for their application in diverse situations.

Examples can be taken from the impact of psychological and sociocultural study. We do not get far in moral judgment by condemning dishonesty, usury, ingratitude, or even injustice. We have to understand the psychological, cultural, and even historical depth of the conduct that is being assessed. Hartshorne and May's *Studies in Deceit*[?] showed the situational character of honesty—for example, we have to know the difference in attitude in the society to taking from an impersonal corporation and from another individual, or the complex attitudes of students to cheating or to copying vs. paraphrasing in preparing essays. Similarly, sociological studies have amply distinguished between the meaning of stealing in the case of juveniles who take a car for a joyride and then abandon it, and in the case of organized appropriation of cars for sale. The usury that was condemned historically consisted in charging high interest in the case of loans for consumption by the unfortunate; as the purposes of borrowing became investment for production, the outcome in moral judgment shifted. Only a failure to distinguish specific content leaves us with the paradox of the "same" act now wrong, right.[3] Even what appears a matter of sentiment—gratitude—turns out deeply rooted. The extremely high value placed on this sentiment in late medieval and early modern literature only makes sense when we realize that it is tied in with a complex social matrix of feudal attitudes and their class-residue, so that ingratitude approximates treason to those upon whom we depend for safety and support. Similarly, justice refers to accepted social principles of distribution, and without reference to them is left only with a vague penumbra of feeling about "desert" or "fairness." Even very specific virtues and vices may turn out to have a complex inner content; for example, in the concept of being shiftless there is a fusion of poor, lazy, and often dirty, that requires for its understanding the specific conditions of a particular era that would tie such disparate elements together.

In general, if we stay on the surface level, we make moral judgments without depth, and with restricted relevance. For we do not see what is packed away in our own moral concepts that makes them capable of application. To unpack the psychological, social, and cultural content in the case of the concepts that we actually employ is to sharpen our criteria for more adequate judgment.

Ordering Concepts: Status Notions of Natural, Universal, Absolute, Normal, and Abnormal

In traditional moral judgment, certain concepts are sometimes applied to values or moral rules or virtues and vices to give them a special status. A value that *expresses man's nature* is felt to have, insofar as it does so, a large measure of justification. A pervasive or *universal* value is thought superior to a merely local one, a moral rule that is *absolute* to one that allows many exceptions. To call a particular value *abnormal* is to devalue, often condemn it. Such concepts traditionally were embedded in metaphysical or theological theses. The idea of the natural belonged in a teleological metaphysics in which a plan was being worked out in human life, and the natural was what conformed to the basic plan; it was what was "inherent" in man and how he would universally behave if not turned aside by special forces. The notion of the absolute was tied to a juridical model of divine command that admitted no disobedience without dire penalty.

In recent times, the sciences, especially the psychological and cultural, have been faced with the problem of interpreting such concepts and winnowing out possible scientific content. It was the evolutionary view that precipitated the intellectual crisis, for it made impossible earlier assumptions that, for example, the natural had to be good because it expressed a divine plan. The natural had now, so to speak, to stand on its own; it was the structure that had survived because it happened to further past adaptations, but it might turn out to be a present hindrance.

If we regard these concepts in terms of their impact in moral judgment, they can be translated as directives, resting on assumptions about human nature and the human situation. To regard something as part of human nature is to say that it is unavoidable, that we cannot hope to change it, that it is a specific force with which we have to reckon. The moral reckoning can thus be an increase in worth if it is a positive value (corresponding to its assured stability), and to be resigned to it or find it less harmful expression, if it is a negative value. Thus, if aggression should be part of human nature, we could attune ourselves to expect it, and seek a "moral equivalent of war,'' as William James suggested. The scientific part is to establish or disprove the unavoidability and to envisage the range of alternative expressions.

Unavoidability has a varying range. In human nature as such, the scientific search for the unavoidable has taken the form of an account of instincts and drives, or in more complicated fashion, an account of basic needs. We have already seen how the notion of needs builds in the values which would be thwarted by nonsatisfaction of need. For major bodily needs these are the agreed upon general values of survival and basic preservation of the individual and the group. In so-called psychological needs, the undesirable element in frustrating them may be varying forms of an ability to function, or even a major loss of abilities which are pervasively necessary for the minimal social cooperation required for group survival. The recognition of the moral import of discovering basic needs has been widespread;[4] philosophical caution is needed to make sure that ethnocentric values are not incorporated into them—for example a basic need for novelty disguised under "growth" may simply represent a restless aspect of our culture. On the other hand, perhaps not enough has been done with the moral import of basic human *processes*, as distinguished from needs. Thus, if survival requires clear perception of the world around us, and if extreme frustration of certain psychological needs impedes the successful development of the ego-function, then the moral role of this need-satisfaction is strengthened by its indirect tie-in to the survival mechanism. It is by the detailed exploration of complex unavoidable tie-ins that the values initially granted become the basis of justifying other more specific value-patterns. A scrupulous scientific exploration here can strengthen the base of sounder moral judgment.

Unavoidability in a narrower sense than human nature applies to cultural values. A sound anthropological science is needed in addition to the psychological to help us distinguish what is a natural given and what is a particular cultural form—for example, in the ideal of success, the question of the regard of one's fellows from the question of accumulating wealth as the major mark. It would also tell us how far what is culturally established is realistically changeable, under what conditions—from the restoration of a no longer living language, in the interests of national cohesion, to the reconstruction of familial or urban patterns long entrenched but inimical to currently social welfare.

Finally, there is unavoidability in personal life too—not in the sense of what could not have been changed, but what may no longer be changeable. A recent complex illustration is seen in the claim coming before the higher court that repeated arrest and punishment for chronic alcoholics is "cruel and unusual punishment" and so unconstitutional. This rests on the view that the conduct is no longer changeable by these persons, and its moral effect would be to force communities to abandon punishment for this large class of offenses and construct treatment facilities.

Universality has in part been dealt with implicitly, since the universal is one mark of the natural. As a directive for moral judgment, to find a value universal is to instruct that wherever it appears within a moral reckoning it can be given a plus. Universality does not affirm the size of the plus, nor how readily it may be outweighed by other values. The search for such universality or generality, once it is divorced from theoretical inferences about human nature, loses its force. It may be an isolated universal of no particular significance. The universals that are important are the ones—like affiliative needs and values—which would help solve other human problems. But universal aggressive tendencies, guaranteed by a "death instinct," would heighten human problems.

For a moral rule to be absolute implies that its mandate cannot be outweighed or ruled out by any other rule or value whatsoever. This obviously is a privileged position, but it need not be tied to an unquestioned command. It can be given a variety of scientific meanings in terms of the possibility that a line of action may always be of overwhelming importance in terms of its impact on the whole of life or the whole structure of personality. There may be absolutes arising from the nature of specific situations ("Never smoke while the plane takes off or lands"), absolutes that express the resultant of a complex reckoning ("Here stand I, I cannot otherwise"), and even absolutes that are given a place in the very structure of morality ("Never treat a man purely as a means"). The justification of these may take scientific form in part: "Because the danger of the plane exploding is too great," "Because the whole integration of my self in its basic value pattern will be shot to pieces," "Because the quality of human interrelations is impoverished and because historical experience has shown that exploitation brings innumerable evils." These justifications would rest on the data of physical, psychological, and historical science. And they are justifications of absolutes—otherwise Luther's stance would be regarded as a confession of paralysis! In a recent case (reported in *The New York Times*, March 9, 1968, p. 1), a career officer who refused to train a pilot in connection with the Vietnam war on grounds of conscience was asked whether he could have acted otherwise, to determine whether his religious beliefs about this war constituted a kind of psychological block. He answered that he could have, if he could have put aside all his beliefs! This is one fairly good index of what is felt as a moral absolute.

The concept of the normal and abnormal requires a more complex analysis than the preceding notions, for it already presupposes the natural in some of its aspects. There has been, of course, considerable controversy over the idea of abnormality.[5] If a purely statistical concept is used, then a judgment of abnormality, as a moral assessment, disguises a demand for conformity; this is a clear case of value smuggling. But other interpretations of normality are found in which it is more akin to the natural; it assumes universal value-tendencies that take place unless there is a counter- or distortion-process. Such an assumption is question-begging unless there is a clear scientific account of what is going on and what kind of interference there has been and what values are thwarted as a result and why those values are superior to the ones brought into play in the distorting process. If, for example, extreme alcoholism is shown to involve overwhelming anxieties or guilt feelings that simply cannot be endured—remember the quip about the superego being that part of the self that is soluble in alcohol—then the judgment of abnormality can probably be ratified quite readily. If a child refuses to grow up, invokes his devotion to childish activities, yet is seen to have desires for increased areas of more adult activity but at the same time to curb them through fears of assuming responsibility, then no ideologies of "Peter-pantheism"can avoid the picture of abnormality. An important controversial case is homosexuality. Many psychiatrists tend to regard it as abnormal, but articulate homosexuals often defend it as simply a viable alternative value set, only made to seem distorted by the cultural prejudice against it. The mere possibility that previous heterosexual capacities and tendencies fell by the way is not seen as more significant than, say, musical capacities or mathematical capacities that were lost through nondevelopment or the presence of incompatible interests. The central issue thus becomes the character of the actual homosexual activity—whether it is a stable and definite value-preference or an unstable compromise between sexual desire and some distorting fear of the opposite sex. It is, of course, quite possible that there are different kinds of homosexuality, only some of which are abnormal, as some types of heterosexuality also may be. Without the deeper scientific knowledge, the judgment of abnormality is merely emotive or expressive of attitude. The growth of the scientific knowledge alone can make it a judgment capable of being shown correct or incorrect.[6]

Ordering Concepts: Criteria of Authenticity

In moral judgment we speak at times of *unrealistic* aims, *real* as against *spurious* values, *true* as against *illusory* happiness, *authentic* as against *unauthentic* values. Such notions are presumed to embody criteria for differentiating a form of the better as against a form of the worse. As in the case of the natural and the normal, these concepts arose in metaphysical contexts, and when the supporting metaphysics gave way they were left with a purely emotive status—at best, terms of praise and at worst terms of abuse. But again, as in the case of the status concepts, a scientific reinterpretation is made possible to the extent that dependable knowledge grows. The search is prompted by the common-sense fact that not all similar behaviors issue from the same psychological background or express the same motivation, that not all experiences require the same interpretation. It is not surprising that the elaboration of the notion of *authenticity* to cover criteria for moral judgment along these lines has come largely from psychological exploration of personality in depth.

The lessons of depth psychology may be applied along the whole range of constituents of a morality, using the same pattern of analysis that was seen in the theory of needs and the judgments of abnormality, amplified by the study of particular areas of personality formation and human emotion. Given an understanding of particular mechanisms of defense, we can distinguish between behaviorally similar "virtuous" conduct—for example, a realistic industriousness and an anxiety-ridden or driven keeping busy, though both get things done. Similarly, criteria of authenticity can be established for feelings and for goals. For example, both in individual cases and as types of phenomena, a distinction can be drawn between the love which is a mutual care and affection and the love which is an insecure dependence; or again, between the power sought as an instrument to aims deemed worthwhile and the power goal that expresses a craving for reassurance. So too, the older, rather indiscriminate treatment of pleasure and happiness in the hedonistic moral philosophies, or the honorific attacks on the "unreal" pleasures of the body in favor of the "real" pleasures of the intellect, sometimes yield to distinctions of pleasures which are relief from unstable tensions, and those which issue from vitality and energy, or to even more careful and detailed classifications such as Fromm's.[7] Again, in the traditional domain of conscience, there is no longer a need for simple belief or else skepticism of its reliability; it can be studied as a psychological process to furnish at least some indices of overrigidity and some marks of sensitivity, and even to distinguish types in relation to the cultural conditions of development, for example, Fromm's distinction between the authoritarian and the humanistic conscience.[8] In the case of sanctions, just as knowledge removed fears of blindness and of leprosy as automatic penalties for immorality, so it can help us remove the fears we project on others, or diminish the aggressive or sadistic in punishment. And in the case of problems generally, we can come to distinguish the illusory problems that rest on our anxieties from the real ones that require practical solutions. Finally, we can come to understand which parts of our morality are running effectively and smoothly and which are breeding frustration. And in general, the increased knowledge is of aid in knowing ourselves, in feeling which of our aims and purposes have depth in our own needs and desires, and which are adventitious, adopted on grounds of conformity, misunderstanding, and unnecessary compromise.

Many of these processes have social analogues as well, as we learn the social functions of our moral ideas. We can thus carry out evaluations of the functions performed and of the performance, distinguishing between the *ideological* and the *realistic*. Many of our concepts of distribution of gains and burdens in a socially complex economic system are still hampered by older ideologies of charity and of work, and of guilt for relaxation and enjoyment. [9]

I have presented the problems of authenticity in the piecemeal fashion in which we find them on the contemporary scientific scene. If we succeed in getting a comprehensive theory of development that integrates the psychological and sociocultural insights of these contemporary disciplines, we may be able to elaborate a systematic theory of the growth of the self to unify in its standards the manifold criteria of authenticity.

Ordering Criteria on a Developmental Theory: The Concept of Moral Maturity

To the pattern of thought that makes possible criteria of basic needs and abnormality, and ordering criteria of authenticity, the concept of maturity brings in addition the idea of a developmental process. This demands a complex expansion of knowledge, without which judgments in terms of "mature" and "immature" remain at most expressive or emotive. We have to ask what kind of knowledge this is.

The idea of a moral development—the outcome of which is a morally mature person—covers a number of different possibilities, indicating different research patterns.

We might begin with a concept of *fully moral* and look for successive steps in the achievement bit by bit of its parts; or else we might think in terms of *more and less moral* in some other way, and plot the growth from less to more. To take oversimplified examples: suppose morality consisted of following ten rules, and we found that as the child grew up he mastered a rule a year (always in the same order) beginning at age 4; then he would be morally mature at age 13, if nothing interfered with his development. Or, in the second case, suppose the notion of morality consisted of a set of ten rules, God-given, and followed out of some emotional attitude toward God. We might then set up a matrix combining indices of rules followed and type of emotion according to moral quality (from fear, through desire to please, to love), and correlate this "degree of morality" with a number of age-spans. In both approaches we would be starting with a notion of "partially to fully moral" and showing that some degree order in it correlated with some order of age-periods. This would be a discovery of a moral development, since obviously it is logically conceivable that the development might follow some different path—for example, the simpler first type might appear in full quite early, then almost wholly lapse in the middle period (a kind of moral latency period), and reappear full-blown in the later period studied. Or no developmental pattern at all might be discovered.

We might think of a moral development instead not as a process in which parts or degrees of morality were actualized successively, but as the acquisition in due order of necessary conditions for the appearance of morality. Here it would not be so much that morality developed, as that some other development in the child made the appearance of morality possible. It would be analogous to the process which ends with the ability to walk—there is first expression of physical energy, then pushing oneself about, then crawling, then staggering, and finally walking. The precursor stages are not stages of walking, but successive addition of abilities—to move, to move the body, to move the body with a measure of organization, to move the body with a measure of organization on one's feet, to move the body with a measure of organization on one's feet systematically in the special way called walking. If we define morality as an internally motivated adherence (not through physical punishment or fear of it) to a set of rules with an understanding of their purpose in the whole of life, then morality appears only when all the conditions are present, but precursor forms may show a cumulative succession or else a systematic replacement up to the threshold of morality.

We might have a complex combination in which a distinctively moral development was related to a distinctively psychological development—either of emotional or intellectual powers or some configuration of the two. Questions might then arise about causal relations and interactions, and also about causal conditions affecting both developments (e.g., parental warmth, or influence of peer groups). They might also, it is important to note, begin to touch the concept of morality itself. For example, if it had not included the emotional attitude with which one obeyed rules, the appearance of an emotional development would almost certainly lead to a broadening of the concept of morality to embrace an appropriate ground of obedience, say respect rather than prudential desires. Again, if the examination of the intellectual development shifted from an ability to understand the laws and their implications to a way of handling or generating laws, the concept of morality might be led into less legalistic channels.

Obviously then, there is no one type of moral development theory that is "fated" at this stage of human knowledge of man and morality. Let us take as contrasting illustrations the developmental accounts given by Kohlberg[10] and Erikson,[11] and consider their implications for the problem of ordering criteria of moral maturity.

Kohlberg treats the stages as types of thought, each representing a structure that emerges from interaction of the child with his social environment, as active processes of the child's ordering of his social world. His three levels, each with two types, are: the premoral, first with a punishment and obedience orientation, then naive instrumental hedonism; a morality of conventional role-conformity, with a good-boy morality resting on the approval of others, and then as authority-maintaining morality; and finally, a morality of self-accepted principles, first as a morality of contract and individual rights and democratically accepted law, and then as a morality of individual principles of conscience. Each of the six types of moral orientation is studied in relation to many different aspects of morality, such as motivation for obedience, mechanisms for resolving conflict, treatment of rules, basis of the moral worth of human life, and so on. Kohlberg looks for formulations that would yield a clear developmental picture. We cannot, of course, go into the rich detail of the studies. They lead him to reject the commonsense notion of a "moral character" implanted by habituation, and even the belief in a self-sufficient internalization process whose degree of success could be measured in emotional terms; and to conclude that the developmental process relevant to moral maturity is that of intellectual growth in general, with the powers it brings of foresight, capacity for weighing alternatives, empathetic ability to see things from different standpoints, maintenance of attention, and so on. We shall see later the concept of morality which these studies involve.

Erikson's sketch in his *Childhood and Society* of eight stages of man is not initially a sketch of moral development. It is presenting the course of psychological development that issues in a mature self-identity. But it can be seen as entering into a view of moral development since, as we shall observe later, the psychoanalytic approach does give its accounts of morality in such a setting. In fact, each of Erikson's stages becomes aligned with character attitudes or personality modes that have a distinctly moral tone. Perhaps it is simplest to itemize them: trust vs. basic mistrust, autonomy vs. shame and doubt, initiative vs. guilt, industry vs. inferiority, identity vs. role diffusion, intimacy vs. isolation, generativity vs. stagnation, and ego-integrity vs. despair. They are tied to successive challenges in typical situations from infancy to old age—invariably they have placed hurdles in the course of the child's growing up. Thus, trust begins in the infant's social achievement of a willingness to let his mother out of sight without undue anxiety or rage, and ego-integrity at the end involves the ultimate acceptance of one's life cycle as inevitable without fear of death. Every stage builds on the previous ones and each in turn adds a new criterion of increasing strength. While some of the concepts may require sharpening, and the role of cultural factors in setting the tasks at various stages requires analysis, the framework as a whole is very suggestive for exploration of basic moral development.

The different accounts of moral development available today go in different directions but they may get closer together in outcome. Kohlberg, for example, finds a certain kinship with the concept of ego-strength. In any case, even if they constitute only partial explorations of the complex field, they are still able to furnish criteria for maturity that go beyond merely the distinctions of authentic and unauthentic discussed above.

It should be stressed again that there is no question of a value-free inquiry into development. The criteria of maturity are not moral rabbits pulled out of a fact-hat. There are explicit values in the guiding concept of the moral and the more moral. In any case, there are values built into the picture of psychological development. Their analysis would follow that which we have seen for the normal and abnormal and authentic, cast in a more dynamic temporal mode, with the properties of the later stage constituting more successful solutions to the problems unavoidably generated out of the earlier stage. A frequent form of the abnormal in this context would be a "fixation" on a lower level, which carries the idea of distorting forces keeping one from moving on or destroying the necessary conditions of advance. If the psychological development is seen in intellectual terms, the increase of intellectual powers would carry obvious implications of increased mastery as a value. But, of course, the values with which a successful theory of development ends in its criteria of the mature are not simply the ones with which the inquiry began. They have become more specific and make possible an ordering which was not achieved before, and the specificity has come from the knowledge gained and wedded to the values initially put in.

That the knowledge brought by different theories of development is partial at present, and goes in different directions, means that we have not yet established a systematic account of maturity, only partial criteria.

Ordering Criteria in Terms of Which Judgment Is Rendered:
Moral and Immoral, Right and Wrong.

Such terms as these are much broader than the ones we have so far considered, and in addition to any other uses they may have in the course of a moral reckoning, they are also the terms in which the decision itself is rendered. (They are not, of course, the only ones; judgment may be rendered in terms of good and bad, ought and ought not, responsible and not responsible, and many more specific ones.) Now recent philosophical ethics have tended to focus on the volitional or performatory side of the use of such terms, rather than simply seeing them as emotive or expressive acts. But if we follow the pattern of thought worked out in the previous discussion, we have to ask whether there is a kind of scientific knowledge which would do the same job of furnishing criteria for correct application of these notions that depth psychology seems to have done for judgments of authenticity and the study of development tries to do for judgments of maturity. Let us focus on the most general notion, that of moral and immoral.

The kind of knowledge involved would have to come from an investigation into the nature and functioning of a morality in the life of a people. As scientific knowledge it would be descriptive and explanatory, but here we find ourselves in a dilemma. For the question of the nature of a morality is settled by philosophers in the philosophical analysis of the concept of morality, and is presupposed in any particular scientific investigation by the concept of morality that is implicitly employed. Even more, a proposed account of the nature and functions of a morality can itself be evaluated, and cannot without smuggling in a demand for conformity be turned into criteria for approving as moral or condemning as immoral. These judgmental concepts would seem to require understanding by reference to basic volitional commitments rather than scientific learning of men's ways. We are thus caught in the general philosophical problem of the relation of philosophical analysis to scientific description and explanation, and of philosophical evaluation to both. This is obviously a larger problem than we can here explore, but perhaps an answer consonant with the pattern of thought I have been advancing can be suggested by a few remarks and a brief case study.

The remarks concern the philosophical analysis of the concept of morality. It is no secret that philosophers have had different ideas about the meaning of "moral" and the mark of the distinctively moral, as well as about the domain of morality.[12] It is precisely this disagreement which led to regarding the concept as emotive or expressive in character. It is true that some contemporary philosophers have thought that philosophical analysis alone could establish distinctive marks, but the results often turn out to rest on a particular selection of uses of moral terms, or reflect the particular pattern of a particular era or culture. The history of ethical theory is alone sufficient to show diversity. Some model their concept of the moral in terms of universal law, others in terms of aspiration to a good, and still others in terms of sympathetic reactions in interpersonal relations. The Utilitarians[13] fashion a mark of concern with general welfare, while the organic approach to the totality, such as Bradley's,[14] emphasizes system in the growth of the self. McBeath,[15] after a study of some forty cultures, chooses a concern with the whole of life as the mark of the moral.

How are we to explain these differences? In part it is because they focus on different domains of human life—one on what goes on within the individual in the play of his desires, another on small-scale interpersonal relations, a third on institutional structures and problems. If so, then a scientific study of the various aspects in the functioning of morality should help broaden and integrate the philosophic understanding by revealing unnoticed relations and implications. Or again, the different philosophical conceptions of morality may embody different value content—one may be elevating a stress on justice and universality into a concept of the moral, another on happiness and well-being, a third on love and mutual concern. In that case, a concept of morality is not a value-free form, as distinct from content, but in part at least it has built-in moral or value or purposive components. Scientific research into the nature and functioning of morality could make clear this moral content and what in the operations of the culture and its institutions supports this elevation of specific values into organizational or structural roles. Finally, the differences may embody different understanding of human powers and human nature. For example, a concept of morality with a cognitive slant or emotive slant or some combination, may rest on particular psychological schools of interpretation. If emotion is seen as a vague precursor of intellectual clarity, then not merely will emotional attitudes hold the lower rungs in a picture of moral development, but the concept of morality will be cast in intellectual terms. If intellectuality is seen as calling for a culmination in disciplined emotion, or in partial action, one of the latter two will hold a central place in the concept of morality. If intellect and emotion are sharply divided, each with its own laws of operation, we get something like the current conflict of cognitivist and anticognitivist in analyzing the concept of morality. Here advances in the psychological understanding of the emotions and the intellect can bring a corresponding refinement in the philosophical concept of morality.

That philosophy is not alone in being faced with all these complexities can be seen if we take as our brief case study Kohlberg's treatment of the concept of morality as a basis for his scientific investigations. He requires a conception of the moral in order to map the stages of moral development. Quite explicitly, he takes justice to be the core or essence of morality.[16] In one sense, justice is seen by him as itself a set of moral values. Liberty and justice are taken to be culturally universal values. On the other hand, he attempts to make out of this a culturally universal definition of morality, by thinking of morality as the form instead of the content of moral judgments. And for corroboration he appeals to the characteristics which philosophers allegedly agree on that make a judgment a genuine moral judgment: moral judgments tend to be universal, inclusive, consistent, and grounded on objective, impersonal or ideal grounds. And the summit of moral development is then seen to consist in the child's ability to make moral judgments and formulate moral principles as his own rather than to conform to others.

The fact that justice is thus being made to serve in two capacities—as a set of moral values of the libertarian-individualistic-democratic sort, *and* as a formal characterization of distinctively moral judgment—shows that there is a more intimate relation of the content and the form than the sharp separation of the notions would imply. From the point of view of the history of ethics, and in the light of the functional interpretation of the relation of form and content indicated above, Kohlberg's concept of the moral would be seen as falling within one specific tradition—the Kantian, which has elevated a morality of justice, and the emphases on universality, equality, and impartiality that it involves—into a universalistic structure to be definitive of morality (as contrasted with the elevation of happiness or love or self-realization). The emergence of such a structuring in modern times reflects the sociocultural breakdown of less than universal groupings (family, class, nation) in a growingly interdependent world, and the counterpart emergence of a democratic individualist conception of the person. And this is not without serious implications for Kohlberg's research. Not merely can we ask how his research would have been different if he had begun with some other conception of the moral, but we can ask how far his special cultural emphases may affect his interpretation. Thus, when he finds that the last two stages in his developmental picture are not found in village peasants or tribal groups, the question may be raised whether these stages should not be regarded as cultural specializations under determinate conditions, and in fact, whether the criteria in the whole third level—with an emphasis on contract, individual rights, democracy, and culminating in individual conscience—do not have a stress characteristic of the intense individualism of Western European culture in the last two centuries.

The conclusion of these brief remarks and the glance at Kohlberg's material is that a concept of morality, whether the product of philosophical analysis of current preanalytic ideas or the construct offered or implicit in scientific research, is never in fact a pure conception nor a creation by sheer volition or commitment. It is a philosophical product, to be sure, but it has built into it elements from the moral content of the period and assumptions about human nature and its operation, as well as elements from the linguistic and intellectual tradition. There is accordingly no mystery in the expectation that increased scientific knowledge about the nature and functioning of morality in the life of a people should improve the existent concept of morality, for it helps furnish a critique of the built-in elements. Nor again is there any mystery in the improvement of the concept in rendering it clearer, refining it, and seeing more minutely how the values that were put into it originally, together with the added knowledge that has come in, stabilizes other values in a central position—that is, how a concept of morality can itself furnish criteria for moral judgment in spite of the fact that its intellectual status is that of a definition. Finally, there is no mystery in the fact that growing knowledge makes possible reconstruction of the concept of morality itself. For this happens to any concept when we gain a wider experience of the material it applies to, and appraise the concept in the light of the lessons of experience and the purposes the concept was designed to serve.

There is thus no contravention of the nature of philosophy and science nor any confusions of the intellectual enterprises of description, explanation, analysis, and evaluation, in the hope that further scientific exploration of the nature and functioning of morality in human life will furnish more definite criteria for correct use of "moral" in moral judgment.

This completes my argument. But it is important, I think, in a Conference such as ours, that philosophical research be not merely programmatic. Accordingly, the second part of this study will serve as a kind of appendix to the first, by going on to show some of the concepts of morality that have emerged in the history of ethics and seeing how they have fared as a result of the scientific work that embodied them.

Some Models of Morality

I shall first present four distinct conceptualizations of morality. These were labelled at the outset as the *goal-seeking model*, the *juridical model*, the *self-development model*, and the *decisional model*. The first three are clear in the history of ethics, the fourth I offer as a construction to interpret some recent developments. They are all abstract models in a double sense. They do not give the content of a morality but only indicate its conceptual structure. And they are applicable in diverse types of enterprise—the structuring of moral judgment, the descriptive mapping out of different moralities, the scientific projection of research on morality, and so on. In the presentation of each model I shall pay some attention to the conditions under which it would be applicable—that is, what man and his world or the human situation have to be like so that it would be a viable model. Whether these are the same conditions that have historically called attention to the model or determined its emergence is a separate question, but they are conditions for its utility. After the presentation of the models I shall take each in turn and consider what kind of scientific research concerning morality seems to be employing the model or falls within its reach. Then I shall consider what the lessons of that research are for the model. In this way, I hope that the cooperation of philosophy and science in the construction and evaluation of models will emerge clearly.

The *goal-seeking model* thinks in terms of the good life. It assumes an appetitive or desiderative structure to human life and the good as the goal or end. Knowledge of the end enables men to order their lives appropriately in its pursuit. This order finds expression as a system of means, principles or rules, desirable character formation or virtues, and methods of decision and application; but what holds it together is the vision or understanding of the good life.

This model dominated ancient ethics, especially in its Aristotelian form. The Utilitarian and naturalistic theories of modern times carried it on, furnishing from the state of current thought their account of the goal which is the good—for example, the hedonistic reduction of the goal to pleasure, or the biological reduction of it to group survival values. The favored concept in this model is the good; the concept of obligation if it occurs at all and is not replaced by some other notion, is reduced to application of the good in decision about particular contexts.

The *juridical model* thinks in terms of a system of laws or rules enjoined on men. It presupposes a structured capacity in men for discernment or reception of these laws, and an organization or sentiment for respecting or obeying them. Conceptions of good character and legitimate goals are framed in terms of the requirements and content of the moral law. The model thus pushes the concept of obligation into central place and, especially since Kant, the formidable "ought" and "duty." The juridical model dominates the western religious tradition, and its great strength in modern ethical theory comes from Kant. Kantian residues in contemporary ethics give it a position of primacy among the philosophers.

The *self-development model* thinks in terms of the self and its qualities and development. It presupposes a sufficient independence, stability, or internally unified operation of a self to make sense of its basic focus. Static forms of the model, such as the ancient Stoics who stressed self-control by the inner self, have given central place to the notion of virtue and vice. Dynamic forms, that conceived of the self as developing, have stressed self-realization and growth. Goals of striving and rules of living in morality have a derivative place in this model, expressing the basic stand on the self and its qualities.

The *decisional model* focuses on the act of choice or decision. Ends, laws, character traits or stable qualities of the self are pushed into the background; they furnish materials which generate the problems for choice, but are not determinative of the conceptual framework of the moral, as in the previous models. There are moderate and extreme forms of the decisional model. In a moderate one, like Dewey's,[17] there is an assumption of a high enough rate of change in human life so that problem situations will constantly be arising, but they will have a definite enough structure to generate criteria for adequacy of solution. Methods of solution generalized for coping with life's problems will determine what is moral in men's aims, and what are desirable modes of interpersonal and social regulation and character formation. Rationality or intelligent handling of decision is the key concept. In an extreme form of the decisional model, such as Sartre's,[18] the instability is greater. The focus is more narrowly on the act of decision and it is freedom rather than rationality lying in established guiding principles which is the key notion.

These four models are not uniquely aligned along the usual controversial dimensions of religious-secular, monist-pluralist, individual-social. For example, men's ends may be God-given or biologically fixated, and the moral law may emanate from divine command or from human reason. Again, a goal-seeking model may be monistic with a fixed all-species plan, or hedonistic with pleasure as a single end, or it may be pluralistic with a variety of culturally determined goal patterns. Again, with respect to individual and social orientations, there may be a common virtue-pattern determined for man as man, or an individual self that each man fashions in his decisions, and so on. What the models do differ on, in their presuppositions about man and the world, is the degree of stability of structure and the selection of area in which the structure is manifest. The goal-seeking model finds basic stability in the area of appetite and striving, the juridical may see the desiderative as a welter of conflicts with stability only in man's capacity for discerning the moral law. The self-development model assumes stability in the character of the self, though aims may be in flux and the human situation too complex for laws. The decisional model may find at most that there are well-structured problems, with all else undergoing constant transformation. In this respect, each of the models is open to the influence of science as it discovers the range and forms of order in the world and man. There is a sense in which the models are clearly at the mercy of the findings. I turn then to our last task, to examine how the models have fared under some of the work that has been done in their framework.

Goal Seeking Model

The human appetite or striving to which the goal-seeking model sends us, is usually conceived in terms of the categories dictated by the psychology of the day, whether it be in terms of an ordered means-ends hierarchy, or a universal pursuit of pleasure and avoidance of pain, or some pattern of impulse expression. How diverse the directions of research can go may be seen if we look within one school of interpretation of human nature and its processes that has been rich in implications for research about morality—the psychoanalytic school.

Although the psychoanalytic approach might accept initially—at least in Freud's view of conscience [19] —the traditional juridical picture of morality as a phenomenon, its own explanatory account has been clearly in the goal-seeking model. It gives some type of genetic picture of initial instinctual energies modified into specific appetitive and desiderative forms by interaction patterns that themselves operate in the service of those energies. The nature and role of morality is cast in terms of the specific theory adopted from the development of those energies. In the structural account of the Id operating on the pleasure principle, the Ego developing modes of reality adjustment by postponement of gratification, and the Superego transmuting fear into anxiety and guilt feeling and rendering internalization possible, morality was of course identified with the operations of the Superego. Hypotheses about moral development would thus attempt correlation between degree of internalization measured in terms of anxiety and specific guilt feeling and developmental stages (e.g., pre-oedipal, oedipal, latency period, etc.). When distinctions sometimes came to be made between the Superego and the Ego-Ideal, comparable distinctions were involved between a morality sanctioned by guilt as a transmutation of fear of retaliation and a morality sanctioned by a threat of withdrawal of love; the latter could be seen as giving more room for the aspiration elements and the pursuit-of-the-ideal elements in morality. In Freud's later conception of the death instinct, [20] morality was seen as largely the attempt to curb aggression in the need for group bonds. It will be recalled that much of Freud's social pessimism was expressed as the belief that widening repression was required for the widening social group in the advance of civilization. When psychoanalytic theory fixed attention on the developmental sequence of oral, anal, and genital stages, and related character formations to each or to the kind of child-parent interactions that characterized each, moral development was seen in terms of a scheme of tasks or hurdles and modes of successful resolution or fixation, each with its specific "morality," and with moral maturity the outcome of a successful running of the course, as in Erikson's stages discussed above.

In brief, the complexity of research on morality within this framework stems from the complexities that have characterized the framework itself. Its insights and resources for research have by no means been tapped adequately by study of internalization in a general sense with general marks of anxiety or guilt as indices. On the whole, the psychoanalytic approaches, by their very plunging on, have helped break up a narrow concept of goal-seeking, cast in purposive terms (though increasingly refined in behavioral studies) or else in traditional hedonistic terms.

A comparable broadening of a quite different sort was effected by the growth of anthropological value studies. Anthropologists have by no means been uniform in interpretation of morality.[21] Some have simply referred to "moral bonds" as social ties not reducible to prudential reckoning. Others have linked morality with religious conceptions of the sacred. Sometimes the reference has been to ideals of achievement, sometimes to stringent rules of conduct in familiar fields of social life, such as sexual relations or distribution patterns. Recent value studies have operated with a much more generalized conception of value in which the moral is only one sector. Kluckhohn's preliminary definition[22] was used in many American anthropological studies: "A value is a conception, explicit or implicit, distinctive of an individual or characteristic of a group, of the desirable which influences the selection from available modes, means, and ends of action." (He adds that it is not merely a preference, but felt or considered to be justified.)

A great deal of the study of moral themes is merged with the general study of cultural value-orientations, or with the delineation of specific patterns such as modes of acquisition and incentives in distribution, or with the study of religious beliefs and world views. Particularly significant has been the alliance of comparative study of cultural transmission with the psychology of personality development, to relate the kind of conscience that is developed to the mode of handling the child, or to general cultural development. Unified studies of moral configurations as a whole have also been found. And there have been some anthropologically inspired studies of specific moralities by philosophers.[23]

The broad sense of value, on the whole, tends toward the goal-seeking model. It thinks in terms of cultural aims and goals, and modes of organization to achieve them. Rules tend to be given a functional interpretation in terms of the aims they support, and patterns of desirable character are seen as expressions of needs, and situations as well as traditions. There are, however, some tendencies toward a self-development model in some of the personality and culture studies, where the organization of cultural values is seen in terms of a character-value pattern which, while historically developed, is assumed to be self-stabilizing and transmitted by the general cultural-educational pattern in which it is embedded.[24] Again, there are some indications that a decisional model of a moderate sort may emerge. The field of value is so broad and so variegated that attempts to find generic features and generic methods yield little more than bare or abstract formulations. Yet at the same time the concentration on the broader field seems to have taken away the pressure for a special delineation of the more limited moral area. Even philosophical discussion here tends to shift to "evaluation" as a general process. This corresponds to the notion of what is desirable, in Kluckhohn's definition. If the marks of the desirable as a key feature in value are sought in a context of selection and decision—

toward which the comparison of differences easily gravitates — the anthropological study of value might readily make a transition to the study of decision modes and decision bases.

The sociohistorical study of morality — the fourth of the influential approaches listed at the outset [25] — has the same effect as the anthropological in being at the same time within a goal-seeking model and yet breaking its bounds. Such study is free — in taking morality as a historical phenomenon, a set of principles, codes, virtues, goals, operative in a given society — to follow the phenomena as it finds them. On the other hand, the precise model employed may depend on the historian's philosophy of history.

What emerges from the sociohistorical treatments of morality is the general recognition that moralities serve social and historical functions, are subject in part to sociohistorical determinants, and to some degree serve as determinants as well. The accent is on change in response to changing needs and problems, whether this response be a dependent reaction, as in the Marxian view of morality as part of the superstructure resting on economic conditions and changes, or whether it be men refining their morality with a growing conscious adaptation, or both at different points and to different extents.

The contemporary outcome of the psychoanalytical, anthropological, and sociohistorical work that has an import on the study of morality, is that the goal-seeking model, while serving as a quite satisfactory conceptual starting point, is expanded or extended almost beyond recognition by the results of the study. No neat picture of human goals remains. It is given in quite different terms with all the variety and sense of intimate cultural penetration that contemporary sophisticated anthropology has brought. In the process there are conceptual swings toward the self-development and the decisional models, but on the whole little or none toward the juridical.

The Juridical Model

The juridical model has had a complex history intimately related to scientific work on morality. While this may seem surprising because of its traditional religious background, its very prevalence made it the assumed account of the meaning of the moral. As a focal phenomenon, the juridical model sends us to commanding and obeying and the inner struggles to secure obedience. In the study of the psychology of morality, this furnished a ready foothold for initial probing.

The effective concepts in the juridical model can be seen by keeping an eye on both the traditional religious form and the Kantian ethic. Morality is the system of authoritative rules. In the traditional view, the authority lies outside the individual in some divine or overarching source, whose demands are capable in some measure of rational interpretation. In Kant,[26] the rules are generated through a categorical imperative that expresses the autonomous legislative activity of the rational agent. In both, the law takes a hold on man's emotional structure. The religious view is compatible with a rich variety of attitudes by the receptive human to the divine commands. In Kant, man feels respect for the law, and the sense of obligation is the constraint felt by the agent when the law thwarts his contrary inclinations. (Kant does not regard these as ordinary sentiments of a psychological variety, which would cause us to obey; for that would contradict the autonomous character of morality.) In both the traditional and Kantian forms of the juridical model, good and virtue are interpreted in terms of this picture of law and duty. Not natural goods or happiness, but the moral will which obeys or acts from a consciousness of obedience to the moral law is the central point here. Goodness is goodness of character, and virture is identified by Kant with strength in obeying duty. The Kantian changes are thus primarily the shift from an outside authority to autonomy of the individual will in a universal community of wills, and secondarily a movement away from a system of laws to a principle of generating laws.

Durkheim[27] takes it as the lesson of his sociological investigation that the source of the outside character of morality, its objectivity, is the social group; and it is this social group that is not merely the source but also the object of respect. This shift sets in motion a host of readjustments. Obligation is no longer to be limited to curbing our sensory nature, but is also a constraint on our intellectual nature. Therefore autonomy requires reinterpretation—it is not a free legislation by the agent, but an understanding brought by the science of morality, which leads to voluntary obedience. This enlightened assent, based on understanding of group demands, he finds distinctive of a secular morality. I need not here enter into a critical consideration of the problems Durkheim gets into concerning the relations of the individual and the social group. The significant point is to see the way presumed sociological evidence about man leads to reinterpretation of elements in the juridical model, yet how at the same time he stays within the conceptual scheme with which he began his inquiries.

The lesson of Piaget's work is the same, much more complex as it is. He too sees morality as a system of rules, and the essence of morality to be sought in the respect the individual acquires for these rules.[28] His intensive investigation of rule genesis and rule functioning in the game context leads him to distinguish two forms of respect (unilateral respect, associated with authority of the elders, and mutual respect, in the milieu of the equality of one's peers), and to propose stages in the development of the consciousness of rules. The notion of mutual respect is tied in with the idea of justice, and stages in the development of the latter are related to stages in the development of the consciousness of rules. Piaget has thus carried further the empirical inquiry into the elements of the juridical model, for the rational consciousness in its unfolding is related to forms of social life. What is more, he has taken hold of the Kantian movement away from rules to principles of generating them; the idea of the sacredness of rules is now seen as an earlier stage preceding emergence of autonomy. Finally, as to the notion of good, while Piaget makes occasional reference to the contrast of duty and aspiration or the feeling of desirability, he seems to limit the theory of good to the autonomy of conscience—in short, to the quality of a person's moral judgment.[29] Autonomy is tied, not as in Durkheim with an understanding of authority as emanating from the group, but with mutual respect which constitutes the good.[30] In their conceptualization of morality both Durkheim and Piaget stay within the problems set by the Kantian conceptual scheme, that is, within the confines of a juridical model.

Kohlberg's concept of morality has already been considered above. As pictured here, he has carried further the departure from the initial model. When a morality was conceived as a system of rules, the stages of development were investigated as stages in learning or internalizing the rules. Durkheim had stayed within the sacred character of rules but given a secular interpretation. Piaget had shifted from rules to modes of handling rules, but not wholly, for his criterion of a sense of justice involving reciprocity and equality is itself a highly general rule or principle. (If is does not give us the specific content, neither does "Pursue happiness" in the Utilitarian tradition, or "Love thy neighbor" in the western religious tradition.) When the relativity of moral content to cultural tradition came to be increasingly stressed, the study of moral development required a sharper decision; how far "more moral" meant a greater possession of allegedly invariant content, or how far it meant instead some principles of handling moral content. We have already seen how Kohlberg still tried to make justice do both jobs. Nevertheless, his work carries him farther and farther away from the juridical model—in virtue of its results rather than its initial conceptualization. The element of a system of rules is minimized; it is not rules nor system but the way of handling them that eventually emerges as most significant. The relegation of good and well-being remains: good is understood as the good of action, that is, the quality of the actor, and, as in Piaget, this is identified with autonomy. Good in the sense of well-being is left to the variable content for the anthropological approaches to play with. Finally, autonomy itself, as developed rational powers, is cut loose from the other aspects, and identified with decision-making capacities that extend beyond the narrower moral domain. Kohlberg's final definition of moral maturity as "the capacity to make decisions and judgments which are moral (that is, based on internal principle) and to act in accordance with such judgments"[31] has almost passed the bounds of a juridical model and moved to a decisional one.

I need not dwell on the extent to which the psychoanalytic studies of conscience in its various forms[32] has broken open the juridical model, as well as the anthropological study of variation in patterns.[33] The effects have been more devastating than on the goal-seeking model, for both psychoanalysis and anthropology always stood outside the juridical model and from the beginning treated it as a special formation in a broader range of possibilities. Freud's motif—that where Supergo was, there Ego shall be—and anthropological study of the puritan morality as a special constellation[34] are indicative of this.

The Self-Development Model

The self-development model has had, as was suggested above, two different stresses in its ethical theory. One pointed to configurations of character and sets of virtues, the other to a general strength of self and its process of develoment or growth. The notion of separate virtues presupposes the idea of unitary character traits; that of strength of character or of will presupposes a relatively identifiable dynamic system. The general issue for research is whether there is a sufficiently stable or internally unified operation of a self to permit a concept of virtue. The Hartshorne and May tests[35] which showed that honesty was not a unified trait in terms of which predictions about behavior could be made for a given individual, but that it was highly "situational," was taken by many to be the deathblow to a concept of virtue. Any explanation of moral behavior by inner traits—such as that a child did not cheat because he was honest—would seem then to be the equivalent of explaining sleep by the "dormitive principle": but after all, granted that any individual is part of a wider milieu or field, so that explanation of his behavior is in principle a function of the total field, the utility of a concept of internal properties for one element in the field cannot be settled antecedently or *a priori*. It is the psychological theory of personality and its utility as a fruitful concept which is decisive here, with research on moral as well as nonmoral data. Contemporary treatments of the concept still leave a large measure of uncertainty about the direction it will take. On the whole, it seems as if the resurgence of the personality idea depended to a very large degree in contemporary psychology on the weight of the psychoanalytic theory of determinants, with the promise that internal operations are distinctive enough to secure an outcome whatever the external (sociocultural) factors; that is, the promise of psychological constancies to be established across historical-cultural lines.

But current work has brought the study of perception and cognition into relation with the study of personality to yield a growingly richer picture of individual development.[36] The impact of this newer movement on the self-development moral model has not yet been assessed.

Such probing for the basis of the notion of virtue yields at present insufficiently conclusive results. This leaves open the question of whether the concept, resting on commonsense beliefs, will be able to operate successfully. At the present time, when some philosophers have suggested that the dominance of the ought and the good in ethical theory has run its course, and virtue is in for its turn, it is salutary to ask what one must take for granted to restore its role. A great part of ethics could indeed consist of minute analyses of virtues rather than delineation of obligations or framing of goals and ideals; and the complexity of our world, in which obligation patterns may undergo revision in rapid social change, and focal goals require readjustment, encourages a search for inner stability. But the long-run question is whether it will lead ethics into the blind alley of a concept that can guide sentiment but point to no action, or whether it can even guide education which—since we ask what kind of persons we want the educational system to produce—is itself the great social context for a notion of virtue. The utility of the concept is thus tied intimately to the success of the psychology of personality.

The Decisional Model

The decisional model points research to a family of decision modes, of which deliberation is the favored phenomenon. Its cognitive character and its expression in judgment make it a ready field for investigation. And the directions in which it would lead research are significantly different. Suppose a Deweyan approach had been adopted instead of a Kantian in the psychological study of moral development. Dewey[37] rejects the temptation situation as a basically moral one, since it involves already knowing what is right, though being tempted not to do it. This would make less relevant most of the lines of inquiry about cheating. Instead, Dewey identifies the moral situation as one in which a genuine conflict of values and moral beliefs makes decision necessary. He thus stresses principles and modes of judgment over rules, standards of evaluation over goals, and the broad spirit of inquiry over specific virtues. The direction of investigation into moral development would thus go immediately into the growth of rational methods of decision. And since Dewey does not make a sharp division between the methods of so-called factual inquiry and those of value-inquiry, the investigation of moral development would call for concern with the social dimension of conditions favorable to the flowering of rationality as well as the psychological conditions of ego-strength.

This type of model for the research into morality has not been focused on directly, although as we have seen, it has been pointed to by some of the results under the other models; it tends to come forward as a result of instability in the structures which the other models assume. If the world, man, and human life turn out to be inextricably complicated, if goals and desires are in flux, moral rules too broad and clumsy, patterns of self under constant strain, then attention turns to the act of decision, though it is at the same time robbed of much that would guide it. Some of the current existentialist outlooks thus leave morality in the role of anguished decision with total responsibility and little guidance. In general, the decisional model is successfully applied only to the extent that the conditions which support the other models have some existence. Left to itself it becomes abstract, formal, even mathematical. Even the criterion of betterness comes to mean little more than "more" or "more complete." And in its original formulation it has already extended beyond the domain of the moral.

In short, all the traditional models, when submitted to philosophical analysis in the light of what research has been done implicitly or explicitly employing them, are found to have broken down. But it is not the breakdown from poverty, but embarrassment of results that carry them beyond themselves. It is the progress that calls for conceptual reconstruction, not for despair about what one is doing. Let me therefore have a final brief reckoning with the models and their lessons. What is the moral of our study of science and morality?

First, there is the lesson of explicit conceptualization. Scientific research requires a more definite conceptualization of the moral. The concept does more than mark out the area within which the inquiry is to take place. It usually furnishes one of the variables for correlation or discovery of concomitance and difference. It is methodologically desirable, therefore, that research make explicit its initial concept of the moral, recognize that its results are relative to that concept, and be explicit in any proposals that it may want to offer at the end for refining or altering that concept.

The second lesson is that while science brings increasing definiteness, it does not entail that there is only one way of going. Work along each of the models has brought progress in reformulating the problems, if not always the stable results one hoped for. But in fact, the research may even treat the models in different ways or regard them in different lights:

- It may regard their distinctiveness as not decisive and expect in the long run to combine them into a working concept of morality in which they would represent different stresses or different interpretations. We saw above, for example in the scientific career of the juridical model, how it almost changed at the end into a decisional model; but this meant definite internal shifts, such as from law as central to the ways of handling laws as central.

- Instead of treating the models as preliminary definitions of the moral, psychology may want to examine them as different psychological configurations. For example, the juridical model could easily be translated into a law-oriented conscience or a form of authoritarian personality. Similarly, anthropology could try out the models as possibly different cultural orientations.[38]
- The different models could be regarded as different or competing theoretical formulations for handling the explanation of a complex common mass of material. After all, there is a common field of human phenomena of appetition, regulation, appreciation, decision, and so forth, in which there are not the sharp cuts that so many theories seem to demand. The different concepts of morality dip into these phenomena and select among them for their constructions. But what one carries away in such a dipping may have relations to what is left behind, that cannot be anticipated at this stage. For example, the models may select different feelings or emotions for their theoretical use (as guilt in the juridical, or sympathy in some self-development models). But what they come up with depends on what emotions are really like. If one model takes a cognitive direction, and a second an emotive, it may turn out that a theory of the emotions eventually prevails in which emotions are seen as having a purposive structure and so some cognitive elements. The lines of intercrossing may lie hidden in the phenomena themselves.

Whichever use scientific research may make of the models, it must not forget that if it goes on to investigate morality it is itself employing some concept. Its view of the models is thus part of the justification for the conception it is itself adopting.

A third lesson is, as the consideration of the previous point also suggests, that categories should not be allowed to harden, that scientific research into morality should not too readily partition the emotional, the behavioral, and the judgmental, or some similar distinctions of different kinds of theories. They may be merely different starting points. The lessons of the weakness of the old faculty psychology apply with equal force here. Similarly applicable is the general methodological lesson of the philosophy of science, that even major categorial distinctions (such as fact and value) have themselves a tentative character, that they are born in a theory and are not inviolable in the progress of research.

A fourth point is that the question of which is a better model for scientific research on morality is perhaps premature. If one model more quickly yields results while another is inconclusive, it still has to be asked whether the definite results may have only a limited scope, whereas the inconclusive may be grappling with a more complex problem. It is even possible to do worthwhile research without any particular model but, as we saw in the case of value studies, by using an extremely broad conception of the desirable. If the results achieved are not definitely about morality, they are about all sorts of interesting items that come somewhere near or in the field. And in the sociohistorical studies, at times the concept of the moral seems to be explicated only extensionally—e.g., the feudal ethic, the Protestant ethic, the ethical beliefs of a slave society, and so on. Out of the anthropological and historical studies there might emerge even the view that the concept of the moral has no inner unity, but consists in a set of psychological and social functions or jobs which happen to be treated together according to present psychological and social techniques; in principle they might be partitioned and handled quite differently. But of course so radical a solution to the concept of the moral could only come from the failure of attempts to find a unity, and it itself would require a specification of the jobs in its altered conception.

A fifth lesson is that the diverse and complex scientific inquiries into morality can serve a mirror function for philosophy and show the diversity and complexity embraced within its own conception of the moral. Even more, if scientific research on morality presupposes a definite concept of morality, then what is discovered in such research, or the kinds of questions that arise in it, have lessons for the philosophical questions in that concept itself. For example, if the concept of ought is investigated as a Superego phenomenon and that of good as an Ego phenomenon, then one dimension at least in the relation of obligation and goodness may be clarified by the psychological knowledge of the relation of Superego and Ego phenomena. In the same way, if ought implies can, then the theory of responsiblity is to be related in some sense to the growing psychological knowledge of inner conflict and what it renders possible and impossible. That the psychological knowledge may not be sufficiently advanced probably accounts for the prevalent tendency of philosophy to go it alone.

Nearly every theoretical concept in ethics rests, as I suggested in the delineation of models at the outset, on some assumptions about the nature of man and his world, though the precise way in which the assumptions enter into the theory requires philosophical analysis. Whether (in a juridical model) respect is directed to the law or the person became, as we saw, a matter of inquiry in Durkheim and Piaget. Similarly, whether (in a self-development model) criticism against an act is always some kind of indignation against a person (as Westermarck maintained) could be subject to research, as could Nietzsche's more specific view that the demand for equality always involved envy. Even the more general or categorial notions in ethics have their underlying phenomena whose structure may be mirrored in the structure of the category. Thus to understand the notion of an ideal involves research into the phenomena of aspiration; if not, we are reduced to some such expedient as G. E. Moore, when he defined the ideal as simply a very high value.[39] But this is philosophical impoverishment by relying on commonsense or surface notions when deeper ones may be made possible by a more intensive research. The relation of the category of virtue to the results of the psychological study of personality, discussed above, constitutes an important illustration.

A sixth lesson is that from the complexities in the concept of morality in both scientific research and in philosophy, a policy recommendation emerges that a less confined and broader conception of the moral be developed which would at the present stage be acceptable for both scientific and philosophical purposes, and which would not foreclose the results of research or shut out research completely. This would be a more comprehensive working concept which would encourage research on all aspects of a morality, not merely one preferred aspect. It would embrace dominant content of different types (notions of welfare as well as of justice), different conceptual organizations (goals and ideals as well as laws), allow for investigation of moralities of different scope (familial honor moralities as well as universalistic types [40]), different affective stresses (moralities of sympathy as well as moralities of guilt and anxiety), different types of sanctions, different modes of decision, and so on. Such a conception would make better use of the anthropological study of varieties of culture and the permeation of the cultural, without necessarily ruling out any universal or general elements. Refinement in such a broad conception would come from the results of the study of the several aspects and their relations, just as in scientific work generally, initial concepts are redefined periodically in terms of established results.

Finally, it is worth recapitulating the general lesson of this whole paper, that as long as science is barred from the study of morality, what is moral and what is immoral has the character of expressive or emotive utterance, reflecting personal fiat or cultural arbitrariness. But to the extent to which scientific research is successful in establishing the orderliness of the world, and man's makeup and man's predicament, a greater determinateness enters into moral judgment; and criteria for correctness, for relevance and validity, for authenticity and maturity, become available on a wide range in many aspects of morality. And while we are still a long way from tying the general use of "moral" and "immoral" as criteria to conceptions of the nature of morality grounded in scientific research of man and his works and powers, yet the procedure by which we may hope to do this, and the pattern of philosophical analysis involved is no different from that which has proved pertinent in the lesser criteria. Meanwhile, then, in reconstructing our concept of morality, we would do well to use that concept which turns out to have maximal scientific grounding, and to regard as the measure of our progress the extent to which our criteria in moral judgment come to reflect the growth of knowledge in our day.

Notes

1. J.P. Sartre, *Existentialism is a Humanism* (1946). In Walter Kaufmann, ed., *Existentialism from Dostoevsky to Sartre* (New York: Meridian Books, 1956).

2. H. Hartshorne and M.A. May, *Studies in the Nature of Character*; Vol. I, *Studies in Deceit*; Vol. II, *Studies in Self-Control*; Vol. III, *Studies in the Organization of Character* (New York: Macmillan, 1928-30).

3. Cf. Karl Duncker, "Ethical relativity? An Inquiry into the Psychology of Ethics," *Mind* 48 (1939): 39-57.

4. Cf. Abraham Edel, *Ethical Judgment: The Use of Science in Ethics* (Glencoe, Ill.: The Free Press, 1955), ch. VI.

5. Cf. Ruth Benedict, "Anthropology and the Abnormal," *Journal of General Psychology* 10 (1934): 59-80; H.J. Wegrocki, "A Critique of Cultural and Statistical Concepts of Abnormality," *Journal of Abnormal and Social Psychology* 34 (1939): 166-78; G. Devereux, "Normal and Abnormal: The Key Problem of Psychiatric Anthropology," *Some Uses of Anthropology: Theoretical and Applied* (Washington, D. C.: The Anthropological Society of Washington, 1956).

6. For further analysis of the example, with reference to the alteration of its position on homosexuality by the American Psychiatric Association, see below pp. 275-76, 281-82.

7. Erich Fromm, *Man for Himself: An Inquiry into the Psychology of Ethics* (New York: Rinehart and Co., Inc., 1947).

8. Ibid., ch. 4. Cf. J.C. Flugel, *Man, Morals and Society, A Psycho-Analytical Study* (New York: International Universities Press, 1945).

9. M. Wolfenstein, "Fun Morality: An Analysis of Recent American Child-Training Literature," *Journal of Social Issues* VII (1951). (Reprinted in M. Mead and M. Wolfenstein, see note 33.)

10. Lawrence Kohlberg, "The Development of Children's Orientations Toward a Moral Order: I. Sequence in the Development of Moral Thought," *Vita Humana* 6 (1963): 11-33; "Development of Moral Character and Moral Ideology," in M. and L. Hoffman, eds., *Review of Child Development Research: Vol. One* (New York: Russell Sage Foundation, 1964), pp. 382-431; "Moral and Religious Education and the Public Schools: A Developmental View," in T.R. Sizer, ed., *Religion and Public Education* (Boston: Houghton Mifflin, 1967); "Moral Development," in *International Encyclopedia of the Social Sciences* (The Macmillan Co. and The Free Press, 1968), Vol. 10, pp. 483-94.

11. Erik H. Erikson, *Childhood and Society* (New York: W. W. Norton and Co., Inc., 1950).

12. Cf. Abraham Edel, *Method in Ethical Theory* (Indianapolis: The Bobbs-Merrill Co., 1963), ch. 7.

13. E.g., J.S. Mill, *Utilitarianism* (New York: Liberal Arts Press, 1949; originally published 1863).

14. F.H. Bradley, *Ethical Studies*, 2nd ed. (London: Oxford University Press, 1927).

15. A. McBeath, *Experiments in Living* (London: Macmillan and Co., Ltd., 1952).

16. Kohlberg, "Moral and Religious Education and the Public Schools," pp. 165, 169, 178.

17. John Dewey, *Human Nature and Conduct* (New York: The Modern Library, 1930); *Theory of the Moral Life*, ed. Arnold Isenberg (New York: Holt, Rinehart and Winston, 1960). The latter is Dewey's part of Dewey and Tufts, *Ethics*, rev. ed. (New York: Henry Holt and Co., 1932).

18. Sartre, see note 1.

19. Sigmund Freud, *New Introductory Lectures on Psychoanalysis*, trans. W. J.H. Sprott (New York: Norton, 1933).

20. Sigmund Freud, *Civilization and its Discontents*, trans. Joan Riviere (Garden City, N.Y.: Doubleday Anchor Books, 1958; originally published 1930).
21. Cf. Abraham Edel, "Anthropology and Ethics in Common Focus," *The Journal of the Royal Anthropological Institute* 92 (1962): 55-72.
22. Clyde Kluckhohn, et al., "Values and Value-Orientations in the Theory of Action," in T. Parsons and E. A. Shils, eds., *Towards a General Theory of Action* (Cambridge, Mass.: Harvard University Press, 1951).
23. R.B. Brandt, *Hopi Ethics* (Chicago: University of Chicago Press, 1954); John Ladd, *The Structure of a Moral Code* (Cambridge, Mass.: Harvard University Press, 1957). For further studies, see Christoph von Fürer-Haimendorf, *Morals and Merit, A Study of Values and Social Controls in South Asian Societies* (Chicago: University of Chicago Press, 1967); Tore Nordenstam, *Sudanese Ethics* (Uppsala: The Scandinavian Institute of African Studies, 1968); John H. Barnsley, *The Social Reality of Ethics: The Comparative Analysis of Moral Codes* (London and Boston: Routledge & Kegan Paul, 1972).
24. Cf. Ruth Benedict, *Patterns of Culture* (Boston and New York: Houghton, Mifflin and Co., 1934).
25. Cf. Max Weber, *The Protestant Ethic and the Spirit of Capitalism*, trans. T. Parsons (New York: Charles Scribner's Sons, 1930); Friedrich Engels, *Anti-Dühring* (New York: International Publishers; orginally published 1878).
26. See Lewis White Beck's translation of Kant's *Foundation of the Metaphysics of Morals* (New York: Liberal Arts Press, 1959).
27. Emile Durkheim, *Moral Education*, trans. Everett K. Wilson and Herman Schnurer (Glencoe, Ill.: The Free Press, 1961).
28. Jean Piaget, *The Moral Judgment of the Child*, trans. Marjorie Gabain (New York: Collier Books, 1962; originally published 1932).
29. Ibid., p. 349.
30. Ibid., p. 352.
31. L. Kohlberg, "Development of Moral Character and Moral Ideology," p. 425.
32. Cf. Flugel (note 8 above).
33. Cf. Margaret Mead, "Social Change and Cultural Surrogates," *The Journal of Educational Sociology* 14 (1940): 92-109; Margaret Mead and Martha Wolfenstein, *Childhood in Contemporary Cultures* (Chicago: University of Chicago Press, Phoenix Books, 1963); J.W.M. Whiting and I.L. Child, *Child Training and Personality: A Cross-Cultural Study* (New Haven: Yale University Press, 1953); May Edel and Abraham Edel, *Anthropology and Ethics: The Quest for Moral Understanding*, rev. ed. (Cleveland: The Press of Case Western Reserve University, 1968; New Brunswick, N.J.: Transaction Books, 1970), ch. XIII.
34. Cf. W. Goldschmidt, "Ethics and the Structure of Society: An Ethnological Contribution to the Sociology of Knowledge," *American Anthropologist* LIII (1951).
35. Hartshorne and May (note 2 above).
36. Cf. H.A. Witkin, et al., *Psychological Differentiation: Studies of Development* (New York: John Wiley and Sons, Inc., 1962).
37. Dewey, *Theory of the Moral Life*.
38. Cf. Edel and Edel, ch. xiv.
39. G.E. Moore, *Principia Ethica* (Cambridge: Cambridge University Press, 1903), ch. VI.
40. Cf. J.G. Peristiany, ed., *Honour and Shame: The Values of Mediterranean Society* (Chicago: University of Chicago Press, 1966).

8. Six Requirements in Search of a Theory: A Study in the Relation of Law and Morality to Social Change

Thirty annual meetings ago [of the American Catholic Philosophical Association] Ralph Barton Perry presented a paper on "Catholicism and Modern Liberalism." In it he proposed a common front of all groups who were ranged against Nazism on the quest for harmonious human happiness. This goal, he maintained, was found in all the traditions to which he addressed himself. They might differ on metaphysical underpinnings, but these could be set off in esoteric doctrines for their own adherents. They might differ on means, but these could be judged by experience in the quest for happiness. The area of agreement he wanted to bring about reads as follows:

> It would consist of the moral ideal of harmonious human happiness and the political ideal of a community devoted to this moral ideal. It would work towards the enlargement of the spheres in which men choose for themselves and make up their own minds in the light of the evidence. To this end it would encourage men to have minds of their own. It would create a social system designed to spread enlightenment, and to give to all men that self-development and command of material goods which will multiply their range of options.[1]

The year before, he had made a similar appeal with respect to Communism and Democracy; both, he said, used the same standard of maximum satisfaction of human needs by inventive enterprise and widening distribution; both wanted to get rid "not only of poverty, disease, and ignorance, but of unfairness, parasitic privilege and wage slavery. Both seek the freedom of the individual, while differing in the tyrannies they fear."[2]

Perry's appeal seemed to fit the historical agenda of human life. He knew that the world in all its variety was coming into focus and its moral quintessence could no longer be captured in demands for a restoration of the "Judaeo-Christian civilization" that had long dominated the West, nor in a fresh crusade against "godless materialism." Yet, at the historical moment in which he made his appeal, it was doomed to failure, or at least to a long postponement. Every outlook seemed dogmatically pitted against every other. Catholicism spent tremendous energies fighting pragmatism. Liberalism was fighting itself, split into a right and left largely on whether communism was to be regarded as the central enemy or as one historical form taken by desperate peoples in the attempt to rise from poverty. Communist writers soon were to bend their intellectual energies, under the impact of the Cold War, to equating pragmatism with imperialistic opportunism. They scarcely looked to the right, where a new Conservatism was springing up which in its more vociferous mood would brand all liberalism as "twenty years of treason." Meanwhile, in the sociohistorical background were the massive historical phenomena: a world impoverished by global war, shattered empires, and the stirring of long-exploited peoples in Asia and Africa, the surge of a renascent capitalism in the United States, the pervasive hopes of a new life articulated on the international scene in lists of freedoms and rights and programs of world development—and, of course, the atomic bomb.

This is ancient history, and in many respects the scene has changed. Everywhere dialogue has replaced philosophical crusades, while on the practical fronts the breakup of the monolithic, whether in religion, politics or even philosophy, has been a central sociointellectual phenomenon of the last two decades. Nowadays in moral philosophy it is difficult even to think back to the time when, for example, religious philosophies proclaimed that secular ethics had problems while religious ethics had answers. Now it is patent that every form of life and every form of philosophy is full of problems. If any generalization has weight over the period I have been describing it is that the pressing problems of human life in the contemporary world find their way into all philosophic constructions and all philosophic movements, and that much can be learned from a comparative analysis of the shape that the problems take in each tradition and of the way this leads to a reformulation of the problems themselves. Intellectually, Perry was right in calling for meeting crises by seeking the common ground of all-human aims, though the search probably must go into some common ground in metaphysics rather than insulating the different world outlooks one from another. Yet, if his formula was oversimplified, its practical base was the intense struggle then taking place throughout the world. It was his hope that the possibilities of a new world would not be missed. Future historians may indeed decide that we missed a turning-point at that time; perhaps today we have another chance.

This paper is a plea to open the gates of a self-enclosed moral and legal theory to the recognition of the underlying world of change and problems. Such a world is not merely an external domain exerting occasional causal powers, between which episodes we can turn away. Nor does it wait patiently to receive decisions from philosophy when the latter shall have settled its own disputes in its own way. My argument is rather that the now global realm of human practice and its problems calls the tune and that theory in morals and law lacks self-consciousness if it does not see this as a proper penetration into its constructions and its criteria. This is not the subservience of theory to practice; it is the recognition that thought and action are fully human expressions in which questions of subservience and domination can have no more place than they have among friends and fellow philosophers.

I am, you see, asking for much less than did Perry: not a complete common ideal, but a common decision on how our work as philosophers can be fruitful. With this as an aim, and if legal and moral philosophy are indeed to play a part in the intellectual retooling needed to deal with the problems generated by social change, then legal and moral theory must themselves be responsive to those problems. I want here to explore six preliminary requirements for such a task. They are directed toward developing a more realistic view of the legal-moral enterprise and securing a richer interrelation between science, value, theory, and practice. They do not foreclose the shape of the theory that will emerge, though they do constitute constraints. That is why they are still requirements in search of a theory.

A Broadened Conception of Law and Morality

The first requirement is that we shift to a *broad conception of both law and morality*. This sounds innocent enough, but it has very wide ramifications. In the case of law it means that we stop thinking of law simply as a set of norms, whether fixed in a code or discoverable from the sweep of judicial decision. Instead, we must think of the entire legal system including legislation as well as courts, enforcement processes as well as decision processes, constitutional change as well as the eternal founding principles, and conceptions of appropriate judicial and legal attitudes as well as conceptions of justice and other ideals in terms of which the aims of the law are articulated. This should not be difficult after all the "legal realism" that has gone into legal self-consciousness in the twentieth century.

Moreover, sociological jurisprudence has not been without influence in legal theory itself. Roscoe Pound, to take an outstanding example, moved far beyond a merely analytical jurisprudence to an engineering model for the law. Unfortunately, he limited the model somewhat by giving the law the narrow, almost retroactive function of bringing legal decision into line with new postulates that have already worked their way into social acceptance.[3] The judicial spirit of the age, in short, was to remain simply the judicial wing of the owl of Minerva, gaining consciousness after the fact and readjusting its limited domain.

The vital movements of legal theory in the first half of the twentieth century had their roots in and were addressed to the great economic and social changes. In western Europe during the very first decades the central phenomenon was the growth of industrialism. Old laws no longer fitted and revision was imperative. The theories of free judicial decision and judicial discretion which played so large a part in the writings of the time about legal method seem to me to be saying in a barely disguised way: the path of legislation is too slow, too entrenched in the conflicts of older interests; let the refined conscience of the judge readjust and retune the instruments of the law for dealing with the changing conditions in all their detail. In the United States of the 1930s, however, the judiciary were on the other side; they appealed to the entrenched rights of private property to hold back social change needed to meet the impact of an already developed industrialism on the working class and the mass of the people. The response by legislative and executive action during the great depression found itself in head-on conflict with the courts. Legal realism showed the judge to be a pivot of possibly arbitrary power, rather than the impartial revelation of the eternal law. As we get closer to our own time, however, note how the picture changes. The courts of the 1950s face the problem of discrimination in education on the frontier while the legislature plods on; the court of the 1970s faces the problem of abortion and women's liberation. Old rights by interpretation beget new rights while all sorts of struggles still go on in the legislative forum.

Can we learn from this story of legal theory how to develop a broader conception of morality? I think we can. But first, what is the present situation in moral philosophy? It is marked by an obsessive concern with the issue of the right vs. the good. The proponents of the primacy of the good, from Aristotle to the Utilitarians, assess rules of right and wrong broadly in terms of the resulting quality of life. The Kantian tradition, so powerful in modern thought, limits morality narrowly to the internal or to what the individual is ready to will, and in social matters to what obligations he will undertake in a contractual meeting of the wills. Rawls' theory of justice is a good recent example of a theory attempting to present a structure of the right, built up by unanimous consent under a veil of ignorance, as though it were independent of, and prior to, conceptions of the good. The contrast of the right and the good, of the deontological and the teleological, of utility and justice, has become a structural dogma of the teaching of ethics since Frankena fastened it on our classrooms in his analysis of ethics at mid-century.

In calling for a broad conception of morality I am not asking you to endorse the teleological as against the deontological or to take a stand with utilitarianism against a conception of rights and justice. Nor do I ask that we thread our way among the numerous theoretical compromises to find a permanent pattern of their relationship. Catholic ethical theory has always had a finely tuned adjustment of the strands, derived from its assimilation of the Aristotelian focus along with the natural law tradition. Rather, I want to draw the lesson from the legal story and to undercut this battle of the giants in moral philosophy by looking to the whole system in which they are enmeshed and to the way they function in facing the pressing problems of human life.

These problems force jobs upon us which can be undertaken in terms of different possible intellectual constructions. Here, the biology of Aristotle, who had a good sense of the role of the functional in theory, suggests an analogy. To focus on isolated segments of the structures and worry about a decision between them is like asking whether the heart or the brain is primary, without even looking to see what functions each performs, how each is related to the other and to the rest of the system, and what are the many senses of "primary." In many respects, the right and good are analogous to the judicial and the legislative in law. The judicial thinks in terms of a system of rights it is expounding and applying, the legislative thinks in terms of ends and purposes of the electorate for which it is devising rights as means. The primacy of the one or the other is a contextual matter. So, too, in ethics the right and the good have their separation of powers, their compromises and adjustments, and their periodic relative dominance in historical crises.

Think back to the crises of the French Revolution. In France, it was fought under the slogan of the "rights of man." In England, Richard Price developed intuitive rationalistic ethics and delivered the sermon on the traditional rights of Englishmen that so enraged Edmund Burke and sent him hurrying to write his *Reflections on the Revolution in France*. Meanwhile, in the utilitarian camp Archdeacon Paley, who propounded the greatest happiness principle for interpreting God's will for man, issued his pamphlet, "Reasons for Contentment Addressed to the Laboring Part of the British Population" in 1793, just when that part of the population was being stirred by the French Revolution.[4] By the time of John Stuart Mill, all was changed. Utilitarianism was now in the forefront of social reform, articulating the demands of broader segments of the population in the rising industrial developments.

In twentieth century United States, the ethical shifts correspond quite closely to those in legal theory. In the 1930s utilitarianism was again in the forefront against restrictive claims of rights. But after World War II, when colonialism was dying, racialism crushed, and the masses of Asia and Africa demanded their place, justice and rights became the theoretical instruments of human advance. Witness the proliferation of human rights in the United Nations documents, specifically directed to remedying contemporary evils from poverty and unemployment to the removal of colonialism, the liberation of women, the protection of children, and the rights of intellectuals. I suspect that much of the force of Rawls' work comes, not from his proclamation of the primacy of the right over the good (which in any case he does not adhere to), but from his emphasis on the rights of the disadvantaged in his "difference principle" of justice. I suspect too, that much of the attack on utilitarianism in moral philosophy today reflects the fact that utilitarianism today is lazy and smug. It is allied to an establishment which feels that in the United States it has achieved the greatest happiness of the greatest number because it has brought a high standard of living to the presumed majority, the all-encompassing "middle class." On such a showing, the greatest happiness of the greatest number is not enough, certainly not something that the excluded minority could acquiesce in.

The lesson of this analysis is philosophical, not historical. It means that the problem of primacy should be translated into the roles that different moral concepts and intellectual structures play in meeting the human practical-social problems in the life which people are leading. A narrow conception of moral theory looks to the structures alone; a broad conception is functional, both in theoretical and practical relations.

There is another narrowing tendency in contemporary ethics which I have time only to mention. It is the over-sharp partition of individual and social morality, which proceeds as if each were governed by a separate set of principles. There are, of course, moral problems of individual action, of interpersonal relations and of group life; but a consideration of any of them carries us without neat boundaries into all of them. The great moral issues of our time are set by questions of war and peace and violence; discrimination and basic inequality; corporatism or participation; distributive principles inherent in the regulation of taxes and wages, foreign assistance and budgetary proportions for consumption and for industrial development; population growth and the limitation of resources; production and pollution; types of property; alienation and the lack of community, and so on. Even the character of the individual good life today becomes—if it has not always been—dependent on the quality of life of the community, the motivations it cultivates, the concepts of success or happiness it expresses in its institutions. In all this, morality, like language, is a social phenomenon. The individual in his maturity may question parts of it and present social grounds for the part that he accepts or rejects or embodies in his life. But this is not an individual morality as contrasted with a social morality.

Empirical and Value Considerations

The second and third requirements concern the interrelation of the empirical and the valuational with the analytic-logical and with one another. I believe we are now past the era in which the accusation of the naturalistic fallacy was equivalent to condemnation or in which a puristic metaethics boasted of its empirical and valuational emptiness. Quite simply, I think that what transpired was a vast experiment, a program of separating the analytic, the empirical, and the valuational. It was argued as if it were a necessary analytic truth, though it was simply a program that failed to be achieved; it took half a century to see this! Our two requirements are therefore: *always look for the material or empirical considerations presupposed at all levels of legal and moral theory*; and *always look for the purposive or value considerations that permeate all parts of the work of law and morality*. We do not have to be told to look for the analytic; our work usually begins with the analytic structures.

I suspect that legal theory, which is familiar with fictions, always knew that factual assumptions play an important part in legal decision, that concepts of reasonable care, negligence and consent require empirical amplifications. The impact of change shows this most clearly. The concept of property has been profoundly affected by both institutional and technological changes, such as the rise of the corporation and now of the multinational corporation or, to take an example near to a book-writing audience, the implication of xerox for copyright. Similarly, the meaning of liberty or privacy is affected by electronic devices, and the meaning of responsibility by changing psychiatric knowledge. Again, legal philosophy has expended considerable energy in showing the selective role of values in the processes of judicial decision, that is, in finding, interpreting, and applying the law. Legislation is, of course, directly purposive; constitutional provisions, in effect, shape the broad character of the kind of institutions that will develop, and hence the kind and quality of life to be lived.

Moral philosophy has been less clear that many moral concepts are held together by empirical assumptions and by purposive patterns. For example, would the old vice of shiftlessness have had much meaning if it were not held together by the empirical cluster of incapacity, laziness, poverty, and perhaps dirt? It applies neither to the capable unemployed who idle for lack of work, nor to the upper class hippie who makes a principle of neglecting cleanliness and avoiding a success-oriented pursuit. Does the virtue of charity similarly presuppose a distinction between the rich and the poor? Would one worker helping another out in a difficult situation be said to be giving him charity? Stealing presupposes institutions of private property and an ability to distinguish that which is open to public use and that which is governed by certain controls. Even a higher-level construct like merit has an inner relational structure which unfolds as it is unpacked: a person deserves a specific treatment (reward) in virtue of a specific act or ability, which is needed for a specific service that is socially desirable. Hence, judgments of merit presuppose some knowledge of the needs, habits, valuations, and conditions of the community in which they are made; they also presuppose a selective acceptance of the specific purposes involved.

Similarly, we are too often unaware of the extent to which the formulae associated with ideas of right or justice themselves involve such a selective component. In the contemporary world, the various formulae in capitalist, socialist or composite intermediate social systems show that definite valuation is at work, much as it may be disguised in one cultural tradition by the argument that "This is what we mean by 'justice' as seen in our linguistic usage." Comparative anthropological materials sometimes help exhibit the selective element. For example, that a preliterate society uses the same term for both stealing and inhospitality shows the value it attaches to hospitality and, in many cultures, to kinship obligations of support. This problem is seen in magnified form in cultural change where the individualistic patterns brought in by our economic ways come into conflict with the kinship patterns of older ways.

The analysis of empirical and valuational assumptions in legal and moral materials is a complex and technical business. They will not always be located in any one logical place: there may be direct variables that need material or purposive values in the analysis of meaning, or there may be presuppositions that are required to make sense of a concept, or there may be conditions that are necessary for the concept to be applicable, and so on. This problem needs analysis with our best logical tools. Sometimes we see a whole process spread out before us in a period of social change. For example, the concept of privacy is being experimented with today by the courts to cover ruling out eavesdropping, personal decision about one's body and its treatment in medical and even psychiatric matters, and to some extent in the use of one's property. At the same time this concept gets a strong moral tone in reaction against the pressures upon the individual in contemporary governmental and private corporate structures. The extended right of civil disobedience is one evidence of this trend. Some of the more standard contexts can be readily handled by more established concepts of trespass and liberty, but as borderlines shift and new types of contexts emerge it may be that privacy will become established as firmly as was liberty, and with a wide coverage. If this happens, it would require philosophical unpacking to see what empirical and valuational materials served as the unifying cement.

It would be useful to show how a highly abstract theoretical element in law and morality can be analyzed in this way. Take the sharp moral dichotomy of right and wrong, paralleled by the legal demand that a case be decided for or against the defendant. The latter seems to us to be practically necessary, and almost constitutive of what a decision is. In some Mediterranean nomadic societies, however, the value of honor is so high that it would be unthinkable in a conflict to condemn one side as wrong and award the palm of rightness to the other.[5] Ethical skill lies in finding a solution which remedies any perpetrated hurts as much as possible and saves face for all concerned. When we contemplate such varying traditions it becomes clear that our moral concept of right and wrong has a built-in sharpness that stems from and tends to perpetuate a rigorous guilt-oriented conscience. We need not, for our present purposes, go into genetic and sociocultural determinants; it is enough to recognize the reference to underlying conditions and built-in values. It can then be seen that even the legal sharpness of decision is not a logical necessity, but a result of transferring any rival value to some other process. For example, to attempt a fair compromise would be, no judicial decision,[6] but mediation; even arbitration is cast in the decisional rather than the mediational mold. In spite of this, the pattern in international law is more like the honor societies than like our own internal judicial model. Gidon Gottlieb has argued[7] that our own sharp model is itself applicable to group-individual relations, and that where the law has to tackle group-group issues (e.g., government-union, or race-race) its operations begin to resemble more the picture of international law. I think the same kind of analysis shown by this example can be carried out for most of the metaethical theories that have occupied the stage in analytic ethics.

Humility Regarding Concepts

The fourth requirement concerns phenomena of social change. If our legal and moral theory is to be responsive to the problems generated by social change they have to be aware of the effects of change within their own structure of concepts, principles, and methods. The requirement then is simply *to exercise humility with respect to our concepts*. Today, the growth of science is bringing about a conceptual revolution in notions of space, time, and causality; the growth of genetics is rendering senseless many previous formulations regarding the relation of heredity and environment; and even tentative studies of cognitive and emotional development begin to refine the language of self. Why should we expect, then, that the concepts of morality, both the unitary description of acts and the terms in which we formulate moral laws and moral problems, will remain more solid than the rock of Gibraltar?

Take, for example, the moral controversies about contraception. I choose it rather than the more heated contemporary struggle about abortion because the latter is in principle a temporary problem. We can conceive of a situation in which improved contraceptive methods and wider education might end abortion as a means or at least make it a relatively rare issue. Contraception, however, concerns a permanent means of control and family planning. An outside observer taking a macroscopic view of social change might very well say that it is understandable that the pattern of large family and prohibition of contraception should have had a strong moral and legal base under, for example, slower conditions of change, a sense of an open world with unlimited resources, concepts of a fixed natural path for human development, intimate relation of sexuality and procreation, short span of life and few modes of security for old age, and the need for greater labor supply and institutions of male domination associating ideals of masculinity with multiplicity of children. However, one by one these supports have vanished, or are about to go, and have been replaced by rapid technological and social change, a sense of a closed world with dwindling resources, technical means for separating sexuality from procreation and associating it elsewhere (for example, with love and companionship, and the desire to share a common life), greater life expectancy and social security for old age, machine production requiring less labor, recognition of the moral basis for liberation of women, and so on. Even the command to be fruitful and multiply, as one Cardinal put it at Pope John XXIII's Vatican II, was addressed to a population of two per square world and so has a different force today. Can the belief in a fixed natural path for human life itself support the older moral injunction on the question of contraception?

I forego here examining the concept of the natural in the proposition that contraception is unnatural. Its conceptual cluster would have to be broken into components of statistical frequency, inherent inclination and value norm, all of which have drifted apart since Darwin's day. To probe this would carry us into the diversity of metaphysical underpinnings. Even for those who have firm faith in a natural order, however, the recognition of the changes raises the question whether many of the components in the older conceptions were not really instrumental to broader and more general values of the spirit, and whether those values may not today require fresh forms.

Some such assumption is evident in the retreat from an insistence on large families to an acceptance of family planning by sexual restraint and then, once it is a question of means, by the advocacy of a "natural" rhythm method. Should the scientist replace the present pill with one that has great food value as well as contraceptive value, the possibility would be opened for declaring it food *per essentiam* and contraception *per accidens*! Such is the precarious path of relying intellectually on the concept of the natural today. If one were to shift wholly to the matter of intention and motivation of the parties, then an untrammelled Kantian model would have to distinguish well-motivated from ill-motivated contraception and family planning. To make such distinctions would bring the circle full round to questions of population pressures, the quality of life, desirable types of interpersonal relations, and the whole array of pressing social problems of the contemporary world.

The same kind of basic change is going on in most of our institutions, whether political, educational or social. The reassessment of aims in altered conditions poses the need for new concepts and new formulations in law and morality. Functions may be shuffled and reasserted in new institutions. Only half a century ago a sociologist could describe the state as a purely political institution which, therefore, by nature should not have an economic participation. Now everyone looks to its action to provide basic guarantees. New functions may emerge and old ones die out. I need scarcely add that this is not novelty for novelty's sake. Changes are difficult and painful when the institutions are deeply rooted and the changes far-reaching. What is more, we have to exercise the greatest caution that a new formulation may not be leading us astray into posing a question in a way that will limit our answers and impoverish our moral possibilities.

Take, for example, the recent meteoric rise of the concept of *triage* and the talk of a lifeboat ethics. It poses our basic problem of sharing food and resources as one of selecting for help those who are most likely to succeed by being helped, not those who will continue to increase in population and bring doomsday nearer for both themselves and the rest of the world. It, therefore, makes our central moral problem that of selecting those to be saved with us as against those to perish. In some respects these discussions remind me of the intense debates that were provoked several decades ago when shelters against nuclear attack were being promoted: for example, if some contaminated person tried to force his way into your shelter, would you be morally justified in shooting him? The very concentration on these questions, novel as they were, could only turn the mind away from the serious issues of nuclear control and the paths of cooperation for peace.

Similarly, now, the basic issues are more likely to be international cooperation on agriculture and energy than a formula for retrenchment under the doomsday threat and questions of whom to allow into the lifeboat.

There is no simple way of constructing new formulations and new concepts for law and morals. We can tell what has to give way, for example, in the law of how to cross-examine witches on trial when they refuse to confess, or in the law of pinpointing responsibility in automobile accidents when no-fault automobile insurance takes over. To determine in what positive direction to move requires a collective philosophical wisdom, but one that appreciates the fact of change and the problems of change.

A Holistic View

Our fifth requirement turns to the ways of going about attempted solutions. This raises the problem of *how to have a holistic view without closed minds*. Social change comes sometimes on a large scale, sometimes piecemeal. But we cannot know whether it is piecemeal without the full view that enables us to see what is not involved. We need, in brief, a view that is long and full, without being closed. One of the best ways to do this is to take conflicting theories and, without antecedent commitment, turn them into models for research. Let me illustrate this for law and morals.

It used to be thought, as Maine argued about contract,[8] that the movement of progressive societies was from status to contract, that more and more legal obligations became the expression of individual will or mutual agreement. A contrary thesis could argue that we are witnessing the socialization of morality[9] in the increasing replacement of social and institutional instruments to carry the burden that individual decision formerly bore. I shall resist the temptation at this point to argue against the voluntarist and group commitment heresies, as Maritain would undoubtedly have done. It is perhaps better to shift from spectator to agent and, instead of asking what is happening as a matter of drift, to ask in what fields, for what reasons and to what extent the two models might be applicable. Thus, instead of dealing with two theories, in this case of allegedly unavoidable direction, each is turned into a model of a different kind of organization in legal and moral fields. There is then no assumption that one model has to dominate the field. The more models the better, for they now function as proposed modes of organization and hence as different tools in our intellectual and social toolchest, to be used where each is best. We can, in fact, point out that the socialization thesis is quite correct for the moral and legal shift from individual charity to social security, that many such issues are now recognized as more desirably dealt with in a social-institutional organization than left on the shoulders of individuals in a society where economies are so integrated and beyond the individual's control. We have long achieved no-fault unemployment, we are getting to no-fault automobile insurance, and the proposals of collectivized health care and guaranteed minimum income are clearly on the agenda. All these are cases of moving into the legal field, based on new institutional forms, problems which to a great extent were moral. The decision to make the shift is a moral one.

On the other hand, if Maine proved wrong about the development of contract which in the domain of economic life has increasingly become subject to necessary regulation, his model fits what is going on in the domain of sex and marriage. Here, everything from the removal of legal penalties on sexual behavior to the experiments with no-fault divorce suggests that a whole area which was part of the law is being moved out of the law or at least being subsumed under individual liberty within the law. The freedom of the individual in these respects does not mean the absence of standards, but that standards have to come from his moral principles. In short, the movement is, in effect, from law to morality.

Once again, here we cannot go into the question of evaluating these movements. Since they are both seriously advocated movements, not chance drifts, I call them to your attention as examples of the way in which a more open decisional attitude is operative in both law and morals. I suggest that whatever your views on the specific questions, the open approach is the morally sound one in a world in which the role of law and role of morality both need constant moral evaluation. It is obvious too that such an open approach cannot be exercised without a view of the whole field of human life and the particular character of the different fields at the given period. Hence, the holistic approach is directly tied to the open approach, paradoxical as this may sound in the abstract.

Activism in Law and Morals

Our final requirement is the *need for activism* in law and morals. I suppose that in a deeper sense all intellectual theorizing about law and morals is activist, even if the activity is one of reenforcing the established ways. However, by intellectual activism I have in mind a fertile and powerful reconstructive tendency in developing new concepts to meet changed conditions and refined and emerging values, or even to sharpen instruments for achieving long cherished values under new conditions. Today, we have a marked spirit of legal experimentalism for the improvement of human well-being, especially among the young who are flocking into the law, though there is no way of knowing how strong it will remain. Activism in morality means trying to work out a practical and institutional embodiment to support moral attitudes and human goods that are frustrated by established ways. Activism in moral philosophy means applying philosophical energies to the analysis of practical moral problems and devising detailed solutions. Neither in law, nor morals, nor moral philosophy does it entail a hectic crisis mentality or breaking necessarily with traditional values, rather than broadening them, interpreting them for the changed world, and making them more effective.

It is hard to think of a moral alternative to moral activism in so rapidly changing a world. The desirable coefficient of activism is probably a direct function of the steepness of the curve of change. In earlier times, when change was slow, conservatism with its hesitation about change in tradition may have been a viable alternative. Now, although it berates experimentalism in morality, the maintenance of traditional ways under rapidly changing conditions is often the most desperate of experiments.

Let me end, as I began, with Ralph Barton Perry. In his first book on ethics, *The Moral Economy* published in 1909, he formulated morality as "simply the forced choice between suicide and the abundant life." [10] When he spoke in 1943 he was pleading desperately for union against collective suicide. We cannot be as hopeful as is his formula for the abundant life, but we can certainly translate it for our world, with its new and apparently permanent crisis, as the directive that in theory as well as in practice we deal with the problems of all mankind on the pain of common extinction.

Notes

1. *Philosophy in Post-War Reconstruction: Proceedings of the American Catholic Philosophical Association* XIX (1943): 83.

2. Ralph Barton Perry, *Our Side is Right* (Cambridge: Harvard University Press, 1942), ch. V.

3. Roscoe Pound, *Interpretations of Legal History* (New York: The Macmillan Company, 1923), viii; and *Social Control Through Law* (New Haven: Yale University Press, 1942).

4. Paley's arguments remind me of Clement's reinterpretation of Jesus' remark about the difficulty of a rich man getting into heaven: the rich young man is assured that all Jesus meant was that you had to get rid of spiritual attachment to riches, not the riches themselves! See "The Rich Man's Salvation" in *Clement of Alexandria*, trans. G. W. Butterworth (Cambridge: Harvard University Press, 1960), p. 291ff.

5. See J. G. Peristiany, ed., *Honour and Shame: The Values of Mediterranean Society* (Chicago: University of Chicago Press, 1966).

6. For the view that values get built into the very process of judicial decision, see Abraham Edel, "On Locating Values in Judicial Inference," *Legal Reasoning: Proceedings of the World Congress for Legal and Social Philosophy*, ed. Hubert Hubien (Brussels, Etablissements Emile Bruylant, 1971), pp. 31-41. Reprinted in Vol. I, ch. XIV.

7. Gidon Gottlieb, "The Nature of International Law: Toward a Second Concept of Law" in Richard Falk and Cyril Black, *The Future of the International Legal Order* (Princeton: Princeton University Press, 1969), p. iv, ch. ix.
8. H.J.S. Maine, *Ancient Law* (New York: E. P. Dutton and Co., 1917), p. 99f.
9. For the philosophical basis of such a view, cf. Felix S. Cohen, "The Socialization of Morality" in *The Legal Conscience: Selected Papers of Felix Cohen* (New Haven: Yale University Press), pp. 337-49.
10. Ralph Barton Perry, *The Moral Economy* (New York: Charles Scribner's Sons, 1909), p. 14.

9. Toward an Analytic Method For Dealing With Moral Change

This paper explores certain relations between morality and moral philosophy today.[1] Our morality is being transformed under the material and social changes of the twentieth century. It needs reconstruction but it is not well served toward this end by our prevailing way of doing moral philosophy. Our prevailing way has neglected the analysis of moral change and so is unprepared to deal with problems of change. Three features of its dominant method are largely responsible for this: it is ahistorical, its analytic approach is too piecemeal and too isolated from context (historical, social, cultural), and its underlying model is unduly individualistic.

This dominant method saw itself as simply logical analysis applied to ethics. But I want to suggest that in spite of this methodological stance it functioned as a whole program structuring the field of ethics, and can thus properly be assessed (as a program would be) by asking what it accomplished in solving puzzles, advancing inquiries, resolving problems, generating applications, making possible systematic developments in the moral field—briefly, in guiding morality fruitfully. Looked at in this way, the dominant method of analysis has been tried out for some decades and I think it has been found wanting. In any case, I shall indicate in outline what a different program would be like which has a strong historical sense instead of being ahistorical, which substitutes a network analysis sensitive to context for the piecemeal isolated analysis; and as for the undue individualism, I shall try to undercut the controversy of the individual vs. the social model by a neutral distinction between macroethics and microethics. As to the structure of this paper, there are two introductory sections, the first giving a brief interpretation of the present state of morality and the second explaining why an analytic method in ethics is to be regarded as a program; these are followed by the three main sections dealing with the three criticisms and offering constructive alternatives.

The Present State of Morality

It should not be surprising that vast and rapid material-technological and social-cultural changes suggest new answers to older moral problems and pose new moral problems. What is really new is the scope and rapidity of the changes. Hence there is obviously some need for theoretical reconsideration of moral rules in their application to areas of change, whether it is to be limited to change in the application to the altered conditions or extended to change in the moral rules themselves or to be something even more fundamental.[2] Yet it is remarkable how little real analytic work has been done on the idea of moral change thus far. This is all the more surprising in that we are all familiar with many forms in which moral changes of one kind or another have taken place in our experience.

Let us state a few of these changes abstractly. Some valuable end, which hitherto reigned alone, is now in competition with another end that has become feasible for the first time. A rule adhered to as expressing an end is now felt to have been merely a means whose stability came from conditions now swept away. Some character traits pivotal in our morality become suspect; it is felt they may have reflected social and material limitations which are now no longer constraining. Some ways in which we relate to other people come to be seen, as we gain deeper psychological knowledge and insight, to have been compromises on our part rather than the ideal aspirations we thought they were. Some ideals of the local scene have lost their meaning as the local grew into the global. New possibilities of achieving old aims, to an extent hitherto undreamed of, bring fresh complications to other established values. It would be highly illuminating to pursue and illustrate the variety of specific forms of change and distinguish, for example, cases where a rule actually "vanishes" from those in which its jurisdiction becomes limited, the meaning of its terms altered, the number of qualifications and exceptions increased, and so on.[3] Let us here rather glance at a few of the changes that have taken place at the highest moral level.

One of these has been a fundamental shaking of our traditional conception of the good. This was part of the puritan ethic with its emphasis on striving, ambition, success, with concomitant guilt for failure. The counterculture of the 1960s, whatever its excesses, offered the value of present consciousness as a substitute for a purposive future-oriented pattern, and an affiliative interpersonal relation (of love) for driving ambition. It accused the prevalent (success) pattern of lacking authenticity and engendering alienation, and extended the confrontation into an active critique of institutions from the military and economic to the educational and familial.

The theory of rights and justice has also witnessed new movements of liberation, new extensions of equality, and even new ideals of distributive justice. The breakdown of older colonialisms and the liberation movements of Blacks, women, and various disadvantaged groups, shook up established and smug conceptions of liberty and equality, of merit and distributive justice. The extension of new human rights (e.g., in the UN Declaration and in the demands of practical movements) far outpaced philosophical analyses of rights. Philosophers, for example, have yet to give a thorough analysis of presuppositions concerning distributive justice in the Third World's demand for a New Economic Order or the grounds for a sharing of technology and resources. Yet these broad and persistent moral trends have not been without theoretical impact. For example, in his *A Theory of Justice*, Rawls takes it for granted that natural advantages do not of themselves constitute a ground for reward in its various forms. This is a genuinely revolutionary moral shift in the theory of a meritocratic society that limited itself to equality of opportunity. (Conservative critics were quick to point this out and brand Rawls as a "New Equalitarian".)

The challenge of fresh moral problems is best illustrated in the rise of bio-ethics, to which hosts of moral philosophers have turned, with even the familiar issues of abortion and euthanasia altered under added knowledge and new technologies. And there are wholly new problems engendered by such techniques as organ transplants. Other fields show comparable results. The theory of scientific responsibility is intensified by technologies of nuclear power, recombinant genetic engineering, and so forth Legal questions of responsibility change form: for example, new concepts of privacy rise in the wake of new technical modes of electronic invasion and new modes of storing information; refurbished concepts of consent and new concepts of autonomy come to the center of the moral stage as new ways of controlling behavior (surgical, drug, conditioning, added to the power of the mass media) constitute new threats. The list is almost endless. The growing area of the new moral problems affects the traditional area of the older problems for the new often magnifies older proposed solutions that seemed only a slight step in a given direction; now their broader implications are unfolded.

Finally, there are genuinely new features in the picture of our world that come into play with respect to fundamental moral attitudes. For example, the idea of limits (brought to the surface in the impact of population growth, pollution, depletion of resources) has challenged older conceptions of indefinite growth and progress. There have also been shifts in our attitude toward nature, and attempts to move from an attitude of exploiting or conquering nature to cooperating with nature.

The radical changes in the world and its problems and the sharp challenges in contemporary morality suggest that the chief task today is reconstruction of ideas, rules, methods, attitudes, in the light of the revolutionary changes in global conditions, knowledge, human powers, and broadening human aspirations. How far such transformation has to go and what are to be the continuities with the past are themselves problems within that rebuilding. Obviously a refined methodology is required for so complex a task, which moral philosophy might have been expected to provide, but at least up to this point the input of moral philosophy has been minimal. In the past decade some moral philosophers, aware of what was going on, have thrown themselves into the examination of practical moral problems. They brought an important sharpness to the analysis but for the most part they do not seem to operate with a systematic ethical theory that is geared to discussing problems of change. In many cases there is the avowed separation of theory and practice: it is assumed that theoretical reflection simply sharpens intuition which is then more sensitive in practice but brings no special lessons to its work. Perhaps this reflects the sharp distinction that was made for several decades between metaethics as an autonomous analytic pursuit, and normative ethics as a substantive (hortatory?) pursuit; it would explain in part why the present concern with practical problems so often looks like a flight from ethical theorizing to normative ethics, without even any feedback revising theoretical models on the basis of the practical analysis and its results. Often, however, the treatment of practical problems is set in a framework of rights theory, but one in which the theoretical work has been done at a remote higher level and little or nothing on such questions as how particular rights are established, applied, their relevance determined, whether they change, and so on—in short, questions pertaining to change and novel situations. Thus when new questions of depletion of resources arise they are discussed as ingenious questions as to whether the unborn have rights, and the new problems of pollution as whether seals and dolphins and trees have rights. One is reminded of Thomas Kuhn's description, in his *The Structure of Scientific Revolutions* of scientists engaging in "normal science" under older paradigms even when all signs point to the need for a new paradigm.

Programs and Dogmas

Why has the moral philosophy which prevailed in recent decades—World War II is a convenient starting point for recency—proved so unhelpful in rebuilding morality? For one thing, it did not treat its doctrines as programs, but rather as proposed analyses to be judged by their logical correctness and conformity to linguistic uses. Programs have to be tried out, tested, they are corrigible against the procedures in putting them into effect and the consequences of their carrying out. Nothing of this sort faces the products that come through analytic processes when they are declared correct, and yet they impose a conceptual pattern on the moral thought of that time. They are thus doing the work of programs without undergoing the tests that programs should endure.

Take a brief example from early twentieth century analytic ethics. G. E. Moore, in his *Principia Ethica*, argued that since "right" meant productive of greater good in the world (than would be produced by an alternative act) and since every act has countless unknown effects through endless time, we would never have sufficient evidence to warrant changing an accepted moral rule and therefore ought to obey it even if it is really incorrect. Now surely there is someting wrong with such an outcome. If we regarded Moore's initial definitions as constituting a program for analysis, that is as furnishing a model, we would want either to refine the model to see whether it could avoid such an outcome as abiding by rules we strongly felt might be wrong, or else consider alternative models to see what outcomes they might yield. For example, Keynes was stimulated by Moore's very chapter in which this argument occurs to work on the theory of probability, which might make probable judgments of rightness more feasible. [4] In a similar spirit, instead of arguing which of the different patterns of the relation of the right and the good was "correct" one could have modelled what kind of an ethics each pattern would produce and what the moral outcomes would be.

That very little work in such a direction went on—except very indirectly through the conflict of different analyses of moral concepts—was not an oversight; it was rather a principle that the conceptual structure of ethics was to be established in a self-enclosed analytic way. Two central general doctrines cooperate to achieve this result—the theory of the naturalistic fallacy with its sharp separation of fact and value, and the equally sharp distinction between metaethics as a purely analytic process and normative ethics as a practical-emotive (or prescriptive) process. The two were themselves regarded as fundamental and obvious, established on logical or linguistic or metaphysical grounds. There was virtually no attempt to carry through the division into fact and value, or to show how it could be done or work out a method to show in detail what would count as fact and what as value and how large borderline areas would be apportioned. The division was assumed in principle as really the case, not as just a reminder that occasionally we need greater explicitness in value assumptions. [5] Again, it was taken to be a fundamental division, but if one asked for the grounds of its importance—for example, as against other possible dichotomies in the field to which attention was not paid—the answer was usually that it brought increased clarity and that it prevented people from smuggling values into a moral argument under the guise of fact or scientific authority. The latter seemed to open the door to an assessment of the moral consequences of a theoretical structure. But when critics of the sharp fact-value separation argued that this doctrine assigned a value-free character to science and was therefore morally objectionable, this was declared irrelevant to the analytic process. Clarity and its values were within the process, but any moral assessments did not affect it; they were rather substantive or normative judgments having nothing to do with it, expressing the attitudes and feelings of people. Clearly the separation of fact and value was a vast program in twentieth century moral philosophy [6] which insulated itself and operated as a dogma rather than as a program.

The separation of metaethics and normative (or substantive) ethics likewise was programmatic. In effect it proposed analyzing moral language in separation not merely from the historical and social context of morality but also from the moral content itself. It had to show first either that there were distinctively moral terms or distinctively moral uses of terms, and it had to show (or assume) they could be understood by an analytic treatment in isolation. (It was almost as sweeping as the thesis in behavioristic linguistics that the scientific study of language could be carried on without reference to the *meaning* of words.) When critics of the distinction pointed to value-attitudes in metaethical assertions, they were in effect arguing that the distinction could not be successfully carried through. When they complained that substantive ethics was left without guidance they were criticizing the presumed consequences of carrying the program into effect.

Today it is clear that both the absolute separation of value from fact and that of metaethics from normative or substantive ethics (as other than simply relative contextual distinctions) created a deep gulf between the domains of moral philosophy or ethical theorizing and practical moral problems. This distancing meant that intellectual tools were simply not being prepared for the needed reconstruction in morality. We have therefore to examine the specific features of our moral philosophy in the last few decades that account for our "unpreparedness."

The Ahistorical Character of Present Moral Philosophy

Change has been recognized in most disciplines during the last century. [7] Questions are recast in a diachronic framework with the synchronic as a cross section of a process: the present state is understood as a relative equilibrium of forces and movements within the process. In ethics, however, the eternal or timeless remains entrenched. The idea of moral change itself is seldom analyzed. [8] When Frankena classified types of ethical theory at mid-century, the basis of division was epistemological. [9] There was no suggestion that in dealing with actual morality a more important basis might be historical (ancient moralities, medieval moralities, industrial-age moralities, etc.) or psychological (guilt- or sympathy-based moralities, etc.) or even goal-based (happiness- or success-moralities, self-control moralities or war-moralities, etc.).

The timeless aspiration of contemporary ethical theorists is strikingly clear in Rawls' demand for a once-and-for-all decision on the formula of justice. No question of periodic re-reckoning with advancing knowledge arises because decision is geared to a methodological veil of ignorance. Significantly, he does not consider totally different models; the chief battle is contractarian vs. utilitarian, largely a family quarrel.[10] Practically nothing in his book examines Marx or Dewey who, in European and in American philosophy respectively, represent the most serious attempts in ethics to come to grips with temporal processes. Though Dewey and Marx differ widely on both empirical and conceptual issues, both carry the emphasis on change and the need for reconstruction into the theoretical concepts of morality. Not merely moral rules and moral content undergo change, but moral concepts are strained, readjust to changing problems and practical situations and can be understood only in terms of their contextual social and historical uses and relations. The central moral problems of an age express social needs and conflicts, not the individual's struggle with temptations about conforming to the traditional ways, even though the individual's consciousness remains one arena in which historical struggles are eventually fought out or historical problems resolved.[11]

If moral philosophy is to help in contemporary moral rebuilding, it is not enough that historical questions and applications be added to preexistent ethical theories themselves developed in an ahistorical way. Nor of course does recognition of change entail adulation of change. It means sober evaluation of what is possible, what can be controlled in the changes taking place, and what changes in morality are called for. The theoretical task facing moral philosophers is not only to analyze the concept of moral change but to revise their theories so as to give due place in their concepts and methods to facts of time, history, and change.

The Piecemeal and Isolated Character of our Analytic Modes

Analysis is, of course, a critical part of philosophical work, but it can be done in many ways and the partisans of each have too often identified "doing philosophy" with their own way of analysis. This obscures the need to assess the different modes either in general or for some special purpose. A plurality once recognized makes such a monopoly impossible. Some way of justifying the mode of analysis one is using is now called for. Simply to drop one or two of the theses that the analytic schools have held—such as the oversharp distinction of metaethics and normative ethics—and let everything else remain analytic business-as-usual, will not do. Yet what criteria can be used in assessment?

Some clues can be elicited from the criticisms found in the procession of twentieth century modes—positivist formalistic, ordinary language, phenomenological, and so on. Thus positivist modes were charged with being too formalistic, too inclined to disparage ordinary language, too quick to develop a technical language, too insistent on large formal systems; they might be successful in mathematics and physics, they had still to show their use in psychology, and were unlikely to be fruitful in ethics.[12] The implication was that they carried us too far from ordinary experience. Some of these defects were remedied by ordinary language analysis (in its several Oxford forms), and insufficient attention to the qualities of direct experience has been remedied by phenomenological modes. Yet both of these in turn have been criticized on other grounds. Phenomenological analysis is criticized for leading into subjectivity and forgetting the natural world that it bracketed at the outset in order to achieve a more concentrated analysis of direct experience; by not correlating its results with the scientific knowledge of that world, it robs the study of man and ethics of the power that a dynamic integration would bring. Similarly, ordinary language analysis is criticized for giving us—in spite of its great sensitivity to finer shades—the current and local dialect of speech and thought, thus locking us into status quo patterns of viewing our world that have been built into the contemporary language. It is neither sensitive to changes that are going on nor furnishes ways of dealing with change.

If such are the criticisms different modes of analysis make of one another in moral philosophy, then clearly they are being assessed in a broad and minimal sense for their potential contributions to ethics-building: what they make possible and what they hinder, whether they follow the complexities of experience or reduce it to abstractions, whether they narrow human vision and experience or give it wide and responsible scope, whether they focus on the eternal or guide us amid the changing. In the long run, perhaps, there is some circularity since the conception of moral philosophy itself fashions these demands but is itself articulated in the process; this is, however, the familiar nonvicious type of circularity with which philosophers have long learned to cope.

We are most concerned here with linguistic analysis of the informal type—whether ordinary language or some variant form—because this has been prevalent in the moral philosophy of recent decades, and so the criticisms of "piecemeal" and "isolated character" are addressed to it. On the whole, linguistic analysis did not reflect much about its character till quite late in its career. Analysis was taken to be a self-sufficient logical process;

whatever its internal character it was not empirical or inductive and it was not governed by any pursuit of ends beyond logical clarification. Its results were therefore claimed to be clarified concepts, autonomous analytical products that might, when the analysis was over, be applied in empirical-scientific and purposive-value enterprises. Now it is precisely this assumption of autonomy that has been challenged. It would appear rather that empirical and purposive considerations present in the materials that are being analyzed and in the assumptions and objectives of the analyst so far from keeping aloof from the process of analysis actually play a vital role and call the turns during the process itself.[13]

From the criticisms already suggested, a proposed revision in analytic mode begins to emerge. Ordinary language analysis was said not to deal with change; to remedy this means applying the shift from an ahistorical to an historical perspective to the linguistic dimensions of a morality as well. Linguistic analysis was said to involve in its process empirical and purposive considerations. Therefore it cannot be isolated from its material and purposive context and it is an empirical question whether its concepts can be understood each by itself without relation to the rest. Hence its conceptual framework should allow for the possibility of these relations. What is proposed, therefore, instead of the prevailing mode of analysis—on the assumption that the inadequacies considered follow from its piecemeal and isolated character—is a *network analysis* to replace the piecemeal and a *context-regarding analysis* to remedy the isolation.[14]

The usual practice has been to analyze terms separately and in isolation from material content. Thus we get separate studies of liberty and equality and they issue in controversies of liberty vs. equality. (Social conflicts are then seen implicitly as between lovers of liberty and lovers of equality; similarly with conflicts between desires for liberty and desires for security.) And we have separate analyses of right and good (e.g., "right" is a prescriptive word, "good" is an evaluative word) and controversies of the right vs. the good, or justice vs. utility, or the deontological vs. the teleological. In all these cases, the separate linguistic uses of each term and of subterms are explored in instances and counterinstances and conclusions drawn about reportorial or expressive or directive role. Even where the result is some formula it rarely relates a term to the whole complex of other moral terms so as to see what they are doing together, how the labor they share is organized or integrated, whether they shift jobs in some subtle pattern of division of labor, even whether the job one term is doing is possible only because the job another is doing fits in on expected occasions. This piecemeal procedure is coupled with (or probably follows from) the overisolated character of the analysis: we do not relate the concepts to the struggles going on and the problems being faced in the concrete situations. We ignore the whole matrix—material, cultural, sociohistorical —in short, the practical context, which defines the jobs the concepts are engaged on, which furnishes the constraints, which gives not merely the means and the application but the meaning (in a sense that needs careful analysis) of the concepts themselves. Our fragmentary and isolated moral concepts, resident in their abstract intellectual heaven, have nothing left to do but dream how they express, prescribe, incite, and engage in other mild residual traces of full-blooded actions; bored with such an existence they can do nothing more than aggress against one another and claim primacy in some undefined sense.

Suppose, however, we analyzed *networks* instead of fragmentary concepts, and in terms of their *functioning with respect to the moral problems* in their full socio-historical context instead of in isolation. We would then ask different questions: what problems and answers are being channelled through libertarian claims, which through equality and why? What human needs are finding expression through aspirations for the good, and what situations are seeking resolution through demands for rights (when conceivably they could also lay claim to being necessary conditions for the good)? For example, to focus in defining liberty, as so many social philosophers have done, on threats coming from the state, and pay no attention to economic and social coercion, makes little sense in complex modern societies, though it may in simple individualistic economies. We cannot, however, carry out this analysis without seeing the needs and demands operating through other concepts and which might as well have operated through liberty. That social and moral judgments are made under a selected description is a fundamental feature of decision. This description determines what is relevant and the concepts open the way for one rather than another body of principles. For example, nineteenth century slaves in the United States were given *liberty*, but twentieth century Blacks seek *equality*. Could not the demands of the latter have been channelled as a search for *greater liberty*? In the present controversies over affirmative action, appeal is made to ideas of *compensation* or *equality* rather than a claim for some liberties as having greater importance than other liberties. The expansion of higher education in the 1950s and 1960s was seen (by many) as increasing *opportunity* and providing necessary conditions for a better life. It could just as easily have been channelled through *equality*— as providing for the mass of the young what the well-off already had in private colleges.[15] If abortion is defended, is it under the rubric of liberty or of property in one's body? Or are these concepts already so preempted by established patterns that a new concept of privacy has to undertake the job (assisted in its rise by other needs that are not satisfied)?[16]

The significance of these considerations is that a concept has to be understood *in relation to other concepts* operating in the field (i.e., not piecemeal) and *to the channelling of content among them* (i.e., not in isolation). Otherwise we are reduced to impressionistic weighing of abstractions or to dogmatic assertions of primacy. For example, Felix Oppenheim argues that public schools did not increase people's liberties, but contributed to their security; in fact he says they diminished their liberty by compelling attendance. He uses this to argue that "economic freedom" should mean governmental noninterference in private economic matters, and "economic security" (not freedom) be used for governmental activity in raising the standard of living, which requires limiting economic freedom.[17] Granted liberty could be narrowed to freedom from legal compulsion, but also that it could be widened (in Hobbes' definition) to removal of obstacles to desire, which should one do? Isaiah Berlin, in his *Two Concepts of Liberty*, argues for a negative as against a positive definition on the ground that the latter lends itself to outrageous totalitarian abuse. But Mihailo Markovic[18] argues convincingly that equally outrageous abuses can be associated with negative freedom.

It certainly looks as if no satisfactory definition of one concept can be reached if we consider it by itself. We have to see what it is doing to the jurisdictions of other concepts, and work out an integrated or systematic order, developing fresh concepts if necessary, and going on to see how that order operates in the life of the community. Such a procedure only brings us up to what other fields have long taken for granted. Other intellectual fields have not hesitated to revise and integrate their concepts and treat them in a systematic way. Physics did not hesitate to revise its concepts in relation to the development of operations for application, and to unify space and time when their separate treatment proved unsatisfactory. A careful political scientist would study the division of powers in the working of a whole system, not seek theoretical primacy for legislation or courts or executive; he would want to know the role of the different checks and balances, and would not be outraged by the idea of courts changing the law or by a concept of administrative law as straddling two of the divisions, or by the recognition that primacy at one time is in the legislature, at another in the executive. He would study what social needs and movements occasioned the rise of such phenomena. Ecology has shown how attention has to be focused on the whole field in order to see the full consequences of what goes on in any part of the field. It also suggests that there are costs somewhere for benefits elsewhere. Such lessons are familiar enough by this time in law and society. Thus at some points evils are endured rather than legally prohibited simply because the machinery of enforcement would produce greater evils. The costs of maintaining a system of liberties are also explicitly recognized, as are increasingly the costs of removing inequities and discriminations. All such lessons of realistic study and assessment might more readily be applied in moral philosophy if we used a network mode of analysis. The difficulties, however, are great, partly because of the complexity and plurality of value interests and (in the case of assessment) because morality itself seems to constitute the ultimate court of appeal and so there is thought to be nothing in terms of which it can itself be weighed (as the law can be weighed in terms of morality).[19]

Moral philosophy today is preoccupied with utilitarianism vs. justice or some kind of rights conception. But looked at historically the opposition between these two schools is like that of the legislature and the courts. Each rises to do the job the other is neglecting. In the 1930s the President with legislative support sought to make institutional changes that the economic situation required if people were not to starve. The courts for a time tried to hold them back. In the 1950s the legislature did not carry through the liberations that the whole impetus of World War II had promised. The courts began the task and at least put some of them forcibly on the agenda of politics. The changes in the 1930s were carried through under the utilitarian concept; those in the 1960s were fought for under the rights concept. In fact the rights concept has gained such strength (particularly since the UN Declaration of Human Rights) that we are constantly besieged by new rights: the rights of animal species, even of trees, find their way into ecological suits, not merely in imaginative literature or philosophical papers.

There were parallels in the eighteenth and nineteenth centuries. The revolutions of the late eighteenth century were under the aegis of rights. The early utilitarianism (for example, in Paley) was conservatively protective of the established property system.[20] British industrialism of the early nineteenth century found its conceptual formulation in utilitarian ideas. Perhaps the present prevalence of the rights concept is a function of the weakness of the utilitarian concept, so that the entrenched rights have to be fought with new rights.

It might be said that all this is interesting history of ideas or sociology of knowledge, but not philosophical analysis. But there is no conflict here: historical material may focus analysis and improve it. For example, if utilitarianism in the United States has been too smug, believing the country had achieved the greatest happiness of the greatest number because the large middle class (in the slogan of some years ago) "never had it so good," then Rawls' introduction of the difference principle can be seen as asserting that the greatest happiness of the greatest number is not enough; it does not cover the disadvantaged and the oppressed. The utilitarian analogue would be a shift from "the greatest happiness of the greatest number" toward "the greatest happiness of all."[21]

Let us accordingly—to assuage the philosophic conscience—take a more formal problem to suggest how a network mode of analysis, anchored to the practical content, could clarify theoretical issues. We may revert to the Rawlsian treatment of the primacy of right, which is surely a most complex issue on the highest theoretical level. Rawls insists throughout that right (and justice) is prior to good. If we keep our eye on the whole system, we see how this builds in a commitment to the priority of liberty in what it enables him to do. By making the good a posterior matter he is able to allow the widest latitude to individual patterns of life; in effect, absolutely necessary social constraints are already taken care of by establishing justice first as integral to all schemes of the good. Thus the priority of the right enhances liberty in the society.[22] But Rawls has not wholly kept the good a posterior matter. A thin slice of the good is already involved in his conception of primary goods that is used to generate the situation in which, under a veil of ignorance, the principles of justice are determined. Though it is clad in the garb of what all humans would want, whatever their plan of life, the primary goods are at bottom some necessary conditions for any good life, if not life itself (as in the old formula of life, liberty, and the pursuit of happiness). Thus instead of the primacy of the right, what we really have in Rawls' theory is: a thin slice of good, a whole chunk of justice, a thick batch of goods. But the middle part is itself quite complex, as the problems of the lexical priority of liberty and the controversies over the difference principles have shown. Once it is possible to divide the good into the thin and the thick, why not divide justice into parts? And once they are all before us, why not allow a more varied order in priorities? Schematically, we might have, for example: $G^1 - J^1 - G^2 - J^2 - G^3 - J^3 - \ldots$ (In this case, the thick goods could be analyzed into necessary common elements, universally more attractive elements, elements with the widest ramifications along whatever lines proved more fruitful; the result would be a set of components paralleling the inner multiplicity of justice.)

Now while such orders could be created ad hoc in an antecedent way, they make sense only if we take them in a developmental way—whether for individual or society. We are then at bottom trying to see what moral principles are to be introduced in human societies in the light of basic needs and developmental needs, desirable opportunities, desirable areas of individual initiative, social necessities under existent conditions, mature aspirations, and so on. Various segments of this range may be conceptualized in terms of good, right, autonomy, justice, liberty, and the rest. Some concepts constrain the variety of choices more than others; constraint and latitude are functions of what we can learn about the human makeup and the formation and possibilities of social living. If the concepts are kept close to their function, then the struggle of the right and the good may be resolved in some equitable partition, rather than in an imperial primacy.[23]

How a network mode of analysis is to be carried out in so complex a context is no easy task. (It does not have the simplicity of the relation of concepts in an axiomatic system.) It requires as much work in moral philosophy in the next generation as the more limited analysis did in the past three decades of its dominance. But self-consciousness may hasten the tempo and bring moral philosophizing and the resolution of our moral problems closer together.

The Individualistic Model in Moral Philosphy

The individualistic model is generally taken for granted in the moral philosophy of our century. It simply continues the traditional model widespread in all fields—the attempt to construct laws of a field in terms of interaction of individual or atomic units. There is no distinction between utilitarianism and contractarianism here: the one calculates social good from maximization of individual preferences, the other works from the meeting of individual wills in an extension of the older voluntarism (no obligation without individual consent or transindividual agreement). In short, the individualistic model has occupied the seat of power in the leading conception both of the good and of the right. What is more, this individualism is particularly entrenched because of the fear of its opposite. To think about the social good without translating it into or deriving it from individual goods or rights is taken to risk some sort of totalitarian theory.[24]

I do not think this question can be tackled in its familiar formulation. The antithesis of individual and social constitutes a set of hardened categories which obscures its own historical development. It has been chiselled at by psychology attempting to show the social character of individual development, that the individual is socially constituted and not merely socially caused. Sociological study has shown the whole variety of community forms that come between isolated individual and the total group—families, wider kingroups, associations, and affiliations. Both social psychology and philosophy have shown the difference between the group concept and the interpersonal concept (Buber's I-thou is a good example). History has traced the way in which changes in economic and social life in the past few centuries smashed the influence of the intermediate groups and associations and left the individual starkly opposed to the whole.

All this is not enough to break the stereotyped model in moral philosophy, and there are no doubt fundamental causes in our mode of life for this attitude, but this does not give it theoretical justification. It may be the reflection of our alienation. I propose therefore a wholly different approach to a controversy that is as destructive as it seems incapable of resolution in our contemporary atmosphere. I suggest we take a leaf from the book of economists and distinguish two fields of *microethics* and *macroethics*, identifying them by the kinds of problems involved, and leave open for resolution in the light of empirical and analytic study what generalizations we may come to make about their relations, that is whether an integrated model will at some time be feasible.[25]

That there are some questions for morality which can be tackled as macroethics should be clear enough. For example, institution-building is generally not one man's concern. Actual discussion of a constitution that is to be framed is not simply voluntaristic even where individuals end up with voting; it reckons with tradition and with the kind and quality of life of the community living under it. Today, moral questions of family structure, sexual equality, population checks, distributive justice, institutions for dealing with crime, and so on, are clearly macroethical. Jurisdiction is something to be decided on in the light of the conditions and character of human life. A moral question can therefore move from micro- to macro-status (or the reverse)—sometimes only because the conditions of coping with it are large-scale. For example, defense in the frontier was an individual or a family matter, now it is a national matter. Moral questions of distributive justice with relation to unemployment and welfare changed into macroquestions at a given point in our history. On the other hand, a whole range of problems involved in conscientious objection and civil disobedience as well as in proposals for moving sexual decisions by consenting adults out of legal control are either acknowledged questions of microethics or proposed transfers to it.

There are no doubt many points about this proposal that will have to be clarified or refined. For example, it should be stressed that the concept of macroethics need make no metaphysical assumptions about the existence of group minds. The whole controversy in the legal tradition about whether corporations or other groups are persons, or in the tradition of social philosophy about the "reality" of nations, can be bypassed. Similarly, there need be no moral assumption about the value primacy of the group which demands the automatic sacrifice of individuals. Questions of overweighing some interests by other interests or the absolute character of some interests are internal matters in its development and procedures. The main accomplishment is that it will become possible for what is often called "social ethics" to experiment with its own frameworks and its own methods, and see whether they can make sense of our moral judgments about public policy without reducing them to the individualistic model. There may also be consequent contributions to the historical interpretation of ethical theories. For example, Kant's universalization criterion might make more sense as a concern with institution-building than as a method for microethical decision. Again, the frequent suggestion that Benthamism makes sense as an ethics for legislation and not for individual decision points in the same direction; to see it in this light gets rid of the horror stories so frequently invented to disprove it as counterintuitive.[26] Contemporary efforts to get away from summing individual preferences to judgments about "the quality of life", while they are sometimes directed to avoiding certain types of measurement, also will support our proposed distinction.

It seems clear that the source of macroethics is the degree of interconnectedness and complexity of human problems in the world today. It is possible that this has reached the point at which a great part of microethics will require the results of macroethics as a framework. It is, however, possible that some questions will remain quite separate. It is too early to predict where the development of the field along lines of such a distinction will carry us. But it does seem a hopeful way to cope with the impasse which the undue prevalence of the individualistic model has brought to moral philosophy. It will, for example, allow a greater realism in the determination of morally relevant distinctions for the formulation of questions. Thus it may prove that the distinction of social and individual is morally less important today than that between global and national; the latter is pertinent to the vital problem of redistribution of resources and technology, as in the issues of distributive justice raised in the idea of a New Economic Order. On the other hand, the division of social and individual may have greater pertinence in microethics, though even here it might turn out to be the distinction not between me and society but between me and other people!

The three critiques offered in this paper are not separate. It is the significance of an historical approach that makes the isolated character of current analysis inappropriate. If morality had in fact the kind of constancy in which change could be ignored, it might be dealt with in terms of invariant features alone. If parts of morality in fact maintained an independence from other parts, then a piecemeal analysis might be feasible. If human life still had social conditions in which the individual (or the familial group) lived and worked and solved its problems alone, the individualistic model might have some applicability. None of these conditions now hold, if ever they did in any fundamental sense. (Even very slow movement is not absolute rest.) We need not become Heracliteans to do moral philosophy in a way that will make our ethical theories applicable to problems of moral reconstruction today. But surely it is long time to stop being Parmenideans!

Notes

1. The paper amplifies a number of suggestions I made at the Bicentennial Symposium of Philosophy, held under the auspices of The Graduate School, City University of New York, at the Biltmore Hotel, New York City, October 7-10, 1976. The suggestions were presented in comments on a paper by Professor William K. Frankena, entitled "Moral Philosophy and Moral Standards."

2. It is interesting to note that in both Tillich's religious philosophy and Sartre's existentialism seriousness (alias authenticity) inherits morality. Such intellectual phenomena are perhaps the best indications of the intensity of contemporary moral problems—that nothing less in the ladder of moral categories, neither rules nor virtues nor goods, retains basic standing in the ethical theory. "Situation ethics" exhibits a similar phenomena in another way.

3. It is quite likely that such a process in less dramatic form is constantly going on, and that the "reflective equilibrium" with which our morality is operating in "normal times" is constantly subject to small changes. To study equilibration processes in morality itself would open up an unexplored area of analysis, just as the discovery of natural selection eventually focused attention on processes of mutation.

4. Moore's argument is in his *Principa Ethica* (Cambridge: Cambridge University Press, 1903), ch. V. J. M. Keynes' reference to the Moorean stimulus to his extensive work on probability is in his "My Early Beliefs" in *Two Memoirs* (New York: Augustus M. Kelley; London: Rupert Hart-Davis, 1949), p. 95. Interestingly, and somewhat paradoxically, he says (p. 82) that he and his contemporaries, enthralled by Moore, paid no attention to that chapter, but concentrated on Chapter VI—The Ideal—which he speaks of as Moore's religion, as contrasted with his morality!

5. Occasionally it was suggested that the division was untenable. For example, Robert Ackerman ("Normative Explanation," *Philosophy and Phenomenological Research*, XXIV [1964]: 522-29) compares the effort to separate the vocabulary of fact and value to the fruitless effort to make an absolute distinction out of theory and observation. John Searle's "How to Derive 'Ought' from 'Is' " (*Philosophical Review* LXX111 [1964]: 43-58) provoked great discussion about missing premises in his deduction, but its significance seems rather to lie in the attention drawn (in its later part) to institutional facts with the suggestion that they cannot be unravelled into separate fact and value.

6. Not of course in moral philosophy alone. It extended also into the social sciences which it dominated for half a century. In the later years the irresponsibility of science justified by the thesis of its neutrality stands out sharply. But in the earlier years the claim for autonomy in social science and the right to investigate empirically human phenomena claimed by religion and tradition no doubt had a quite different role. In moral philosophy similarly the variety of defenses—linguistic, logical, metaphysical, empirical, valuational—will have to be examined for the intellectual context in which they emerged. Perhaps in the long run, to understand so complex a phenomenon as the dogma of the fact-value separation over the greater part of our century, we will have to see it not in philosophy alone but in intellectual movements as a whole. The categories of a historian may prove a necessary supplement to those of a philosopher.

7. This was obviously prompted by the theory of evolution and its general outlook. In some disciplines, however, there was an occasional hiatus, largely because of excessive claims made in a short-sighted use of evolutionary ideas. For

example, nineteenth and early twentieth century anthropology had its linear stage-theories drawing the line in separate cultural and institutional areas from primitive beginnings to contemporary heights. Similarly in ethics the liberal mind was outraged by Spencerian "Social Darwinism" while the conservative mind was apprehensive of Marxian historical-evolutionary pictures.

8. This is probably not an accident. To distinguish the great variety of points it would involve—as a scientist might distinguish in a scientific change whether it was a theory being abandoned or altered in part, or a mode of application with a change in operations, or a whole principle of analysis in theory-construction, and so on—would involve treating an ethical theory in its relation to its functioning in morality in historical process, and so depart from the isolated character of ethics to be discussed below. Again, the intellectual defenses against historical treatment have been complex and would require separate analysis. Chief among them has been the assumption that any appeal to historical-genetic considerations is a "genetic fallacy," that is, a substitution of history for philosophical assessment of truth or adequacy. This fails to recognize that a study of genetic development of an idea is not merely a study of *origins or causes* but can help clarify *the present meaning* of the idea. The discussion of equality, for example, makes little sense if seen only as an arithmetical formula, but a great deal of sense if we trace the historical shift of content that constitutes the core of equalitarian demands. For a general discussion of the genetic fallacy accusation in this domain, see Abraham Edel, *Method in Ethical Theory* (Indianapolis: Bobbs-Merrill, 1963), ch. X1.

9. W. K. Frankena, "Moral Philosophy at Mid-Century," *The Philosophical Review* LX (1951): 44-55. The differences were on such questions as cognitivism vs. noncognitivism (including emotivism), naturalism vs. nonnaturalism (not in the metaphysical sense but whether ethical terms named a natural characteristic). Frankena's picture, of course, was quite accurate for what was going on.

10. One is reminded of Locke's battle with Filmer, and the neglect of all the rich development in the seventeenth century of political positions (e.g., the Levellers) to the left of Locke. Of course the revived divine right of kings doctrine was relevant to the crisis of 1688. No doubt much of our contemporary parcelling of liberal and conservative, which attempts to penetrate to their "essence" in abstract terms has the same character.

11. Dewey, of course, is the philosopher who has outlined such a position in detail with reference to the variety of schools in the history of ethics and the conflict of schools in the twentieth century ethics. The neglect of Dewey in contemporary ethical theory is thus a serious mistake, since it still remains the chief moral philosophy addressed to the problems that the present state of morality faces. Dewey defines a moral situation primarily in terms not of the struggle to conform to already established principles or rules but in terms of the conflict of rules, claims, obligations, in a context in which one is trying to decide what is right or what ought to be done. Emphasis is thus thrown on the processes of decision and evaluation. This is carried through in detail in refashioning the interpretation of the concepts of the good, of virtue, of right and obligation. The constancies in the concepts and their relations are linked to constancies in the structure of human life and institutions, and throughout there is concern with the way morality is defined and developed with the growth of knowledge. Hence both corrigibility and conceptual change are accommodated in the process. Dewey's clearest account on these matters is in his *Theory of the Moral Life* (the middle part of the revised edition of Dewey and Tufts *Ethics*, New York, Holt 1932). It is worth noting that during recent decades, when Dewey was included in books of readings for classroom use

there was a tendency to choose from his epistemological parts (for example, from his "The Construction of the Good") or other parts that could be discussed in terms of the issue of the desired and desirable, that is, the naturalistic fallacy.

Of course in the nearly half-century of human experience since Dewey developed his moral theory there is much to change, but the reconstructive outlook is definitely central in a way in which even other naturalistic theories of this century rarely achieved. Such significant naturalistic works as R. B. Perry's *General Theory of Value* (New York: Longmans Green, 1926) or Stephen Pepper's *The Sources of Value* (Berkeley and Los Angeles: University of California Press, 1958) seek rather the permanent structure of human life. Dewey is sensitive to the theoretical consequences not only of change but of the rate of change.

12. Even this should not be too hastily concluded. They developed the field of deontic logic and logic of preference which has yet to prove its worth for coping with moral decision; but it is too early to reject it as useless. For a comparative study of analytic modes, see Abraham Edel, Introduction to Volume I.

13. Perhaps such self-consciousness comes only with the dusk, when methods are challenged. One is reminded of Oakeshott's remark that to attempt to build up a conservative philosophy is already to indicate that the conservative rule is slipping. At its height, the type of analysis we are considering was too secure (first in England, then in the United States) to have a sense of slipping. For a detailed argument about the internal role of the factual and valuational within the analytic process, see Abraham Edel, "Analytic Philosophy of Education at the Cross-roads," in *Educational Theory* 22 (1972): 131-152.

14. About the first part there may be no philosophical inhibition today; network approaches in epistemological analysis have strong advocacy, and the question may be rather how it is to be done for ideas of morality and social philosophy. About the second part there may still be strong inhibitions (of the kind indicated in the note about the genetic fallacy above) based on a tradition of drawing a sharp line between an idea and its applications as well as derogatory conceptions of *psychologizing* or *sociologizing* where "logicizing" or analyzing is called for. This is not the place to reopen such old battles, nor to make claims for one or another side. I call attention to these problems here simply to insist that what I am about to discuss is reference to context as part of a mode of analysis itself. Perhaps the expanded studies of meaning and symbolization of the last two decades have carried us past the old controversies.

15. It is interesting to note that if the original system of public schooling in the United States is seen under the ideal of equality—to provide for the mass of the people what the few had—it would be quite parallel to the scheme of a strong affirmative action, with governmental expenditure and coercion (compulsory attendance as well as compulsory taxation to support the schools). Using the public schools so seen as a parallel in today's controversies about affirmative action would prescribe setting up compulsory government workshops to provide jobs for women and Blacks. By comparison all talk of quotas and infringing others' liberty would pale to triviality.

16. The rise of new concepts and their subsequent career should be of special philosophical interest. It is hard to see how they are to be understood apart from quite specific relations to the problems involved in their genesis and the (possibly differing) ones to which they are addressed in successive periods. "Autonomy" is an excellent example. In such cases—and why not in all cases?—the categories of biography (even *floruit*, with a date) appear more appropriate than some of the impressionistic features that cling to the abstract notion. There is also an element of

inspiration in asking of a concept, as one may ask of a word in the language, whether it is U or non-U (that is, used by the upper or lower classes). One could quite warm to this subject. Think of the checkered career of *fraternity*—with its ups and downs—as compared to liberty. It is philosophical folly to treat this as if it had no philosophical relevance.

17. *Dimensions of Freedom* (New York: St. Martin's Press, 1961), p. 123. If the compulsory attendance feature were abolished, would public schools mean greater freedom, or would the argument shift to the compulsory character of supporting taxation? Similarly, if governmental support for the unemployed came from running (and profiting from) new enterprises, like deep sea mining or off-shore oil drilling in as yet unleased locations, could the gains of the unemployed be construed as greater economic freedom, or would the fact of not leasing to private enterprise (and the dream of profits lost) inhibit the use of the term "freedom" in this economic context?

18. Mihailo Marković, in his manuscript, *The Philosophical Problems of Freedom*. I am indebted to him for the opportunity to read some of its chapters.

19. This ultimate character of morality makes it difficult even to give a meaning to the notion of the costs of morality, just as there has always been felt to be a difficulty in the question "Why should I be moral?" A meaning does, however, become possible if there is a wider domain of values of which morality is one province, other provinces being religious, aesthetic, physical, etc. (One could then ask whether morality is beautiful, healthy, or impious in having its autonomous standards, etc.) Again, within morality, there could be a cost to the maintenance of one of its concepts in terms of the application of others: thus there are sacrifices of the good in doing one's duty. It is likely that, apart from the reference to a wider value theory, questions about the cost of morality have to be translated into questions about the cost of enforcing or supporting morality or the cost of some category of morality in relation to some other category of morality (e.g., the effect of emphasizing ends on the attitude to means, or the effect on present action of placing a high value on remote ideals).

20. See William Paley's pamphlet on "Reasons for Contentment, addressed to the labouring part of the British population" (1793), in *The Works of William Paley*, Vol. IV (London: George Cowie and Co., 1837), pp. 391-403.

21. This is a jump over the intervening possibilities justified perhaps for the United States in terms of its high productive power. For South Africa, where the majority is exploited, the formula of the greatest happiness of the greatest number might still have power. Or perhaps it is too late for this even in South Africa: it might sound as hollow as to set up for the United States the greatest happiness of three-fourths or five-sixths. In any case, the utilitarian theory did not mean by the greatest number a majority, as most of the horror stories about deducible consequences of utilitarianism in marginal cases seem to imply. Utilitarians surely meant the greatest number possible. This transfers problems to the meaning of "possibility"—logical, scientific, technical or practical. Perhaps more immediately relevant would be "possible with respect to human nature," "possible with respect to cultural conditions," "possible with respect to existing institutions," "achievable by reconstruction," "achievable through revolution," etc.

22. John Rawls, *A Theory of Justice* (Cambridge, Mass.: The Belknap Press, Harvard University Press, 1971). I do not think we find this in his argument, but it stands out clearly on the whole view. Compare also the way in which Nozick's utopias, in his *Anarchy, State and Utopia* (New York: Basic Books, 1974), can have greater latitude, after he has established his initial constraints.

23. If Prometheus provided a myth for knowledge, Antaeus may serve for moral philosophers. It will be recalled that he kept his great strength as long as he touched the earth; Heracles conquered him by lifting him into the air. Theoretical problems like that illustrated above may be dealt with strongly only by keeping our concepts in touch with the ground. But of course, this does not mean letting them stick in one place. I think it was Dewey who used the analogy of walking, which involves one foot in the air and the other on the ground, in turn, as compared to shuffling to distinguish a scientific from a crude empiricism.

24. There is more than a touch of the Cold War in this mode of thinking, as if each side had appropriated one of conceptual opposites: we-they, free-enslaved, individual-social.

25. The economic use of the distinction is not precise enough to transfer bodily, it only offers a clue. It would simplify things if this were taken, as is sometimes found, to set off the management of individual enterprises from the management of whole societies. But in fact the distinction is drawn in different ways, and much that one might think of as macro- is often included in micro-, perhaps because of hope that the laws or principles involved will prove deducible from the laws of the microsystem.

26. It should be pointed out in this connection that, in the light of current use of horror stories as a technique of refuting an ethical theory by counterinstances, little effort is needed to show that every ethical theory can generate horror stories at its limits. What a horror story proves is that the theory needs further constraints at those limits, not that it is disproved.

Part II
Fact and Value in Practice

10. Prometheus on Trial: Technology and Morality

This paper concerns the relation of technology and morality, both the impact of technology on morality and the constraints and directives that morality imposes on technology. We hope also to reach some conclusions about the *desirable* pattern of their relations. So complex an undertaking involves the combination of several methodological approaches. There is conceptual analysis as we decide what we are to understand by "morality" and "technology," terms whose use has not been without considerable ambiguity and vagueness. There is social and cultural description of their institutional forms, and historical description of changes in both technology and morality. (No post-Darwinian philosophy can blithely seek the "essential" relations of such forms. Philosophy is compelled to trace historical shifts and interactions, and if it wishes to establish essential relations it must do so in the shape of presumed constancies or invariants for all periods and changes.) There are throughout some theoretical presuppositions for explaining change and its determinants, usually tied in with a general philosophy of history and of cultural integration. And in the normative reckoning there is some implicit mode of evaluation. It should be added that these four methodological approaches—analytic, descriptive, explanatory, evaluative—are not to be cut off into separate provinces of operation. The operation of each presupposes the materials and tentative results of the others.

Such complexity should not alarm us, but rather make us cautious as we thread our way. I shall begin by considering more precisely what morality is and what are to be its limits for the purposes of our investigation. This is important here, in part because morality rather than the more general problem of values is our present concern. Second, I shall examine some of the difficulties in the concept of technology, as seen in a comparison of some nineteenth and twentieth century accounts of their interrelation. This will raise also the issue of one-way determination or mutual interaction, especially as so many of the theories assume technology to be the independent variable and morality the dependent variable. Finally—and this is the central part of the task—I shall review what appear to me to be three distinct stages in the recent relations of technology and morality, culminating in normative suggestions.

It is clear enough, whatever one's conceptual formulation, that morality is in some sense a sociocultural institution, like language, religion, law. It is further back among the fundamentals than most, more demanding in its tone, less likely perhaps to be regarded as instrumental to something else. We expect a morality to tell us[1] what human goals are worth aiming at and how they fit into a life that is good; what general lines of action are to be followed or avoided as right or wrong respectively and what obligations and duties and responsibilities they entail, as well as what means are out of bounds; what kind of character to develop in ourselves and our children and what kinds to avoid (the traditional "virtues and vices"); what motivations, sanctions, or feelings to attend to seriously (questions of conscience, shame, guilt, social approval, etc.); how to decide, when we are in deep doubt on moral matters. Out of our morality we can even expect to elicit the ways of reflective justification for our paths of life, though in the contemporary world we are sophisticated enough to know that justifications differ, that some will look to a traditional religion, others to the social needs of men, others to an inner conviction, and so on. On the whole, however, we expect to find a common pool of moral rules and obligations and responsibilities and even aspirations, in spite of different backgrounds of philosophical justification in the conflicting theories of man and the world and its ways.

The notion of values is clearly much wider than that of morality. In philosophical usage, it is often a general grab bag in which we may thrust all our for-or-against attitudes, all our desires, interests, aspirations, aesthetic appreciations, pleasures and pains, all our purposes and preferred patterns of conduct, even all our effective instrumentalities for achieving what we want. Not all values are moral values, and the moral somehow provides us with criteria for selection from the domain of values, whether it uses the language of worth or of obligation.[2]

For the purposes of our present inquiry, let us select out of the picture of morality four central topics. One is the scope of the moral community, the basic group of those who are to count and those who are to have responsibilities. (Once upon a time it was the kin group, at times it has been the limited family, the nation, on a few occasions the race; in the contemporary world it is increasingly thought of in terms of all men everywhere.) The second topic is that of distributive justice—the basic principles for the apportionment of goods and burdens among people. The third is the character of interpersonal relations, how a man regards his fellow man and how he is set to behave toward him. The fourth is the ultimate one of what a man expects out of his life, the general character of his basic aspirations. These four are to be the themes of our discussion, and should provide a varied and representative as well as significant focus for considering the relations of technology and morality.

In its derivation, "technology" is of course the study of "technē" and the latter is the ancient Greek word for skill or art or craft. In the Platonic and Aristotelian writings, the standard illustrations are the craft of the builder, potter, sculptor, doctor, etc. The scope of technē was already then being broadened; the assumption that ruling is a craft underlies Platonic political elitism. The assumption that all life is a craft underlies the teleological or human nature ethics (as in Aristotle), which posits fixed purposes as the goals to be achieved in living. The interplay of morality and technological concepts is thus an old story. Twentieth century existentialism is bucking an entrenched tradition when it argues that man creates his own purposes as he goes along, so that his morality cannot be reduced to a craft with antecedent goals.

The term "technology"—the study of techne—comes to be used for a craft set-up in a society rather than the study of such a set-up. It is a shift in usage of the same type as our speaking of "sociological conditions" in a society, when we mean social conditions. This may be due to slovenly speech habits. But it may instead have a profoundly theoretical import: "the technology of a society" may indicate that actual set-up in the society which is studied by technology as a science! Comparably, the sociological conditions of a society would be those selected by sociology as a science, and would be differently identified accordingly as one defined "sociology" differently—for example, as limited to the study of patterns of groups or as the broad cultural anthropology of complex societies. If there is anything to this suspicion, it would bring us directly to the view that "technology" is a term within a social or philosophical theory, and takes different conceptual shape according to that theory. Ellul in his *The Technological Society* [3] is perhaps wise to speak basically of "technique" or occasionally of "the technical," in spite of his title.

The accordion-like character of the concept of technology is best seen in the comparison of historical classifications and theories of historical change. In the narrowest sense it would signify the fundamental tools and skills for making the material things men need for survival. This is familiar enough in references to Stone Age technology, limited chiefly to stone tools for hunting and cutting, as compared to the later complex tools for mining, agriculture, and manufacturing. The anthropological concept of "material culture" included basic skills in dealing with such tools, but not usually basic habits in bodily posture, such as sitting, though it will include basic modes of manipulating the hand and the fingers, in hoeing or in sewing. Presumably to name an epoch in human development by the use of stone or copper or iron is to imply a theory of widespread differences in the mode of life that such technology supports.

Not all technological or quasi-technological interpretations of human culture need employ this concept directly. Herbert Spencer makes a central point of the shift from a military to an industrial society, with its concomitant shift in virtues and moral rules. [4] Such descriptions of a mode of life are fairly wholesale, fusing technology and purposes, instrumentalities and current practices. Thorstein Veblen isolates the technology from business forms quite sharply. [5] And in dealing with intellectual traits he relates their prevalence to the mode of production specifically—for example, belief in luck to a hunting mode and in causal operation of nature to a manufacturing industrial mode. [6] John Dewey, in an early paper on "Interpretation of the Savage Mind," [7] argues that occupations determine the chief modes of satisfaction, the standards of success and failure, and the working definitions of value. They determine the objects and relations that are important, and so the qualities of significance, and produce the integration of a functioning whole. He illustrates this in detail in an examination of the effects of a hunting life as contrasted with an agricultural life. In a much later book—*Liberalism and Social Action* [8]—technology is seen as a kind of crystallized or congealed social intelligence which is the dynamic factor in producing social change.

Let us compare in somewhat greater detail the conception of technology in the Marxian view, in which the economic base of a society determines the superstructure of institutions and ideas, and Ellul's conception of *technique*, which is perhaps the most extreme among recent views of technological imperatives as taking over the shaping of human life on the globe.

In the Marxian view, [9] the economic base is divided into the forces of production (in effect, what we should think of as the technological set-up) and the relations of production or type of economy, embodying modes of exchange, distribution or control. In the theory, the technology demands certain relations of production, and this whole complex plays a central role in determining the kind of social and cultural life that becomes possible, the superstructure which embraces both social institutions and intellectual and artistic production. As in the long run the technological base undergoes change (cumulative mutations as it were), the older relations of production become a fetter to the expansion of production but are clung to because they preserve the established class structure. And so the class struggle is intensified with its conflict of opposing systems and opposing philosophies and opposing moralities. The outcome, unless it be mutual destruction, is the victory of the newer system demanded by the newer mode of production. The morality of a period thus reflects to some extent the technological base at its particular level of development, to a much larger extent the relations of production, to a definite degree the stage of maturation in the class struggle, and in a lingering way the older superstructural institutions and residues.

In this fashion, a feudal morality gave way to a bourgeois morality, as the production for the growing market took over from the self-enclosed locale producing for itself. Feudal morality had a practical local community, in spite of a theoretically universal religion. Its principles of distributive justice gave clear dominance to aristocrat and lord as against peasant and serf. Its picture of interpersonal relations was a web of preassigned obligations in which every man stayed in his destined place and had his allotted responsibilities, doing his duty with gratitude and loyalty to those above. The aim of life was stated as salvation; the best most men could achieve was resignation, starting as they did with the liablities of original sin. On the other side, bourgeois morality became increasingly universalistic, making the atomic individual the core of its morality. Distributive justice was expressed in a theory of property rights that opened up the world to individual effort and exploitation. Men's relations among themselves were primarily contractual, expressing the individual's will and decision. The world was open, unlimited for possible achievement, and men could properly, under one theory or another, pursue their worldly success and the accumulation of wealth. Although all sorts of traditional elements and local variations entered into the moral shift, the dynamic element was the change in the forces of production and the kind of organization it demanded.

The Marxian theory separates off the technology from the economic and social relations. There are, to be sure, some uncertain borderlines; for example, the organization of production in a piecework cottage industry or in a centralized manufacturing plant appears to be part of the technological set-up. In theory, it might conceivably be shifted to some intermediate category between technology and relations of production, if alternative modes were feasible under the same general conditions. Technological imperatives in any case are permeating, stringent, and determining, and history spells them out. The mode of determination is partly direct causation, partly a selective process on a kind of evolutionary model in which superstructural "mutations" are adopted according to basic fittingness. In this latter sense the Marxian theory is driven to relax its determinism. In part it has to give greater scope to the consciousness of men in fashioning the direction of their social life and its ways. Its concept of leadership, in which there is awareness of emerging directions of history (though itself a causal product of an advanced society), calls for an intellectual-political spearhead in transforming society. Morality, therefore, while hemmed within the broad conditioning picture, is capable of interacting with differential strength depending on the total balance of forces; and within limits it can fashion directions.

There is one element of morality as we construed it which Marxism never questions. This is embodied in its conviction that men's drive for freedom—that is, for greater productive control of the material world—underlies the advance of technology, the rise and fall of classes, and the unavoidable outcome of a classless society on a high productive level. This is the dynamic core in the all-human value system. Whether it is to be postulated as the nature of man or exhibited as an historical invariant, may not always be clear. But this much is clear—Prometheus is never put on trial.

Jacques Ellul, on the contrary, is the Cassandra of our contemporary technological society. Whereas Marx tried to separate technology from its multitude of social and value relations, Ellul's concept of *technique* is all-devouring and wipes out borderlines in its inevitable march. That is, not merely is technique taking over the world, but the concept of technique itself becomes all-encompassing. At some point tied to the machine, it may retain the flavor of the mechanized, but comes to characterize domains in which machines can play no role. Very soon we are told:[10] "In fact, technique is nothing more than *means* and the ensemble of means." Or again, technical operation "takes what was previously tentative, unconscious, and spontaneous and brings it into the realm of clear, voluntary, and reasoned concepts."[11] Reason creates new means and methods and measures efficiency, selecting the means best adapted to the desired end. "Thus the multiplicity of means is reduced to one: the most efficient. And here reason appears clearly in the guise of technique."[12] The conscious search for the best means in every field replaces nature, chance, instinct. The aggregate of such means produces technical civilization. Men become preoccupied with this search, a science of means is developed, and no activity escapes the technical imperative.

In moral terms, Ellul is tracing the growth of consciousness and selfconscious critique in human life. But whereas Marx saw this as growing freedom for achieving human ends, Ellul's warning is in effect that human ends are thrust aside and only means pursued. Yet there is no saving the situation by the merely dialectical argument that he is unduly separating means and ends and describing a situation in which one end—

efficiency—replaces other ends. For if, in Deweyan fashion, we recognize the continual interpenetration of means and ends, the predicament may be even worse—our very ends become altered in the interaction, and technique is itself the all-powerful end, our ordinary human ends being reduced to means to feed it!

Ellul's study attempts to trace the linkages that alone might justify the expansion of the concept of technology to that of technique. These are empirical claims, such as the growing primacy of applied science over pure science and the incorporation of the latter into technique, or the growth of a technical state of mind as an historical phenomenon. He outlines the several factors that explain the growth of technique, both economic and intellectual. And he traces the imperial conquest in one field after another—the management of economics, politics, the social environment, art—with some consideration of its strategy and tactics, up to man's final locking into his artificial creations. In Ellul's constructs and claims Prometheus stands condemned for the very contributions by which he would try to justify himself—and perhaps with Kierkegaard as prosecutor.[13]

How then shall we choose, for the purpose of studying the relations of technology and morality, between a narrowing of the technological concept, as in Marx's procedure, and an expansion of the concept, as in Ellul's notion of technique? In principle, the path is clear: responsible analytic method requires that at the outset at least we make the conceptual boundaries of technology as sharp as possible. The definition may be changed eventually, but only after the empirical claims are established. If the anthropological concept of material culture is now too narrow, a wider sense of technology as the application of science for production of things men need or want or for the doing of jobs men want done (as in medical technology) is probably as far as we can safely venture at present.

The limits of what constitutes technology can be seen if we scan its conceptual boundaries in different directions. On the one side is pure science, whose theoretical knowledge technology draws on, and for whose experimental work technology constructs instruments. (Any claims for specific types of relations have to be independently established.) On the other side is the use of technology, determined by independent human purposes, although the experience in use constitutes technological experience and contributes to technological knowledge. Thus it is one matter to develop a technology for supersonic transport, or for chemical warfare, and quite another to decide whether to use it or not and in what way. It is quite conceivable that men may decide in some fields to develop the technological knowledge and even to apply it to acquire some experience, and then decide not to use it thereafter. Just as in human history there have been bad reasons for not developing a feasible technology (for example, to maintain the profit or control of a group in power), so there may be good reasons for not using technological powers we already have (for example, to keep down a level of pollution). The claim that all inventions are bound to be used is a quite separate historical hypothesis and prediction.

In the relation of technology and society there is also a boundary line to be marked. This is between the technology as a mode of producing things and the sociological organization in which it is embedded. For example, the relation between technology and city life has been a close one in the past, and it may have been an unavoidable one in the growth of large-scale technological use. But the conceptual boundary has to be sharp if men are to envisage other possibilities and improved modes of organization for their lives.[14] Whether these are bound to be idle dreams is an independent question for argument. As we shall see, this kind of question is crucial today in the relation of contemporary technology and large-scale corporate forms.

Finally, there is a boundary to be drawn between technology and the habits of mind that are associated with it. The claim that all technology involves subservience to the single ideal of efficiency as sole end is a conceivable one, but whether it is an unavoidable connection or the product of a particular historical subservience of technology to business interest is precisely a point at issue. It is, for example, possible that the traditional myth of the value-free character of technology as such rests on the usual isolation of different enterprises in which technologists obediently serve, and that a noncompetitive bringing together of enterprises for multiple social ends may break through such habits of mind.

Such sharp boundaries help give a greater conceptual clarity to the notion of technology. They may rescue it from wholesale adulation or wholesale condemnation. In any case, it gains an initial independence so that its relations to science, to social purposes and to social organization, to habits of mind, and to morality require separate and empirical study and evaluation.

Similar considerations apply to the question of whether technology determines morality or whether morality can shape technology. Marxism gives a grudging but growing place to human conscious effort, and so to morality as part of consciousness. Ellul appears to have morality tied hand and foot. If we had time here to canvass other traditional philosophies of history, the variety of theses would be much enlarged. For example, liberal philosophies of history give values a central propulsive role in human life; they are almost the independent variables for human action. In conservative philosophies of history the moral component is even more basic. It is almost the independent variable for explanation as well as for action. In short, no consideration of the relation of technology and morality in the twentieth century, and even more with respect to the onrushing twenty-first century, can make a dogma of one-directional determinism between the two. There is no doubt a large measure of causal operation of technology on morality. But it is still an open question how far men becoming conscious of this causation can affect the direction of subsequent relations in the light of their values.

We can turn at last to the actual interrelations of technology and morality. In the historical vista of the last few decades we can distinguish three fairly clear stages. Let us trace them in retrospective mood.

The first stage was the sudden vista of abundance ushered in by the development of atomic energy, beginning in the 1940s. It was envisaged as a growing one-world economy, developing the resources of a total globe to serve human needs. (If it seemed momentarily to be centered on furnishing incredible weapons of war in a growing cold war atmosphere, this was regarded as a temporary aberration.) The moral impact of the new technology was thus in the first place as a vision of a world in which the means of survival, health and livelihood, with their material resources, at long last were to be achieved, the occasions of human conflict were to be fundamentally reduced, and the opportunities of material and cultural progress open equally to all mankind. The effects upon at least three of our four moral themes were tremendous.

The moral community could be broadened to the whole of mankind; no serious moral theory could now center upon a partial group such as nation, class, and certainly not (in the light of the Nazi experience) race. No basis remained for the major discriminations which had haunted mankind, in which an exploitative aim had always appeared through many rationalizations—though the central discrimination against women had still to be raised to fuller consciousness.

Distributive justice, in a context of potential plenty, could operate with principles of allotting to each according to his need, not just to his work or according to the returns he gave to social well-being, which might stem from an initial chance distribution of skills and gifts; and it could allot burdens according to ability to bear them. Indeed the growth of technology and economy in the second quarter of the century had already steered toward a collectivization of distributive justice in field after field, more openly in socialist countries, more piecemeal in capitalist countries. Obvious examples in the United States beginning with the New Deal of the 1930s were unemployment insurance, social security, welfare provisions, medical care in one or another form; and of course the United States was far behind in most of these respects compared to many of the European countries. In the unleashing of production after World War II and in the growing computerized automation of the late fifties and sixties, it was moved to the point where systematic political programs of guaranteed national income could be offered by all sections of the political spectrum.

The impact of the new dream on the character of human relations was more complicated. Let us look back on this perennial and central problem of morality. The moral ideal at its fullest has always been one of love, mutual help, and shared creative experience. The two major deviations (one might almost think of them as "heresies") have been the voluntarist idolization of individual self-assertiveness in which human relations become purely contractual, and idealist immersion of individuals in causes or group loyalties, in which moral relations stem from the grip of the common ideal. There are some genial moral elements in both deviations, and there were no doubt historical causes for their exaggerated character. Thus the individualist heresy grew up with the rise of the bourgeoisie and justified itself as the path to initiative, competitively forced discovery, and generally beneficial production. The idealist heresy taps the moral side in which men reach beyond their selfish selves.

The character of human relations envisaged in the first stage reflected to some degree the particular historical situation after World War II. The war had been fought, in the consciousness of men, against Nazi brutality and oppression. It had involved widespread cooperation and organization, and it had penetrated almost the whole globe. One of the immediate aftereffects had been the breakdown of colonialism, and the awakening of Asia and Africa. It brought a heightened awareness of different cultures and religions and ways of life requiring cooperation in a new and shared universal tolerance of differences. The vista of technological abundance offered a double hope in human relations—the improvement of quality by a diminution of aggression and the organization for common human purposes in a reinvigorated and extended democracy in which all men would participate. The early hopes for the United Nations fed into this conception.

The impact of the vista of abundance on the fourth moral theme was less revolutionary. The picture of the good life was not affected, so much as the hope of moving toward its achievement. For the level on which the good life was envisaged at this point concerned the basic necessities—bare freedom from the want which had characterized so much of the submerged peoples of the world and so much even of the developed societies under the impact of war. If ideals of liberty, of education and access to knowledge, and opportunity of achievement, entered into the picture, they were formulated in an abstract way—the development of the capacities of all human beings, as against their restriction and destruction.

To the impact on these moral themes may be added a kind of general insight, or hindsight, into people's own moral reactions that the vista of abundance helped bring about. It was a retrospective realization that many of the elements in our traditional morality had been addressed to a world of utter scarcity, and that a thorough canvassing would be needed to expose these (hopefully) outworn components. Thus a historical survey could make clear in ethical writings themselves that the fact of basic scarcity in human life had been reflected in a repressive attitude to basic human desires and in an ideology of their grasping and irrational character. Repression in one's internal order as well as in the social order had become almost the mark of the moral, and to give way to desire was still equated with the temptation to moral evil. The whole fabric of morality had prefaced itself with the conception of man as his own worst enemy, and the whole fabric of politics with the picture of man as a wolf to man. How much of human institutions, moral and social beliefs, had ultimately no other justification than that they served to bring resignation or repression that a now outgoing scarcity had in the past made imperative?[15] An age of moral doubt, and if we were fortunate, of moral reconstruction, was in the offing.

Such was the first impact of the spurt of technology on moral principles—the universalization of the moral community, the collectivization of distributive justice, the humanization and democratization of human relations, and the fulfillment of individuality. It was an old story, but somehow ever new, and now it seemed nearer and more realizable.

The second stage spans barely two decades. On the technological side it is a record of headlong advance—one might have said with breakneck speed if the adjective did not sound so uncomfortably and literally threatening. It encompasses not merely computer technology and automation in industry mentioned above, but space science, electronic refined developments, medical technology, beginnings of biological genetic technology, and area after area of specific advances in all directions. Nevertheless the general impact is not one of continuing the dream of abundance from the first stage. If not abandonment, there is at least a hiatus in the dream. And its moral consequences have been extremely complex—not merely striking a note of despair, but stirring up basic questions especially about the expectations of human life and the content of the good life, and even a moral revulsion against technology itself.

The usual hypotheses to explain these phenomena no doubt carry us part of the way. They are the obvious diversion of resources for needless and immoral warfare (clearest in the Vietnam war), in which inhuman killing and brutalization continue on a technologically improved level; and the wastefulness of a competitive profit-seeking economy in which social well-being is continually sacrificed to individual gain. But correct as these causal analyses no doubt are, the problem has gone far beyond them to question the aims and procedures common to both capitalist and socialist societies in the modern world. Involved in the critique is a realization of the uneven development of technology itself, of the scope of unintended consequences and the extent of "external" costs, as well as the costs of technological organization. The rise to central consciousness of the problems of overpopulation, pollution, war and structural violence, and corporate control over the individual, constitutes a focal point in a general mood of almost total alienation. Let us again explore this in the context of our four moral themes.

The universal moral community has not lost force. On the contrary, the growing technological unification of the globe has continued to render any lesser community obsolete. But new threats have grown on fresh horizons—the overpopulation of the community, the simple danger that there may be too many of us! Will an overcrowded planet make life morally cheap, as in overcrowded countries death has sometimes become a casual event, with the value of the person pushed into fantasy in a myth of promised reincarnation? And meanwhile, moral battles are being fought over population control, over the "naturalness" of birth control or over the legitimacy of abortion.

The collectivization of distributive justice has gained tremendously in momentum in the second stage. It is evident both in the expansion of socialism and in the new forms of property that are arising in such capitalist countries as the United States. Just as in the early part of the century fresh devices of corporate growth brought about in effect legal forms of property through incorporeal instruments—patents, securities, copyrights, and modes of corporate control—so today there is a vast proliferation of tenure rights, job protection, welfare rights, and modes of protection which are quite comparable. The moral collectivization of distributive justice is thus equivalent to a public assumption of a guarantee of social justice. If anything, it has gained in strength through the growing awareness of the so-called external costs of technology. This is the realization that a large part of the actual costs of production, in terms of the pollution resulting and the lowered quality of human life that ensues, has been borne by the public rather than the people who have profited from the production. The new general threat to the moral collectivization of distributive justice comes therefore not from any inner abatement of momentum, but from the general threats of pollution and depletion of resources, that is from the very technological base whose progress made its development possible.

The third theme—the humanization and democratization of human relations—has been the scene of the most violent reactions. The new technology has been taken to unveil itself as the support of a total impersonality and alienation on the one hand, and as an antiindividualist corporate authoritarianism on the other. Hence comes the now familiar picture of a man crushed, not liberated, by technological imperatives. In part this involves the assumption that a large-scale technology is unavoidably accompanied by a large-scale social organization (whether socialist or capitalist) which cannot be other than authoritarian if it is to be effective. But sometimes, even more, the inhuman element is traced back beyond technology into science itself which is said to involve an impersonal aloofness from value concern and to treat human beings as objects, not persons. Insofar as science itself is deeply rooted in men's conception of truth and reality, the growth of science, so far from being seen as a great height in human achievement, becomes seen as preparatory to the self-destruction of humanity. While only a minor theme as yet, this outlook has played into the hands of a growing irrationalism in human affairs.[16]

Such analyses of pervasive alienation and pervasive authoritarianism are particularly crucial because they make these evident phenomena—for there is no doubt about the brutality and the oppression—essential consequences rather than accidental consequences of the very trends from which people had drawn their hopes and dreams.

Most striking, and perhaps most profound, has been the fourth moral theme. Whereas in the first stage, the old ideals of success and achievement were unquestioned—rather their actualization was sought in greater degree for the mass of the people—the new tendency, most marked in the so-called "counterculture" but extending far beyond such groups, is to reject the aim of accumulating material goods, to question the strenuous purposiveness of the moral tradition, to substitute the intrinsic value of present experience, to focus on present satisfying human relations. It is not to be equated with a hedonism of the moment, which in its older forms had an exploitative or egoistic grasping quality, although sometimes the two are confused. It is rather a thorough questioning of an ideal of success and achievement which required postponement and self-repression and enslaved beauty and love and enjoyment to a puritanical concept of success.

Two further components in this basic questioning can be distinguished, one concerning the motivation, the other the object of desires. The motivation of success has, in recent centuries at least, been competitive: to succeed was not merely to find satisfaction in doing or in accomplishment, but in standing out above others. Perhaps such ideals played a part in gearing men to intensive effort in production, but in contemporary retrospect they have an alien and alienating quality. The object of desires is often derided as the accumulation of material goods, perhaps in an overgeneral way failing to distinguish liberating material techniques from the sheer accumulation of the gadgets and tinsel of a plastic age. Perhaps even here, the critique basically concerns again the moral element in the motivations underlying the accumulation—the power and show in large and constantly changing automobile models, the flow of mechanical and plastic toys forced upon our children in such a way as to breed passivity rather than imagination, the stereotyping and isolating effects of television. Such critiques are familiar enough, sometimes as romantic diatribes in a fantasy of an unspoiled nature, but their moral point is clear. They warn us that in the acknowledged interaction of means and ends, it may be that our technological age will produce enough to satisfy our desires; but the desires they satisfy will themselves be the debased products of the technological age, not the exalted ideals with which men entered upon their noble struggles for liberty and equality and opportunity and abundance.[17]

The second stage is still with us, though it seems to be passing. Much in its picture is overreactive, but the questions it raises are deep and abiding. Its contribution to a sophisticated conception of the relation of technology and morality is a considerable one, even if it came in thunderous or apocalyptic form.

The third stage may be conceived of as the growth of *critical awareness*. It is emerging in the midst of the controversies engendered by the brusque impact of the second stage on the dreams of the first stage. Its virtues are sobriety, realism, and a search for limits, but most of all for an awareness of the presuppositions of human directions. The outcome is an openness of outlook which seeks what room there is for moral influence and what influence in turn the limits have on human morality. Let us preface the consideration of our four themes in the third stage with a few of the intellectual lessons that have emerged.

First, there is no more place for a fragmentary approach to the historical material of our concern, whether technological or moral. We cannot rest on a "law of progress" or a "law of development" for either technology or morality treated in isolation. We need nothing less than a full systematic analysis of a totally developing process in the full interaction of its components. This does not mean that we may not distinguish the moral from the technological or that everything gets mixed with everything else into an undifferentiated mass. It is rather that relations and interactions have to be included in the picture of the development of any components.

Second, man is not something outside of the system of the natural world. Just as his bodily processes take place in the natural world and affect other things in the world, so his ideas and values are capable of affecting changes, and are in turn affected by what is going on. In this sense morality just as much as technology is part of the man-nature process.

Third, among the striking events in the whole process will be hard choices that human beings will have to make. The growth of technology makes certain alternatives possible which were not possible before, and much that would in the past have gone on as a matter of course—for example, that certain people die of a certain illness—could now be avoided or postponed, but at a given social cost balanced by other social losses. These are not necessarily new types of decision; in the past, deciding that a war should be waged, and with a given draft system, meant probable life or death to given segments of the population, and in the same way the decision to use certain technologies entailed the probability of a given proportion of accidental deaths. What is now more likely is the necessity for self-conscious choice where before there had been unknown accidental factors at work. There may be a certain moral temptation to wish that nature took away again the human responsibilities of choice— and even the proposal of a randomizing machine to decide who benefits and who loses. But such restoration of nature's blindness can only be justified where it is shown that men are no better off for conscious decision. Its use would thus be a result of responsible decision itself.[18]

Fourth, in a world undergoing so much change, there can no longer be a bias against intervention by human action where it is clear that there is sufficient basis for guidance or control. For the most revolutionary thing at a given point may even be to try to hold back change or to ignore its direction. The old advantage of social conservatism as applied to institutions and social forms is thus obsolete, and conservatism often turns into its opposite by becoming counterrevolution. But in both cases— nonintervention as well as attempted control—one will certainly expect some unintended consequences. Hence both paths require constant watchfulness. It cannot be left to a preprogrammed decision any more than to blind nature.

With these methodological reservations, we can look to the relation of technology and morality in the context of the four themes considered above, though in only a summary but hopefully not too cursory a way.

The universalization of the moral community faces the problems that emerged in stage two, but is not basically overthrown. The ancient moral dreams of the brotherhood of man (or a slogan more palatable to Women's Lib—the siblinghood of persons) are basically strengthened by the technological interconnection of human interests and communication on a global scale. It is conceivable that fresh developments in technology could make possible a world of small communities wholly independent for their support, just as some scientists once dreamed that the development of light metals from universally available energy and clay might make every nation self-sufficient and end the older wars for coal and iron.[19] But it is likely that the technological substructure on a global base required for such isolation would work against this. The realistic problems for the universal moral community lie in the dangers of overpopulation. Here the swift reaction in a great part of contemporary morality has been toward planned population control. It is amazing how swiftly moral attitudes to birth control have been transformed under technological development of methods. And it is quite likely that further developments here will undercut the traditional religious arguments about what is natural and unnatural in such techniques. The abortion struggle in morality today is a quite different issue. It is tied in with conceptions of killing, with moral rights of women to their own bodies, and so forth. But the whole abortion controversy represents a relatively short-range issue which may be superseded with more effective contraceptive technology and more effective education. The collateral question of sexual abstinence or sexual gratification and its relations to reproduction is also a quite different matter. The so-called moral revolution in sexual attitudes has turned from a reaction against older repression to a search for new moral forms of interpersonal relations. Its chief distinguishing mark at present is its mistrust of the older justifying bases for repression. In this respect it is an excellent illustration of the tremendous consequences of technological innovation in precipitating moral change. The next great area of hard decision here will probably be the looming possibilities of genetic control in reproduction.

As suggested with reference to stage two, the collectivization of distributive justice is not likely to be reversed, and there are excellent grounds for its enhancement if technological processes carry through the promise of abundance. Even if such growth comes to an end, there are good moral reasons for maintaining the moral achievements. For in a period of stabilization or material retrenchment there is always the danger that aggressiveness in the seizure of scarce resources will recur unless a deliberate pattern of social decision on the lower level of material conditions is maintained. The problem is no different in principle from the lessons about rationing and black markets learned as early as World War II.

The current reactions against the idea of economic growth are perhaps extreme, because so wholesale in character. As in the question of pollution, we are in danger of reverting to romantic conceptions of unspoiled nature and of idolized primitive conditions, but there is no unspoiled nature in this sense. The sober question is rather the tolerance level for specific types of changes with specific choices within specific limits. There has probably been some intellectual gain in looking at the world as a space ship within which a total systematic approach is needed to maintain balance. And the criticisms of economic growth have revealed how narrow is its conception of a gross national product, as diagnostic even of our economic well-being. (It might be helped, but perhaps not much, by a concept of a gross national happiness product, if the concept of happiness were not itself already captured in Madison Avenue advertising.) But in the long run the question of whether the world we live in is a closed system or a relatively open one may depend on science and technology itself. If we were to develop a clean source of tremendous energy—whether solar energy or some combination of diverse sources—then the present tendency to see our world as a closed system might be reversed. And insofar as numerous moral rules hang on such a presupposition, the direction of a segment of our morality depends on the path of technological success.

In fact, however, the reverse determination is beginning to play a significant part, and it is not simply in the speculative terms of a moral choice of resigned moderation as against expansive happiness which then inhibits scientific research and technological endeavor. What we find is a withdrawal of the assumption that technological expansion is of itself beneficial; there is a growing critique in terms of human well-being directed at the consequences of its various projects. Thus we are witnessing proposals even among scientists to phase out the nuclear industry; fears are voiced about the development of recombinant genetics, and the impact on the natural environment of the petrochemical industry (leading even to some suggestions that parts of it be phased out). The concept of "intermediate technology" for countries of the Third World has been offered not as a stopgap until "big technology" becomes feasible there, but as a kind of technology more suitable for a qualitatively different and possibly richer life. Such phenomena may not be decisive against large technology itself, only against its direction of development and employment in its present alliance with corporate business; but the moral basis of judgment is nevertheless becoming primary.

The third theme—the humanization and democratization of human relations—poses the most uncertain questions. Do impersonality and alienation, bureaucratic subjugation and loss of dignity, follow unavoidably from our modern technology? Or are they the heritage of the system of administration that has accompanied the growth of industry, or the victim of the conflicts between economic systems, or the expression of traditional authoritarianism in our social structure? Certainly if we look at logical possibilities and make sharp boundaries between technology and social organization, between technology and the purposes that guide its use, there would seem to be ample ground for combining technology with conceivably less dismal alternatives. There are attempts at decentralization with more democratic controls. Necessary centralization can be associated with democratic control of basic uses. Imaginative speculation can certainly work out relationships far different from the ones that exist. But the pivotal question is feasibility. Today much of the theoretical criticism of large-scale authoritarian structures, whether socialist governmental ones or private corporate ones, comes from the residues of the older anarchist social philosophies which keep springing up phoenix-like in fresh forms. This thought has shown a greater vitality than one would have expected, since the time when Marx charged anarchism with being the remains of a lower bourgeois ideology of private property. Its strength today comes largely from the revolt against alienation, dehumanization, and authoritarianism. Its stress on mutual aid, on cooperative community, and the naturally affiliative in human relations gives it the ideal moral cast that other social philosophies often appear to lack. Its achievement in contemporary thought has been largely to prompt the search for alternative paths in the very midst of the battles of socialism and capitalism. And it is quite conceivable that this contribution will effectively keep alive the wider range of possibilities in working toward the future, a contribution that in the past has sometimes come from the anthropological presentation of varied social structures or from the historical study of changing forms.

In any case, the current breakthrough of a moral individualism almost like the older voluntarist heresy but in a new setting and with a new direction is a remarkable phenomenon in our day. It is in many respects like the liberalism of John Stuart Mill a century ago. Each man's life is his own, let him make his own decisions and work out his own pattern, joining freely with those of his fellows he wishes, and do his own thing. The move for democratized decision in all areas, while far from successful, shows the depths of human resistance that authoritarianism begets. At times it is left merely with the power to destroy. But the theoretical lesson for social philosophy is the superiority of actual men over their institutional creations. In a world in which more and more becomes a matter of human decision, the ideals of moral autonomy, democratic decision, and maximal freedom of decision point to fundamental moral directions, however much they have appeared to run against the technological grain. It is oversimple to plot the curve of technological organization as going from the small-scale to the large-scale, and dismiss all individualism as myth. Technology itself might do better to recognize that it has been entrapped in a system of social alliances with authoritarian and corporate structures, to disentangle itself and set to work on the technology of freedom and democratic action.

Reference was made earlier to the contemporary existentialist revolt against the view that life is a craft or a set enterprise whose rules may be discovered and permanently enshrined as morality. This is not in fact simply an existentialist revolt. It is found in a number of different philosophies of man of varied sort—the religious concept of the dignity of man which is not reducible to its roles or functions, the materialist belief in the evolution of matter producing fresh qualitative levels each with its own laws and necessities, the Deweyan type of pragmatism which recognized the pervasiveness of change and human beings as the center of decision in the succession of novel problems in fresh contexts of life. Common to them all is a presupposition of man as the source of crafts or enterprises, as the creator of purposes. If this converging view of man is not without foundation, then Ellul's view that to propose a direction for technique is to deny technique [20] cannot be conclusive. Whether his pessimism is well-grounded is a separate question, but it cannot rest on an inability of human beings, once they become conscious of the purposes underlying and serviced by existent technique, to criticize, reorient, and struggle for an altered set of purposes.

As for the fourth of our themes, the re-reckoning of the human good, this clearly cannot be denied human beings, especially when they look back upon the procession of human goals at different stages of human development and come to realize that their goals are their own, made by themselves not arbitrarily but in a context of human conditions in the natural world. It is a mark of the depth of this growing consciousness that the attack on technology has had to work all the way back to an attack on consciousness. For, as we noted earlier, the very act of scientific inquiry is seen as putting an undue distance between man and his object of inquiry, and so breeding a cold impersonality that is most marked when men are themselves the object of study. [21] But such "distancing" is the very mark of consciousness, not of science as such. All consciousness, as post-evolutionary theories of mind have made clear, arises originally in some frustration or block to the steady discharge of impulse, it produces a situation in which tentative trial in imagination is substituted for action, and in the long run it becomes the distinctively human mode of reconstructing the world and experience. It is always, of course, possible to argue that consciousness should be barred from some areas in the name of instinct or direct immersion in the immediate. But such arguments after a long career have had little to recommend their embrace. The re-reckoning of the human good is, on the contrary, one of the most glorious of human tasks. Consciousness of this quest, in spite of all its dangers, is the one way in which a serious sense of alternative possibilities can be developed. Nor can the spirit of inquiry in moral matters be stilled once it has arisen, especially in troubled times. After Socrates, it is impossible to be pre-Socratic, even though Socrates has been put to death. The vastness of changes in the twentieth century has created a moral turmoil. [22] Morality is in the midst of bringing itself into the contemporary world, not in the sense of adjustment to what technology has made of the world, but in the sense of re-evaluating what is possible and what is desirable. It is too early for a complete report, but it is already clear that a great many of what appeared to be its fundamentals—in rules, in attitudes and character, in selected goals to give shape to human life—were really secondary or even tertiary features evolved in response to conditions of life that are in fact gone by.[23] Morality in shedding such elements and altering its shape today may thus be getting closer to what it has always struggled to give expression. Or else it may be in the creative throes of bringing human beings to an advanced point transcending the past. Whichever it be, morality in the firm revolts that it has been waging against its past, against the easy ways of technological development, is not involved in a breakdown but in massive reconstruction. A partnership of technology and morality in critical appraisal and in the setting of new directions is an enormous power for the human good.

Notes

1. Cf. May Edel and Abraham Edel, *Anthropology and Ethics: The Quest for Moral Understanding* (Cleveland: The Press of Case Western Reserve University, 1968, rev. ed.; New Brunswick, N.J.: Transaction Books, 1970).

2. While this is a common philosophical usage, the sociological is sometimes different; it uses "value"for the selective element and "norms" for what the philosophic usage employs "values."

3. Jacques Ellul, *The Technological Society* (New York: Vintage Books, 1967). Original French publication 1952.

4. See, for example, Herbert Spencer's *Principles of Ethics*, Vol. I (New York: D. Appleton and Co., 1896).

5. Thorstein Veblen, *The Theory of Business Enterprise* (New York: Charles Scribner's Sons, 1932).

6. Thorstein Veblen, *The Theory of the Leisure Class* (New York: Modern Library, 1934).

7. John Dewey, "Interpretation of the Savage Mind," reprinted in Joseph Ratner, ed., *John Dewey: Philosophy, Psychology and Social Practice* (New York: Capricorn Books, 1965).

8. John Dewey, *Liberalism and Social Action* (New York: G. P. Putnam's Sons, 1935).

9. E.g. Frederick Engels, *Anti-Dühring* (New York: International Publishers). Part III. See also Karl Marx, *Capital*, Vol. I (New York: Modern Library).

10. Ellul, *The Technological Society,* p. 19.

11. Ibid., p. 20.

12. Ibid., p. 21.

13. Ibid., p. 55.

14. Kenneth E. Boulding, *The Meaning of the 20th Century: The Great Transition* (New York: Harper Colophon Books, 1965).

15. Cf. Abraham Edel, "Scarcity and Abundance in Ethical Theory" in *Method in Ethical Theory* (Indianapolis: Bobbs-Merrill, 1963); also, Herbert Marcuse, *Eros and Civilization* (Boston: Beacon Press, 1955).

16. Cf. Theodore Roszak, *The Making of a Counter Culture* (New York: Doubleday, Anchor Books, 1969), ch. VII.

17. Aldous Huxley's *Brave New World* (1932) is perhaps the best example of the literature foreshadowing this point of view.

18. For a sober view of costs and benefits, see James A. Michener, "One and a Half Cheers for Progress," in *The New York Times Magazine*, September 5, 1971.

19. See, for example, Launcelot Hogben, *Retreat from Reason* (New York: Random House, 1938).
20. Ellul, *The Technological Society,* p. 97.
21. See Roszak, note 16 above.
22. For a consideration of the obligation this imposes on scientists, see chapter 11, below.
23. For a striking illustration, compare the remark of a Cardinal at Pope John's earth-shaking assembly, that the moral injunction to "be fruitful and multiply" was addressed to a population of two per square world!

11. The Scientific Enterprise and Social Conscience

The four papers that follow, written at different times over a half dozen years, deal with a common issue—the social responsibility of scientists or more generally of those who have special knowledge. The first paper carries out a critique of the abstract ways the topic has generally been handled and proposes a more empirical approach. The second paper looks more specifically at the issue for scientists and engineers. The third explores the relations of theory and practice, especially in connection with a number of the professions. The last gives a detailed analysis of the problem in a controversy which involved many sciences.

Collectively the papers offer a critique of traditional approaches insofar as they have tried to settle questions of the social responsibility of scientists by appeal to one or another sweeping principle. Complete freedom of inquiry is demanded as a matter of right without regard to consequences since truth is there begging to be discovered and truth is intrinsically worthwhile. Or else it is asserted that no human activity is without social responsibility since it is born and bred in the matrix of social life, fed with the intellectual tools of a mother culture, shaped and cultivated by a social inheritance, given the free gift of the past, and looked to by the future. The first comes with an occasional twinge of guilt, the second with a lingering regret for the freedom that is being abandoned.

The argument continued, but the scientists did not worry very much for they were firmly entrenched. Science was basic to human progress and as long as there was confidence in progress the argument seemed more confined to "theory." But in our world, events overtake our debates. With the argument still unresolved, science grew to vast proportions, technology remade the face of civilization and its consequences began to multiply problems. At first these were not sufficiently pressing, or else sufficiently recognized, to raise a hue and cry. Some voices might, and did, condemn Prometheus. Yet most were content to have science run a repair department, like Sears and Roebuck, for its products. But the first half of the 1970s witnessed a turning point. One after another, the *costs* of our technological advances as carried out had been unfolded—the dangers of pollution and the unintended deleterious consequences of our very improvements (of which DDT was an early paradigm). Some of these may be charged to technology itself, many no doubt to the intransigence of our social institutions. Now a general attack on the ideal of progress was launched. Models of doom (beginning with the Club of Rome's *Limits to Growth* in 1972) made inroads on assumptions of endless growth. Concepts of intermediate technology were offered to replace large-scale technology. Such attacks and proposals differed from the older irrationalist attacks on science in that their arguments were themselves scientific and did not appeal to some anti-scientific methodology. Whether the result was to suggest a red light rather than a green light for technology and science, or only a yellow light, made no difference in effect. The challenge was a practical one and could no longer be ignored. The problem of the responsibility of science was placed on the immediate agenda. It might even become a matter of legislation. It could no longer, in a favorite metaphor, be kept "on the back burner."

The emerging lesson is that the responsibility of science or of all knowledge is not to be settled by sweeping principles which affirm or deny but that it is a function of the conditions of life, the extent of our knowledge of the world (including what we know of these conditions), the part that science is actually playing in the world, the part that it could realistically play at a given time, the extent to which knowledge can be expanded, what institutions are available for these tasks, and what the costs are and how they are to be borne. Thus included in all these conditions are our moral principles that specify our goals and conceptions of justice. Hence freedom of inquiry and social responsibility can both be among our principles, but the actual responsibility of science is not uniquely determined by either of them. Its determination requires almost a whole practical science of social responsibility, and that is, a distinctive body of knowledge oriented to determining responsibility in different fields for different projects as an ongoing matter—a science as distinctive as a science of the weather or a science of navigation or a science of meeting health emergencies. Perhaps we should call it a discipline rather than a science. I use the term "science" for its shock value, to indicate that it requires a whole body of empirical knowledge, not a set of *a priori* principles or theorems deduced from them, but that there has to be constant reference to conditions of living that undergo change, that it therefore involves continual assessment, not a finished batch of maxims or a code of conclusions.

The development of this thesis in its historical-philosophical perspective is the task of the paper that follows.

* * *

Nearly every group in our society—from business to policemen to teachers—finds no difficulty in talking about social responsibilities. Why has there been confusion about it in the case of the scientific enterprise? At least one reason, apart from historical and sociological ones, has been the conception of science—the strange mixture of the timorous and the lordly stance—which attempted to give a single answer in terms of a particular conception of the lone scientist as intrinsically a truth-seeker. This is both narrow and wide—narrow because it ignores the understanding of the scientific enterprise in terms of its changing historical relations, and wide because the very pursuit of truth itself, if faced as a path in the contemporary world, carries its practitioner much farther than he may think on empirical and historical grounds.

In a time of crisis the problem of the responsibilities of scientists as scientists should constantly recur. What is surprising is that it should keep coming back in the same old terms and with the same old dichotomies, and—in spite of twentieth century philosophy having been characterized as an Age of Analysis—without a clearer analysis of the questions themselves and their presuppositions. It is as if we started with some fixed definition of the scientist, whether the layman's image of a father-figure in a white coat or the philosopher's fallibilistic doubter, and set it against some equally fixed concept of social conscience embodying the usual hard-line division between individual and social. It is as if we simply held up and compared the two pictures and reported that there was or was not a path between the two conceptions.

There is something very wrong with this procedure and its results. Science is a historically changing enterprise and its responsibilities do not flow simply from its perennial features but from its place in a given time and a given level of development. A sense of social responsibility does not take its character from the perennial features of the human conscience alone, but from the whole sociocultural complex in historical development. Hence any picture of the relations of the two rests on assumed pictures of the whole of the human world in its operations in our age. Once we realize this we can see how complex is the scope of our thesis. I shall therefore begin with the changing character of the scientific enterprise, go on to the changing character of social conscience, and finally draw conclusions about the contemporary social responsibilities of science. The sections are labelled: "The Scientific Enterprise," "Social Conscience," and ". . . And . . ."

The Scientific Enterprise

The material and sociological changes in the scientific enterprise (or enterprises) are familiar enough. There are said to be more scientists now alive and at work than the past total in all human history. Science is more systematically organized in its pursuit, though still fragmented at many points. It has large resources. Basic research is now encouraged and subsidized, not merely applied research. Sociologically, science is not a self-determining field. In spite of the occasional dreams of technocracy or the occasional entry of a scientist into the directory of ruling classes, science is the servant, not the Platonic guardian. It has many masters and many strings by which it is pulled, even in the atmosphere of the freer university, and its practitioners can soothe themselves with a truth-for-truth's-sake ideology. We know that some scientists have even shifted from physics to theoretical biology, despairing of any physical research as out of military reach. By comparison to such uses the relation of the scientific enterprise to business and the search for profits seemed almost benign until sociology elaborated the concept of the military-industrial complex. Now, I want to explore the internal changes in the scientific enterprise—both in the theory of knowledge and in the practical attitudes of men—that the progress of science has brought about; it is these changes, I shall argue, that determine the responsibility of scientists in the contemporary world. I want to distinguish four such changes:

- A shift in the view of human interference in the course of events.
- The growth of science to the point where we no longer set theoretical barriers to its possible scope.
- The development of what may be called an "ecological mode of thought."
- An apparent change in the relation of practice to theory in the scientific enterprise.

The Model of Human Interference

Men have always wanted to extend their control of the world and themselves, and in primitive societies we find magical endeavors. But the ideal of science has not always been associated with that of control. In ancient Greek philosophy the ideal of science was the intellectual grasp of the eternal, what could not be otherwise; the purer the science, therefore, the less the extent of human control. The idea of knowledge as power had a slow growth. We can see this clearly in the human attitudes to crisis. There is a kind of weather-model; you wait for the crisis to blow over and hope for the best. We do this whenever we are helpless; economic crises were treated in this way till quite recently. There is an intervention-model: you intervene but only to remove obstacles and hindrances, so that nature can take its course. This was the medical model under the older teleological approach that nature works for the best. With the Cartesian view of the body as a machine the idea of fashioning, interfering to control came to the fore; now some see the body as a mechanism with replaceable, even improvable parts. In the social field, resistance to intervention is old. We may recall Aristotle's story, in his *Politics*, of the society in which a person who moved a change in the laws had a halter put around his neck, and if his motion lost, he was hanged on the spot. The adulation of tradition, whether in the British conception of the common law as a slow unconscious growth or in a Burkean conservatism, has an almost parallel character—to try consciously to remake, to plan the whole, is to exhibit the height of folly. The control-model permeates contemporary attitudes in all fields. This reflects not only the vast expansion of science and the hope of designing improvements, but also the desperate state of many of our problems, in which a weather-model would mean the acceptance of disasters. Of course even the attitude to the weather is changing: the next generation may think such a name for a resignation-model rather queer and inaccurate.

The Growing Scope of the Scientific Enterprise

It was barely yesterday that arguments were still popular about the inherent limitations of science. First a sharp line was drawn between the physical and the human-social, and the latter declared out of bounds because it involved the particular, the subjective, the free will, and the qualitative. Then parts of psychology and the social were surrendered, but the cultural and the historical were ideographic, empathetic, value-ridden. We need not track down all the barriers that were thrust aside. Of course, the conception of science changed in the process: it ceased to be universalistic, mechanical and quantitative in the nineteenth century sense; probability and statistics made their sweep into the human field while generalized and refined mathematical conceptions of order upset the sharp distinction of quantity and quality. The outcome is that the whole domain of knowledge lies open to the attempts of science. To attempt is not, of course, to succeed, but *a priori* and metaphysical limitations seem to be a thing of the past. The domain of ignorance is and will be indefinitely vast. But from a practical point of view it can no longer be used as an *a priori* veto on attempts at knowledge and control. In more stable days, it could be said that no experimental ventures should be made in human life which involved a plunge into the unknown, because disasters might result. Now the same argument often can be urged against foregoing experimental ventures; for the consequences of continuing in the old ways in a rapidly changing world may be quite as unknowable and quite as disastrous. This argument does not justify recklessness in experiment; we are learning how reckless we have been, but it also underlines the recklessness of conservatism too. In short, the emphasis falls on responsible inquiry and responsible attempts at control. The burden of responsibility falls with increasing weight on the scientific enterprise.

The Ecological Mode of Thought

Part of the recklessness has come not from ignorance but from neglect of knowledge in other fields that either already exists or could be acquired. A changed mode of thought is arising which we may call "ecological" because it is sharply illustrated in ecological studies. We have become very sensitive to the way in which attempts at control in one direction have upset the balance of nature in others, as in the case of insecticides and the disposal of industrial wastes, so we now demand that the application of knowledge be carried out in terms of the whole range of relevant knowledge available. In a column in the *New York Times* [1] James Reston quotes Prime Minister Clement Atlee's remark that when he concurred in President Truman's decision to drop the atomic bomb on Hiroshima, they knew nothing about the genetic effects of fallout, though in fact, as Reston points out, H. J. Muller had won the Nobel Prize in 1927 for his evidence of the genetic effects of radiation. Another aspect of the shift in outlook is a demand that one-sided evaluation should not dominate policy; for example, when oil leaks from off-shore drilling, the oil industry may worry about the seepage of salt water into oil, and the seashore population about dispersing the oil lest it cover the beaches. But the chemical used to disperse the oil may have a more deleterious effect on marine life than the oil itself.

On the theoretical side, an ecological mode of thought involves a systems approach, in which there is not only a meeting of different sciences in relation to a particular problem, but there may be a recasting of formulations in the hitherto isolated sciences. This approach may in part constitute a critique of isolated abstract formulation of knowledge itself in an unduly narrowed domain—the fallacy, for example, of the presidential candidate in the late 1920s who argued that the American economy was in fine shape but something happened abroad and it spread to cause the Great Depression. He failed to realize that the very description of an economy in the modern world should be as part of a world-system.

It is probably space research which most dramatized the need for a full picture which combines the work of many sciences. When this is applied to the whole of human living we begin to think of the planet itself as a space ship, a relatively closed system in which the cyclical processes maintaining a balance have to be known and reckoned with if disaster is to be avoided.

The Relation of Practice to Theory in the Scientific Enterprise

Practical questions are playing a greater role in science today. Experiment itself requires more extended use and organization of resources. Developed sciences experiment over a broader field: nuclear tests involve a wide geographic area; medical experiments may require a large population of subjects; economic and political experiments have to take place in the on-going life of a society. The very tools of testing and observation become large and complex technological achievements, whether it be the telescopes of astronomy or the standardization of tests in psychology and the use of computers in behavioral science generally. The field of practical application may itself be furnishing a test in experience which if not a controlled experimental design may nevertheless add weight for or against a theoretical position. For example, the collapse of a bridge brings to a test the strength of the materials, the appearance of side-effects tests the safety of a drug, the day-by-day sessions of the psychoanalyst constitute some kind of check on the theory of therapy. And the subject matter of experiment, especially in social science, may itself be the practical issues of human well-being, so that the experiment itself is one of how effectively to diminish crime or use of drugs or to achieve fewer family breakups. In fact this is so widespread that some theoretical attempts have been made to redefine the social sciences by human objectives, for instance economics as the science of securing high productivity and wide distribution without depressions.

In much of this, where the scientific study is of human beings, the integration of practical application and experiment becomes so close that it seems to be almost two different ways of saying the same thing. Thus in medicine the line grows thin between the medical efficacy of a drug and its experimental effect. In governmental hearings on the contraceptive pill, one outcome was a recommendation that every doctor regard every use of it by a patient as an experiment.

One could draw an interesting parallel between the integration of the empirical element in science and the integration of the practical element, whose symptoms are now evident. In the early history of the sciences it took a long time until experience got built into the notion of the scientific enterprise; before that the model of science was wholly mathematical-conceptual, as in ancient philosophy, and experience had merely an outside suggestive role. In time, the areas that were "merely empirical" achieved respectability as fit subjects for science, and scientific knowledge about our world became regarded as "empirical science." Practical application too has been traditionally conceived as having merely an illustrative role or a facilitating role. But its closer relations seem to bring it near to occupying the place of an insider in the scientific enterprise. The integration of practical application within the complex of theory and experience may mean that the concept of the scientific enterprise is itself being refashioned.

If these four tendencies outlined describe a significant trend in our understanding of the scientific enterprise, there will as a consequence be serious inroads on the traditional picture of science as value-free, admitting of individual devotion to the ideal of truth, but having only external relations to values, social policy, practice—in short, on the view that the scientist as individual or the scientist as citizen may have social responsibilities, but not as scientist. How is such a view of science possible in a world in which the scientific enterprise has come increasingly to take a control-stance, to range over the whole of human life, to adopt an ecological mode of thought, and to bring practical application within the scope of its work? Does not such an emerging view of the scientific enterprise itself demand a social conscience?

It is possible to invoke the metaphysical dogma of the sharp separation of value and fact as an *a priori* barrier to this demand. But as we have seen (ch. 2), it must not be assumed that science is equivalent to fact in such a dichotomy: science may very well involve some facts and some values, no matter how strongly the dogma be held to. In any case, if the scientific enterprise is allowed an internal value of the pursuit of truth, it becomes an empirical matter how far into the value domain this carries the scientist; for he is committed to defending the pursuit of truth as scientist, not merely as citizen or individual. And if the picture of the world should happen to be such that only a particular political policy will preserve the pursuit of truth, and all others will subvert it, he may find himself as scientist committed to political action. Of course, there is the possibility of drawing back. It might be said that while the scientific enterprise, as a human affair, involves values, science as an ideal type of activity which has a place in the enterprise does not. But this argument is, I suspect, a desperate move. It will end up by holding that the aim of science is not the discovery of truth, but only the discovery of theoretical systems to fit accumulated data—that the aim is not even to yield warranted beliefs, but only to show which theoretical formulations are assigned what degrees of probability on the basis of what evidence. This can, I think, be worked out with some refinement. But the result will bear little resemblance to what we think of as the scientific enterprise; it is rather a particular redefinition of science, which has the burden of justification on its shoulders. And it would be circular to argue that it is justified because the value-free character of science is thereby preserved.

Social Conscience

The social conscience of a society can be described as a pattern of assumed and felt responsibility for others as well as a concern for the well-being of people and for the solution of dominant social problems of the age. Every society has some such pattern, as it has a specific social structure and specific social institutions. Individuals may differ in the extent and intensity in which they exhibit social responsibility. But the scope of social conscience, its mode of expression, and the kinds of topics to which it is directed, are historically variable and can be seen as sociocultural formations. The only way really to understand the present character of our social conscience is to see it as the outcome of a historical development of the last few centuries.

By the seventeenth century, a new pattern of conscience was in the making. We need not enter into the background of the emerging economic order in which an acquisitive individualism became dominant, nor the religious break with the older authoritarian church as a result of which the lone individual directly faced his God. Soon the individual was no longer enmeshed in the guilt of original sin with its weight of obligations and hopeless struggles. He became increasingly an atomic will, exercising his choice and recognizing no obligation that did not issue from his will. This *moral voluntarism* or, in interpersonal and social relations, *moral contractualism*, became enshrined as an individualistic pattern of obligations and responsibilities. It is clearly marked in political, legal, and moral theory. In politics the very state was conceived as contractual in origin: atomic individuals entered with an initial capital of natural rights, and took on burdens only by consent, for the effective maintenance and expression of their rights. In law, the field of contract increasingly took over human relations that had been the subject of institutional regulation; in Maine's familiar phrase, the movement of progressive societies was from status to contract. In the theory of tort and crime, men went far toward shedding fault and responsibility for anything that could not be traced by direct connection to their will-acts or by indirect connection to their negligence.

It is perhaps the abstract regions of ethical theory that show most starkly the character of the shift. The older pattern of duties imposed on men by God's will and applied by derivation from natural law, without consulting individual will or consent, gives way to a primary dichotomy between self and other. In the "other" are telescoped all the intermediate kinds of ties—family, kin, small group, and society at large. Moral philosophers in the eighteenth century, faced with Hobbes' stern egoism, attempted to justify benevolence, that is, to persuade the individual sitting on his rights and interests to stretch his hand toward others, his non-self. They seemed to think, as Hobbes himself had done, that a sober rationality would take a man beyond himself, if only to protect himself, and a greater wisdom would find an identity of interest with others, that beneficence would be a good investment yielding appropriate return, or that private profit pursued would redound to public well-being through greater productivity.

These roundabout routes for mustering a social conscience are familiar enough. Nor were they questions of abstract theory alone, for their anxious character reflected the breakdown in traditional ways of handling widespread poverty, suffering, and social displacement. The career of parish relief and poor laws in England, supplemented by Dickens' novels and the bitter history of trade union organizational struggles, are evidence enough. When the twentieth century outburst of industrial progress faced men with the familiar dislocation—industrial accidents, unemployment, poverty, social insecurity—the intellectual equipment for social responsibility was utterly inadequate, and justification for what was socially unavoidable and socially desirable had to be fashioned almost afresh.

I need not recapitulate the familiar story of the growth of social responsibility and the struggles, both theoretical and practical, that were waged to secure workmen's compensation, unemployment insurance, social security, welfare support, medical care, extension of educational opportunity, and so on. The general character of the moral shift was that something formerly conceived as a matter of individual responsibility or individual hardship (to be unable to get a job used to be felt as a personal failing and to have an accident in industry was one's own tough luck) became conceived of as a burden to be socially carried, paid for either by contribution of those who stood to gain by the work or though the general social tax fund. It is a sombre paradox that humane treatment was often argued for not on the ground that a man was a fellow man, but that he was a factor in production whose depreciation should be borne by those who gain from using him, just as they had to stand the losses in the wear and tear of machinery. But of course this presupposed that men, unlike worn-out machines, could not simply be thrown on the scrap heap, or would not endure being so thrown. Nowadays even the literal scrap heap has become a problem of social responsibility and the debate goes on whether pollution is to be faced as a social problem met through the tax fund or through throwing the burden as "external costs" on the enterprises that produced the pollution, as a normal part of their operation. But perhaps the best example of how far we have moved in developing a pattern of social responsibility is the current consideration of a guaranteed minimum annual income. About half a century ago, Bertrand Russell, in his *Roads to Freedom*, advanced the idea of a "vagabond wage," a minimum support everyone should be given. ² Russell assumed enough people would want more than that, and so keep the wheels of industry going, and his justification was that those who were content with little because they wanted leisure and philosophy and the pursuit of the impractical should be allowed this option. Compare this today with Milton Friedman's advocacy of a "negative income tax" to assure a minimum income.[3] The grounds are quite different—an attempt to cut through the welfare system and increase individual control over his own spending, a realization that the economy has to give at least minimal support to people and can afford to do so. Friedman's advocacy is of course from the premises of the political right in American economic thought; the center and the left have other grounds. But all three are moving close to some form of guaranteed annual income. However diverse or murky the roots, the scope of social responsibility is clearly expanding.

The growth of a social conscience in all these ways did not, however, spell the end of the individualistic tradition in morality. Strangely enough, it is becoming more, not less powerful, and taking over provinces hitherto marked as social. Perhaps the most extreme form of individualist reconstruction is seen in the rise of individual responsibility *against* authority and the state, as contrasted with the older social conception of patriotism and obedience and loyalty. A number of diverse forces had fed this reliance on individual judgment. One is no doubt the weakening of patriotism as a dominant binding relation in the development of the wider loyalties of a growingly unified humanity. A great share of causal responsibility goes to the discrediting of the mystique of the state in the evidence of Hitlerism and its deeds; this is best seen in the outcome of the Nuremberg trials of the Nazi leaders, in which even disobedience to military commands is enjoined where basically immoral action is commanded. Konvitz, in *Religious Liberty and Conscience*,[4] called attention to the fact that this principle of the Nuremberg trials is now established in international law, that the Universal Declaration of Human Rights by the United Nations includes recognition of conscience apart from its relation to religion. Even in Catholicism with its doctrine of papal infallibility there has been the recognition of conscience in Vatican Council II, not as a new right but as a continuing traditional doctrine. Konvitz concludes that the case for constitutional recognition of conscience in the United States is even stronger than that which supports freedom of association. I think that a third factor elevating individual judgment lies in the lessons of experience with intellectual repression—for example, such impositions of ideological dogmatism in the Soviet Union as the notorious Lysenko affair and its domination of genetics, or our own experience of the drive for conformity in political dogma in the so-called McCarthy period of the 1950s. Writing at the opening of that decade, in his *The Fear of Freedom*,[5] Francis Biddle, who had served as Attorney General and surveyed the growing hysteria, was led to question the whole idea of disloyalty to the government. If government is the servant of the people, how can a man be expected to be loyal to the government rather than the government as servant loyal to the master? Summoning Josiah Royce's conception of loyalty in his *Philosophy of Loyalty*, [6] Biddle concluded that men are loyal to their ideals, and that ideals cannot be dictated but are the individual's own choice.

In the 1960s a fourth factor was added in our experience—the Civil Rights movement, in which legality was on the side of discrimination, and later, the opposition to the Vietnam war, often extra-legal in its tactics. The growth of civil disobedience as a technique of social change has thus been rapid, as well as the movement to give greater legal scope to conscience—for example, to allow conscientious opposition to a particular war, not merely to war itself, as a ground for draft-exemption.

The ambivalent attitude to individual judgment in contemporary society seems to reflect conflicting forces. On the one hand the growth of corporate enterprise and large-scale organization presses for conformity. On the other hand the complexity of the technological and social organization and the weight of problems and the rapidity of change in all fields of life demand a high degree of inventiveness, individual initiative and a constant stream of new ideas. And so we have almost the paradox of nonconformity becoming a conformist demand. Probably never before has the weight of individual decision and the multiplication of occasions requiring it been accompanied with so little social guidance.

In its theory as well, the individualistic form of morality has been growing rather than receding. The treatment of morality as *autonomous decision* was already central to Kantian ethics. He gave it the special form of universal legislation by the free individual for the community of rational beings. Since then the legislative aspect has moved into the background, but the decisional element has become more and more pronounced. This is not a feature of any one type of ethical theory, but fairly common to diverse types. Thus a naturalistic ethics like Dewey's, with an integral stress on the scientific and the social, pinpoints a really moral problem not as one in which a man is fighting temptation to do what is moral, but one in which he is moved by opposing principles or values and has to decide what is the moral thing to do; and traditional ethical theory is recast by him in a methodological vein to provide the best available mode of decision in a flux of human, social and individual, problem situations, under changing historical conditions. And a subjectivistic ethics like Sartre's, postulating complete human freedom, also focuses on the momentousness of present decision, with no reliance on a God, a human nature, a past trend of choice, a dependence on others' advice; to reach out to any of these is itself a choice. Sometimes even analytic ethics—Hare, for example—finds every invocation of a principle to be a prescriptive decision, as sharply as in Sartre's view.

Yet, though individual decision and individual responsibility are the central focus of theoretical development, this is no longer the old individualism of the atomic self, cut off by initial stipulation from society as its opposite. Dewey's individualism rather proposes individuality, the rich development of the person as a social goal for education and morals and social institutions. And Sartre's intense focus on this individual has him assume responsibility for all that is immoral around him. A man cannot, says Sartre, shift off responsibility for a war that he had no part in making; he could always be asked what he has done to stop it. For the individual conscience in the moral philosophies of today social responsibility is central: it is no longer peripheral or simply a good business transaction. What is happening is a long overdue breakdown of the individual-social dichotomy. Both the growth of our knowledge of man and the development of our complex interrelated modern life make this dichotomy less significant for understanding what a man is, what kind of self he develops and what his obligations and responsibilities are. The dichotomy is recognized more and more as the historical cleavage of a particular type of life and society which is becoming outmoded. It is not yet clear what kind of categories will emerge as central in ethics and in thought. At the present time that of the active or creative, as against the static, looms large, but this too may be reflecting the intensity of change. Yet it does contain the permanent lesson that man's self-knowledge is an active point of self-reconstruction rather than a learning of what is already fixed by nature. This lesson was already clear in the nineteenth century. In historical terms it is found in the Marxian conception of freedom as the growth of human awareness of the laws of the world and man which enables man to make greater progress in attaining his human values. In individualistic subjective terms it is clearly stated by Kierkegaard when in *Either/Or* he contrasts the Socratic moral maxim of "Know Thyself" with his own maxim of "Choose Thyself."

Morality is self-making and society-making and there is no cut between the two. The growth of social conscience in the contemporary world represents a profound transformation in the life of men, breaking into their consciousness and reshaping thought and sentiment, and creating the opportunity for a freer reconstruction. Whatever be the precise historical and social forces that have brought it about, it has a growing firmness which imparts to it the voice of judgment. It is therefore with this conscience and its demands that the scientist must reckon as he attempts to shape—whether to expand or limit—the responsibilities of his profession.

... And ...

On this view of conscience, what role should the scientist take, what responsibilites should he accept and assume? On the one hand, the scope of scientific knowledge suggests a greater share in the social conscience; on the other, the high standards of evidence and the disinterested character of scientific inquiry suggest distinguishing sharply between the scientist and the citizen and assigning responsibility to a man as citizen or as individual, not as scientist. In the latter case, too, he might even as citizen plead draft-exemption from social activism on grounds of occupation.

There are two ways to deal with this line of argument. One is basically revolutionary, upsetting the categories and dichotomies in terms of which the question is framed. Thus it may be said that the role-playing which distinguishes the man as scientist, as citizen, as individual, is becoming an increasingly meaningless game, that it will go the way of the older distinctions between the economic and the moral man or the self as individual and the self as social. There are particular moral problems of conflict in virtue of different relationships, but no general partitioning of the person and his responsibility; man and human life are becoming too integrated for that, and even in the past such distinctions were never more than relative isolation of systems and practices in a basically unified human life. The second path is less drastic: even if one wishes to preserve the distinctions between the various roles, the decision about what social responsibilities fall within which role is itself a scientific one, contextual rather than general. I think in the long run the first is the more profound: yet to be more than a general insight it would have to work out its detailed modes of assigning responsibilities. The second path, however, is the one I want to pursue here.

Suppose then that the scientist argues against taking a policy stand on social matters because as scientist he is aware of the vast amount of justifying evidence needed in authoritative judgment; one has fewer cognitive responsibilities if he judges social matters as a citizen or as an individual, since it permits more subjective judgment. The difficulty is, however, that on many questions the scientist knows the central evidence only as a scientist—the genetic effects of nuclear fallout as a biologist, the inflationary effects of the Vietnam war as an economist, and the psychological effects of ghetto life as a social psychologist. As a private citizen, he might have had aberrant notions. Of course, part of the evidence may come from other scientific fields, not his own. And part may indeed be just his belief as a layman. If these scruples stand in the way of expressing a scientific social judgment, the scientific thing to do is not to plead subjectivity and individual bias, but to be more precise about the extent of his evidence and specify its credentials. Thus a particular social stand by biologists might be advanced with the addendum: seventy percent as biologist, ten percent as relying on economists, twelve percent as general intellectual (all intellectuals presumably having a more sharpened sense of evidence or relevance), five percent as citizen (in terms of accepted social obligations), and three percent as individual subjective conviction.

Think of the generally educative effect of such pronouncements. If a classification were developed for social judgments, think of the height of sophistication if the public could respond to a flaming headline—"Political Scientists Issue Grade B Condemnation of Federal Pollution Policy; Ecologists Support with D Resolution."

Sometimes I have the impression that the scientists' plea for exemption from social judgment as scientists is a normative judgment quite parallel to an occupation's plea for automatic exemption on the ground of its social importance. Scientists are too busy for political activism, or incipient rebellion. Yet here again, the answer is unfortunately not open to antecedent determination. Whether or how much rebellion is involved is an empirical matter and depends on the state of the country and the character of the issues. In Nazi Germany, to make a biological assertion about lack of evidence for Aryan superiority, was probably equivalent to revolt. And in the Oppenheimer case, as we recall, it was a scientific hesitation about the feasibility of the hydrogen bomb that played some part, as well as moral consideration of the consequences of pushing on with its development. But a large part of social action that can fall into the province of scientists is scarcely of this dramatic kind. Many social questions are not a matter of introducing new and revolutionary categories, but of shifting some area among mutually acceptable categories. Thus if ecologists want a nationally directed water policy or economists and sociologists want a governmental housing industry, they need not be voting on socialism versus capitalism. The categories exist within our society; for example, our army is a collective institution—we do not advertise a war to be waged by the lowest private bidder. (The postoffice is a more customary example.) Nor was the suggestion made some time ago in New York that subway rides should be free, an anarchistic-collectivist aspiration: it was simply saying that subways should be the same kind of municipal service as garbage collection. Certainly these are social-science issues in large part. I am reminded of the clarity with which, if I recall a newspaper account rightly, Milton Friedman, when he was testifying for the negative income-tax, cut through the remark of a Senator to the effect that at least people who got public money in this way should forfeit their right to vote. Friedman replied that if putting one's hands in the public trough warranted loss of the vote, business men would be the first to lose it.

As to types of social responsibility for the scientific enterprise, a number of different ones may be distinguished. They are considered along two dimensions: distance from the center of scientific work, and locus of responsibility—whether they would fall on individual scientists or be more effectively carried out by scientists in associated groups.

There are, first, obligations that arise in the pursuit of the scientific work itself—not simply the moral obligations of truth and scrupulous evidence, but obligations in the professional and public milieu with respect to the work itself. For example, Bentley Glass [8] lists such obligations as: to publish one's methods and results in such a way that another may confirm and extend the results; to see that one's work is properly abstracted and indexed; to write critical reviews in the field; to communicate to the general public the new great revelations of science; to transmit the knowledge to the succeeding generation. Such obligations follow from the state of the field as well as the general objectives of the enterprise; thus proper indexing rises to importance because given the stream of present contributions the dangers of work being lost in plenty are very real. Again, the obligation to ensure communication to the general public reflects the tremendous importance of a wide base of public understanding if the lessons of science are to play a part in the advance of culture and social life; this obligation is distorted if scientists think of it only as a way of ensuring financial support for science. Every scientist need not be busy on all these fronts. Some of the obligations can be carried out in an organized professional or even institutional way—for example, the rise of scientific journalism as a profession itself rather than an an additional burden to a scientist who may not be gifted in this respect. While there is no scientific obligation to be polemical about conflicting theories and approaches, the obligation to do critical reviews seems to suggest not only the wider purview of the field but the participation in sharpening theoretical approaches.

There are, in the second place, direct social responsibilities to others who are involved in the work or come within its ambit. Types of such responsibilities are extremely varied. Familiar examples of such responsibilities are those of medical researchers to subjects; psychological experiments which involve lying to or misleading the subjects (e.g., the extreme case of the Milgram electric shock tests, in which the subject is told to increase an alleged electric shock in order to see where he will revolt and draw the line as he watches a faked tortured response); relations of anthropologists to informants in native villages whose ordinary relations may be quite upset after the researcher's departure; questions of invasion of privacy of informants in modes of research and modes of publication; participant observation as a technique and its effects.

There is further the general responsibility of maintaining the conditions under which science can be continued. This may become a matter of direct political participation where the general freedom of inquiry is threatened.

Other issues may have a comparable status. For example, the imposition of secrecy on research projects where they are connected with military or political applications has been much opposed by scientists as a hindrance to the free flow of scientific communication. The imposition of political qualifications on scientists as a condition of engaging in research is often seen as disruptive of the community of science and its professional criteria. There is no advance way of knowing what kind of conditions may turn out to interfere with scientific work and progress, but when scientists individually or in organized fashion oppose these conditions, it is as a scientific responsibility or an exercise of a scientific social conscience.

Moving gradually into the social context of scientific work, it would seem to be a scientist's responsibility to know or be aware of the various social relations of his scientific work—how it is supported and financed, what practical purposes motivate the support and the work, what applications are likely to be made of it and who will benefit and who be affected in what way. So far I speak merely of the obligation not to remain in the dark on these matters. Many such questions have often been raised—for example, whether certain psychological work is primarily for increasing the efficacy of advertising by finding the depth hold of certain symbols; whether in a given period British anthropological study might have been furnishing the knowledge for the maintenance of empire over native peoples; whether Project Camelot was not (in spite of the latitude for disinterested research it would allow to reputable social scientists) primarily a preparation for preventing and suppressing often needed revolutionary changes; whether particular research in some areas might not be sponsored by industrial interests in order to secure patents and hold back marketing the products to avoid competition with present processes.

Knowledge about one's scientific work and its context would seem to carry some responsibility for decision—whether to abandon the research under these conditions, to do it but make public or agitate against its intended applications or to work out alternative ways of carrying it on. With the development of such large-scale problems in our scientific culture, paradigms may well be established in the ethical code of the profession. For example, research in biological warfare might well have been banned by scientists even before its partial rejection by national edict. Many fuzzy borderlines still remain to be dealt with. It is not inconceivable that a union of engineers should include in its bargaining with a given corporation provision for disposal of waste that should not add to pollution of the environment. Just so a teachers' union may include in its bargaining the provision of school breakfasts or lunches for children—in part because of the help this gives to the educational process, in part because of the general obligation for the welfare of those affected.

Where there are crucial problems affecting the whole life of the society, it may well be a responsibility of all intellectual, scientific and cultural leadership in the community to ask itself what it can do to help face the problems. Thus in our contemporary world one could pinpoint the problems of war, discrimination in its various forms, overpopulation, and the pollution of the environment and exhaustion of natural resources as the four great threats to mankind. Hence there would be no question about the scope of social conscience in general with respect to them and about the obligation of scientists to ask themselves what their fields could do to ameliorate the situation. In fact, the obligation of scientists here is greater because of the part science has played in generating the situation, even where its action was directly beneficial as, for example, in increasing life expectancy by reducing infant mortality, and much more so where it gave the instruments to blindly acquisitive business institutions. Two excellent examples of the exercise of this obligation of science in crucial problems and threats are: the reaction of anthropologists in the 1930s to Nazi racialism and the agitation by atomic scientists in the 1940s and 1950s for controls of nuclear power against war uses. Geneticists nowadays are much worried about the breakthrough in their field and the questions of control over human biological development it raises.

Let me conclude with a few reflections on the modes of action a sense of social responsibility among scientists may call for. Again there is no simple answer. We may distinguish individual action, informal group action, and action in structured associational groups. Individual action may take the form of public criticism or withdrawal from a field of work, or engagement in some form of political action. Informal group action has tended to be ad hoc; it is a familiar feature of our society to see advertisements of scientists on the question of the Vietnam war, or on overpopulation, or occasionally even on some particular flagrant injustice.

Organized group action is less developed. Sometimes there is the exercise of negative fighting functions, parallel to strikes by unions for specific demands; this has not been employed very much by organized scientists, but is quite conceivable in the present state of things. There is also the exercise of what may be called a ferment-function, to generate all sorts of new ideas and plans and intensify consciousness of the problems and possible solutions (for example even the minimal educative function of showing how decisions are actually come by). Thirdly there is what we may think of as institution-making, which has in some sense been more common than we may think. Thus the development of insurance as an idea was a mathematical discovery which underlies vast social transformation in modern societies, though not directly applied by scientists themselves. Group medicine was an invention of medical practitioners. The development of clinics for psychotherapy and the growth of schools for mentally ill children arose from the work of professionals and readily passed into government programs. Recent attempts to organize the poor for taking part in a concerted pursuit of their own welfare also had professional origins. There is nothing implausible in current suggestions that organized scientists market their own discoveries for public welfare, for example, in drugs or even in certain industries. We may compare the fostering of housing and banks by certain unions, or even the suggestion that Harold Ickes made after World War II that what the government had built up for industry during war production should be turned over to a corporation with all veterans as shareholders instead of being sold at a cheap price to industrial corporations. Of course such suggestions run up against the realities of basic power, but the amount of free play in our society would be tested by social experiments along these lines. There are large avenues for the legitimate exercise of the social conscience of scientists, far beyond the mere expression of a collective voice where there is one. There could very well be sections of the scientific societies (or of the whole AAAS) on institution-making, and on international scientific cooperation, for example on implementing the abolition of biological warfare. It may not even be too early to think about the possibilities of international citizenship for scientists. But that would add international responsibilities.

Notes

1. *New York Times*, Sunday, January 4, 1970.
2. Bertrand Russell, *Roads to Freedom* (London: George Allen and Unwin, 1918, 2nd rev. ed. 1919), p. 179. The background of his proposal is discussed in chapter IV.
3. Milton Friedman, *Capitalism and Freedom* (Chicago: University of Chicago Press, 1962; Phoenix Books, 1963), ch. XII.
4. Milton R. Konvitz, *Religious Liberty and Conscience: A Constitutional Inquiry* (New York: Viking Press, 1968).
5. Francis Biddle, *The Fear of Freedom* (Garden City, N.Y.: Doubleday and Co., 1951).
6. Josiah Royce, *The Philosophy of Loyalty* (New York: Macmillan, 1908).
7. Soren Kierkegaard, *Either/Or*, Vol. II, trans. David Lowrie (Garden City, N.Y.: Doubleday, Anchor Books, 1959), p. 263.
8. Bentley Glass, *Science and Ethical Values* (Chapel Hill, N.C.: University of North Carolina Press, 1965).

12. The Social Responsibility of Scientists and Engineers

Most professions have traditional patterns of responsibility embedded in practice, morality, and law. To inquire into them at this point is to ask how systematic or well-structured they are, how they are justified, or again—in the light of changing contemporary attitudes or the emergence of fresh obligations—how they should be altered. We need not draw sharp lines between these aspects. The present world is undergoing revolutionary transformations in all directions, and it is natural enough as well as pressing for scientists and engineers to ask what is expected of them and even more what they are to expect of themselves.

That to be an engineer is to occupy a social role and so to have social responsibilities would be questioned by no one. If there is a momentary twinge of doubt about the scientist it is because the earlier image of the scientist had him outside of the standardized social roles. He was an independent inquirer wandering wholly on his own in the pastures of knowledge, nibbling at what attracts him, and responsible to no one for his intellectual decisions. No doubt even then an anthropological or sociological conception of social role would have little difficulty in discerning a definite pattern, but that is beside the point. Today such a romantic image would have to be cast as a proposed answer to the question of responsibilities in the social role, not as an antecedent denial of a social role. Certainly society today educates the scientist, supports scientific work, anticipates the continuing contribution of scientists and is more than irked—indeed threatened at its foundations—if the scientific contribution and the stream of scientists is reduced to a trickle. If scientists turn out to be in a marked degree more independent than is the case in many social roles, it is because such an independence is required for the success of their work. And if their social responsibilities should turn out to be minimal—and what they are is the object of our inquiry—that too would be dictated by the conditions of their functioning. The social requirements, not the individual desires alone, here call the tune. The history of many fields has followed the same pattern of the growth of consciousness. First earlier beliefs in absolute individual rights, then grudging concessions of social responsibilities accompanying such rights, finally there is a recognition that the role itself is a social one in a web of social responsibilities, and that the apparent individual niche is in fact a social office held under social conditions. The history of the idea of property is perhaps the most striking example of this progress. The scientist, the artist, the philosopher have been the most recalcitrant in facing their own social roots.

Responsibility in any vocation is determined by at least three interrelated factors. A first is the extent of accumulated knowledge and particularly the degree to which it is systematic knowledge. It is always those who know more who bear the greater responsibility, whose judgment we rely on in situations of risk or danger. A second is the pattern of purposes or aims which generates or supports the work of the vocation, just as the goal of health underlies the doctor's work, sanitation and an adequate water supply the plumber's, or protection the locksmith's. So too the needs for fresh knowledge to improve our human lot and solve our pressing problems underlie the scientist's work. This does not entail gearing the scientist into a "knowledge industry" for it may very well be that far from assimilating the pursuit of knowledge to the practices of industry we would do well to assimilate the conditions of industry to those in the pursuit of knowledge. A third factor determining responsibilities is the state of the field to which the knowledge is being applied in the light of the pattern of underlying aims. For example, the work of the politician depends not only on the principles of politics, however authentic they may be, but also on the character of the people he is attempting to lead—for instance whether there is a cultural homogeneity or ethnic variety, whether there is a high or low level of employment, etc. The work of an economic advisor is far different for a country with vast resources than for one wholly dependent on supplies of energy from outside, though the same economic theory is employed and same aims prevail. Of course the three factors I have specified are not to be sharply cut off from one another. Knowledge is required in interpreting the state of the field as well as in understanding the purposes at work, human purposes permeate the pursuit of knowledge and the development of the field, and the state of the field advances or hinders the development of knowledge and the advancement as well as the very articulation of purposes. It is an interdeveloping whole at every point, not an encounter of distinct forces.

An important initial consequence of this brief analysis is that social responsibilities of both science and engineering are not to be captured in an eternal formula, for all three contributing factors have undergone great change even since the times of Gallileo and Newton. The growth of systematic knowledge certainly has, not only in physics itself but in the emergence of biological, psychological and social science. The world to which knowledge is to be applied is itself far different in almost all its aspects. And even if one would claim that human aims should be taken as constant, the forms of their expression in the interplay of factors have altered considerably. The consequent role of science (including applied science) has moved from the peripheral to the central; science supports not only the present character of our lives but the very means of maintaining life itself. One could almost use as a paradigm the reversed relations of city and country in the dramatic changes in agriculture. Once rural agriculture occupied the majority of the population and cities depended on the country. Now in the United States less than five percent of the population are directly engaged in agriculture. Instead of the city depending on the country, agriculture itself depends on power, fertilizer, marketing, shipping, and in some areas on seasonal population movement. Central to contemporary agriculture is the work of science and engineering; new strains of seeds and irrigation are only the more striking features. The very sense of security which enables agriculture to put, as it were, all its eggs so frequently in one basket, planting a single crop over limitless areas without thinking about possible mishaps that older generations constantly faced by varied plantings, is a practical vote of confidence in science. It is taken for granted that if some new beetle turns up to threaten the crop, scientists will be on the spot to track it down and stop it before incalculable damage is done. These are only a few ways in which the structural foundations of life today depend on science and technology.

When we survey the whole global scene in this light, when we see that in a reliance on science and technology whole cultures are changing their form, what a strange contrast is the nonchalant avowals of social irresponsibility that have—though in very different ways—characterized both scientists and engineers. Science and engineering have taken the easy way. Scientists have shed all other responsibilities by proclaiming allegiance to the pursuit of truth, irrespective of the consequences on human life, by denying even the obligation to trace and reckon such consequences. If this noble quest, stricter than the strictest monogamy, has characterized the scientists, the engineers on the other hand have simply transferred responsibility to their employers, foregoing the right to have ends of their own. Let us consider each one separately.

Scientists have used different arguments in confronting those who would assign them social responsibilities. One is the separation of fact and value. Taking their cue from a dominant intellectual trend of the first half of our century they have accepted the value neutrality of science and limited themselves to the domain of the allegedly factual, leaving value to either arbitrary personal commitment or outside authoritative imposition. A second argument has appealed to the division of labor and the limited character of expertise in each of the fields of knowledge. A third, long-standing argument began by trusting to progress to take care of things and then flipped to the opposite view that since we cannot predict what the consequences will be in any case, the conditions for a judgment of responsibility are nonexistent.

A philosophical and sociological analysis of the belief in the value neutrality of science would see it as a grand experiment proclaimed in the early part of our century consequent upon a presumed gap between fact and value or the *is* and the *ought*. It arose in the emancipation of ethics from traditional theological and metaphysical binding and its enunciation of moral autonomy should have meant the right to reassess our moral and social obligations in the light of our growing scientific knowledge of the world and the changed character of our institutions. But in fact it overshot the mark and was directed against science as well. Hence value was cut from *all* existence and all understanding of the factual world. So confident was this hidden metaphysics that it asserted value and fact to be utterly different in their nature or their logic or their essence. It thereby assumed that what was really a proposed program of separation was already entrenched as a success. The grounds for rejecting this half century experiment as unsuccessful are complex. Let me simply assert that its failures are by now manifest in many pathways of philosophical inquiry, and it is now ignored rather than argued about. Sociologically, the dogma served on the one hand to relieve the scientist of responsibility for the social consequences of scientific finds and applications, while on the other hand it allowed others to dictate the directions and uses of scientific work. It is little wonder that science was dominantly guided by the demands of war and power and profit, while the work on problems that would improve the quality of life for the mass of the population was left to chance.

Appeals to the division of labor as a ground for not assuming responsibility recognize rightly that value judgments about a scientific discovery or an engineering accomplishment depend on more than the work of a particular scientist or a particular field. Even in a classic case such as the introduction of the automobile, one would have had to be a prophet to guess the results that actually followed for the character of life. An economist might have hazarded an optimistic view about increased employment, but could a political theorist have foreseen the power of the auto industry in national politics, or a sociologist or psychologist the disruption in traditional sexual patterns, or geographers and biological scientists and chemists the spread of pollution in the atmosphere? What was a mere engineer responsible for as he gazed on the motor that he had wrought? Granted that at the time we did not as a people have sufficient self-consciousness to inquire into social consequences as we plunged ahead, does it follow that we can learn nothing from our own history? To appeal for absolution from responsibility because cooperation would be required to conjecture the consequences is simply to throw responsibility on others. If a man can be saved from drowning by three men one of whom has a boat, the second a rope, and the third can swim, it is wanton neglect for each to maintain that he has no responsibility if the man drowns. So too, if scientists can exercise responsibility only by cooperation, let them work out ways of cooperating. This becomes more urgent as the consequences of today's scientific accomplishments are comparably much more dangerous, as the many disputes about the safety of nuclear installations or about the possible effects of supersonic travel on the ozone layer of the atmosphere have made clear. Dangers penetrate even to the earlier stages of research, they no longer wait to become issues about the effects of the finished product. Thus there has recently been public discussion of the fact that a committee of biological scientists warned fellow scientists to defer a line of pure research because gene transplantation of the kind envisaged in that line might create drug-resistant germs or new types of cancer-causing viruses. This was treated in the press as a first in public recognition that scientists have the obligation of considering possible dangers throughout their work and should move ahead only when it is sure that disastrous consequnces wiil not ensue. [1]

It should be noted in fairness that many individual scientists have felt responsibility for the social applications and consequences of their work. Notable examples are Nobel worried about the uses of dynamite, Banting about the social availability of insulin, Norbert Wiener lecturing to labor groups to make them aware of the dangers of unemployment arising from computerization, and most clearly the Federation of Atomic Scientists working to secure civilian control of atomic energy. And of course legions of scientists have taken on responsibilities as citizens while refusing them as scientists. But as the role of science in our lives has increased it becomes necessary to work out new forms of alerting people and new ways of exercising responsibility, as well as to relieve scientists of this scientist-citizen schizophrenia.

Problems arising from the division of labor concern not only the larger social issues of responsibility but also the specific responsibilities for a complex job and even the workaday relations of apparently technical issues. For example, the interlocking of architect, engineer, contractor and job foreman may be so complex that it will be uncertain whose responsibility it is if critical defects appear in the finished structure. Who should have foreseen a buckling or a crash? Who should have known what would happen to elevators in a fire high up a high-rise building? If a plastics engineer warns of the fire potentials of his materials but continues to supervise production though he knows it goes eventually into children's toys, can he shrug off responsibility to the business department? Similarly for automobile engineers and the cost decisions that produce an eventual "built-in obsolescence" in the cars rolling off the assembly line. Whose responsibility is it to press for new and better models rather than patch up the established modes of production? To move from the socially sublime to the organizationally ridiculous, I recently had an experience which I cannot help but regard as paradigmatic for our problem. A bathtub had a clogged drain and, it turned out, also a worn-out faucet. Two different groups in the plumbing establishment came to repair and replace. Each did its work but no one worried about the coordination and a functioning bathtub. I ended up with a clear drain and a working faucet, but no way to stopper the tub. The question of responsibility, after this was repaired, no doubt became an in-house matter and I do not know how it was settled apart from the mutual curses of the two teams. (It could scarcely be called an act of God.)

The belief in the inevitability of progress on the whole played its part in the general view that the individual could go on with his narrow purposes in the pursuit of enlightened self-interest (or its more exalted analogue, personal intellectual interest) and history would balance things out. When this optimistic thesis was found inadequate, it gave way to the opposing pessimistic thesis that it was no use trying to figure out the consequences in any change. But nowadays, given the rapidity of change, there are equally unintended consequences in holding on to the present form of things. We could, of course, simply cultivate a fatalistic resignation to whatever comes to pass. We could even draw upon a traditional religious attitude and assume responsibility for everything based on a concept of divine punishment and original sin. But both these attitudes, which in the past have been found in people's reactions to earthquakes and other natural disasters, hardly seem appropriate for happenings in which human activity and human planning have played so large a role; nor are they attitudes with which the scientific temper can be comfortable. There is really no alternative, in spite of the fallibility of knowledge and prediction, to a discriminating responsibility in the light of the best knowledge we can muster and the continued inquiry into those problems that are relevant but obscure. It is one thing to live with probabilities and run their risks. It is quite another to remain wilfully blind.

I conclude that the three kinds of arguments most commonly used to shrug off responsibility are at least in the contemporary setting no longer appropriate. The only path left for scientists is therefore to canvass scientifically the forms of responsibility appropriate to them in the present stage of the character of scientific work and the present general state of our civilization. I need scarcely add at this point that the result will be a whole area of inquiries, not a simplistic principle or a few commandments.

Before suggesting forms of responsibility, let us look back to the engineer. We intimated earlier that lacking a cohesive ideal (as medicine has the ideal of health) engineering becomes the servant with business (in our system) or government in the role of master. This is perhaps oversimplified. An alternative to a cohesive ideal is a pluralistic pattern of specific ideals with well-demarcated fields of application. Thus agricultural engineering (or the whole of agriculture viewed as a human technology) has a quite clear and stable goal of healthy and more plentiful food. Certainly it has value choices to make, for example in the impact of the green revolution on population growth and cultural habits, but its human aims are never in doubt. It is not therefore necessary that engineering have as unified a goal as medicine does. Nor will having one itself ensure social responsibility. The medical responsibility for health has not ensured the profession's allegiance to the widespread provision of health care although it has made plausible criticisms of the profession in the light of its goal. Similarly, the central ideal of justice has not rescued the contemporary legal profession from its current ethical self-reproaches, though it has made this self-criticism possible. The trouble in engineering has been on the whole—with a few areas excepted—that it has acquiesced in a concern with efficient means and has been able to put off specific responsibilities on outsiders who call its tune. A simple linguistic observation will show how deeply this is embedded in our ways of thought: a doctor who helped the construction of Hitlerian gas chambers would generally be regarded as violating his medical responsibilities—he is a "bad doctor." An engineer who prepared and supervised the use of napalm in attacking a civilian population could even be praised for his engineering skill—he could be a "good engineer." The language tolerates an ambiguity in "good" for the engineer which it does not permit for the doctor.

The results of limiting engineering responsibilities so preponderantly to means have been large-scale engineering participation in activities that yield military destruction, environmental pollution, ugly urban sprawl, spreading concrete parkways and parking lots, surveillance devices destructive of privacy, nuclear hazards, and so on in an indefinitely extended list. Not only the utopian visions of technological progress, but the perfectly realistic projects of utility and beauty that technology at its present stage could bring—and of which it gives us occasional tantalizing glimpses—are thwarted so that todays's youth is turned off from engineering as well as science, and technology is increasingly assailed as the enemy of mankind. Even the ordinary comforts, for which ages have longed as they made their way with the sweat of the brow, are now labelled as "crass materialism." I suspect that in self-defense, if for no other reason, engineers will be driven to face their social responsibilities. In any case, technology today exerts too much power in the actual operations of our lives to have it left to the laissez-faire attitude long gone by in economics itself.

Reconstructing Responsibilities

In turning to lines of possible responsibility in science and engineering, I want to repeat the central focus of this study. I am not primarily inquiring here into whether science and engineering always *had* a pattern of responsibility which they ignored. I am dealing with the tasks of reconstruction in a world that is undergoing massive transformation in all its institutions and traditions, and asking in what directions the responsibilities of scientists and engineers should move today in the part they play in determining and supporting the character and prospects of human life. Our concern is thus with problems that are new or at least transformed, not new answers to earlier questions. This is worth stressing. If we had more time, we could scan the intellectual horizons and see the new views of man that have emerged in the psychological, and social, and philosophic disciplines. We could trace the shake-up that is taking place in all fields of thought and in all professions, compelled by the novel developments and the novel problems and tasks of our century. It is no wonder that new obligations and responsibilities will have to be hammered out, not simply old and neglected ones revised. Indeed, effort is required creatively to determine what these responsibilities themselves are to be. By way of suggestion I want therefore to outline briefly:

- The basic moral responsibility to assume responsibilities.
- Some fairly clear intellectual responsibilities, and some more difficult problems of moral attitude.
- The responsibility to develop a scientific *community*.
- How far science and engineering should broaden their traditional list of aims.
- Instrumentalities and agencies.

The Assumption of Responsibilities

It is clear from the argument of this paper that the responsibility to assume responsibilites is not morally trivial. It is a recognition by scientists and engineers that, however much there may be disagreement about what the responsibilities should be, scientists and engineers have a social role which demands the exploration and assumption of social responsibilities.

For many this will be a break, by no means an easy one, with traditional attitudes. It is not merely a moral set of will but rests on analysis of their work in the contemporary world—as has been indicated sufficiently.

Intellectual and Moral Attitudes

Several questions arise about intellectual and moral attitudes. Since the borderlines are not wholly clear, I shall simply treat them seriatim:

The assumption of social responsibilities does not mean abandoning the traditional values of the scientific enterprise. For example, if—as some have argued—the assumption of social responsibilities for scientists may inhibit free-ranging discovery, then devices would have to be worked out to counter such an effect. Available strategies are another matter. For example, they could be organizational, such as institutes of basic research with adequate control over their own findings and their possible use; or educational, such as the consolidation of a public attitude that not everything a scientist discovers to be possible should be immediately turned over to engineering construction, or that not everything that can be produced by the engineer should automatically be put into public production for general use. (The issues about the SST are a case in point).

In dealing with the consequences of their work on human life, scientists will have to cultivate a kind of holistic mode of thought which may run counter to the intensive specialization of scientific fields and problems. This was noted above in discussing the need for cooperation. The consequences of nuclear discoveries are biological, psychological, political and social, as well as physical. Certainly the engineer has to some extent developed this mode of thought in facing practical problems. But too often we still find scientists proposing biological solutions for social problems with an obvious disregard of social and cultural components—for example, in problems of behavior control with respect to crime and delinquency. Ecological studies constitute a breakthrough into a mode of thought which when generalized, takes into account the constant interaction of all factors in an on-going process. We begin to see, for example, that the costs of pollution are a part of the whole cost of industrial production—like raw materials, labor, insurance, etc.—instead of seeing them as external matters, someone else's business. Of course the problem of how costs will be borne is a separable one, but somebody has to pay the cost and the picture is incomplete without it. The interrelatedness of our problems is the realistic basis for the holistic approach, not an intellectual preference. We are not surprised that dealing with inflation will require an integrated program in the same way that the project of reaching the moon did. Sometimes caricature hits closer to the truth than economics. For example, current discussions of the effects of inflation on unemployment often take for granted that unemployment will have to go up to stop inflation and assume that some help for the unemployed will be found elsewhere. Russell Baker in a column in the *New York Times* plays with the question of who shall be added to the unemployed. If these additions are part of the battle against inflation, why should we not use the new principle of a volunteer army as against a draft of the unwilling? Why not invite volunteers for the army of the unemployed, people who consent to leave their jobs and go on relief till the inflation is over. (For the academic world this would be like inviting professors who are eligible to take sabbaticals at half pay when the college is in financial difficulties.) Such caricatures have the effect of looking at the whole scene in terms of what the end picture would be like rather than remaining trapped in the categories and rationalizations that serve to dump the costs on a particular part of the community.

A holistic view in the contemporary interconnected globe leads directly into a global or all-human point of view. Atmospheric nuclear tests had to be assessed for their results through the whole global atmosphere, not simply for what they did to the one country. Scientists would have the requisite realism to explore whether a global agricultural policy is now required rather than the familiar nationalistic and rear-guard actions. Biologists have set the even further problems of effects on the whole evolutionary process—for instance the problem of the trade-off between increased deleterious mutations and increased nuclear power.

Difficult questions will arise when potentially disastrous consequences go beyond the more readily agreed-on matters of food and health and safety to psychological, economic, cultural and even political effects. Here an attitude of scientific realism as against ideology with its dangers of imposing the values of special interests becomes paramount in importance. We cannot expect scientists and engineers to be without national or special cultural interests, but it is clearly their responsibility to face where these enter their thinking and not to confuse ideology with science.

Such a realism should assume a responsibility for facing, not obscuring a conflict of values. For example, the military proponents of Project Camelot no doubt regarded as disastrous consequences any revolution in South America that would disrupt the existing order of American interests. Social scientists in their reactions to the project took quite different interpretations of the situation. In part, they objected to being used for ideological purposes. In part, they were unwilling to subsume the efforts of Latin American liberals to bring their countries up from feudalism to a contemporary democracy under the rubric of communism. In part they had different conceptions of American interests. In any case, in estimating the effects that their research would have had in the project, they had to take a stand on one set of values as against another.[2] To assume social responsibilities is clearly to assume an obligation to face issues of possible value conflict.

In assuming social responsibilites, scientists and engineers will have to work with a sophisticated concept of locating control. It is not just the question of whether the scientist is to be responsible for himself or controlled by his peers or associations or by law and regulative governmental agencies. Obviously there are different areas and problems to which different modes will apply. A greater refinement and richer development of alternatives and the meshing of responsibilities is required. This is part of the responsibility of facing responsibilities.

A Scientific Community

Scientists and engineers constitute a distinctive group in many ways, and are capable of becoming a community. They could share a scientific temper, a respect for knowledge together with a sense of its tentative, self-corrective and cumulative character, the readiness to incorporate new knowledge for practice and to follow out its implications. No wonder that an occasional enthusiastic philosopher (Dewey, for example, or Russell at times or Singer at the University of Pennsylvania) thought that all that was needed was to extend the scientific outlook into the social and moral fields to solve the major human problems. Although this proved too optimistic, there is yet the sufficient basis for common interests of inestimable value to mankind to justify developing a communal cohesion in scientists. This need not be modelled directly on what has been done in the medical and legal profession. It could work out its own desirable forms.

Broadening Traditional Aims

That engineering should go beyond its traditonal aim of efficiency and avoiding waste is clear enough by now. The question is whether the increase of responsibility can take determinate shape. In the contêmporary large-scale technology, certainly safety is being added. Engineers might very well be held responsible if they acquiesce in the course of their work in unsafe structures and unsafe products. Their refusal to do so—whether in nuclear plants, in mining, in automobiles, or in the designing of flammable garments or of dangerous toys for children—might very well be protected and encouraged by legal devices. (How about triple damages for dismissal of an engineer who condemns a product as dangerous?) Whether to some extent beauty should be added to the list of aims is a more difficult question. In some areas—highway engineering, for example—perhaps it should. But the direction of this broadening of responsibilities is clear. It is an attempt to free the engineer from servility, so that he can participate in the judgment of human values with respect to the products of his work.

The broadening of aims for the scientist is a more difficult question. Perhaps scientists could stablize a list of *great problems of mankind* to which, insofar as they became relevant to scientific interest, the direction of solving rather than intensifying them should become paramount. I do not think there will be much difficulty in specifying the problems—sources of energy, optimal population, adequate food supply, adequate transportation and improved communication, avoidance of war, reduced pollution and more adequate preservation of resources. In large measure, much scientific effort already incorporates some of these aims specifically. For example, there is the search for a "super plant" which will combine drought tolerance, resistance to disease, self-fertilizing root, and high yield.[3] The touchy points will be rather the protection of scientists who take issue with national or business policies that run counter to the solution of these problems. There is the grim reminder of what happened to Oppenheimer when he opposed the hydrogen bomb. There are the current efforts to turn back the advances made in the fight against pollution. There are the complex problems that emerged in the UN conference on population, such as the fears on the part of underdeveloped countries that population control was being urged as a substitute for assisting them in economic development. In many areas, especially in the social sciences, the proper direction of science may itself be highly controversial. The assumption of scientific responsibility will entail the scientific treatment of such controversy itself.

Instrumentalities and Agencies

If there is appropriate growth of scientific community to match the extent of present-day social responsibility, a number of instrumentalities for carrying them out may be suggested.

Systematic monitoring is one option. Independent scientific agencies for monitoring rather than sole reliance on governmental or business agencies would be very imporant in many fields—whether protecting the general public or safeguarding consumers. American experience has been that too often the governmental groups designed for this end fall under the influence of the very interests whose regulation is the point at issue.

The whole field of control of patents in the social interest by participation of scientific groups—not necessarily the actual inventor—is well worth exploring. At the minimum it is another area of monitoring, tied to the scientific community by the fact that its members were the creators.

But it is possible that stronger participation than monitoring—perhaps involving some censorship of monopolistic uses adverse to social interest—could be developed.

Medicine and law have procedures of disbarment against individuals who grossly violate professional standards. Engineering could make use of such a tool against gross violators of social responsibilities.

The union weapons of strike, boycott, and the like, could also be made available for scientists both to maintain freedom of science and to meet gross violations of public interest. They could have been used earlier in the battles against biological warfare, and in problems of nuclear warfare. Of course, such weapons presuppose a high degree of organization, and require responsible care.

Selective institution-making, it has sometimes been suggested, could help strengthen the scientific community in dealing with social responsibilities. There is no incongruity in a Mathematics Association running an insurance enterprise—after all, mathematicians provided the theoretical basis of insurance. Some areas of manufacture practically call for the independent scientific touch: manufacture of medicinal drugs is the obvious example. And there is no reason why important areas of research that business has neglected because of its present investment in competing areas should not be undertaken by associated scientists and where important discoveries are made productive enterprises could be instituted and controlled by scientists in associated forms. This is neither capitalism nor socialism, but more along the lines of the traditional cooperative movement, operating in areas of public interest.

Finally, the suggestion sometimes arises that creative scientists belong to mankind rather than to one nation. In the rapid interdevelopment of the globe, a universal citizenship for certain classes of scientists might be a weapon of as yet untried strength for human welfare. This has a sound moral basis, for the intellectual enterprises of scientific inquiry—letting go the vicissitudes of specific institutions of science and temperaments of scientists—does constitute an intellectual habit and demands conditions of work which gather its practitioners into an ideal community. It is as if nature and truth conspired to fix the brotherhood of scientists (or shall I say the siblinghood of scientists?) as more universal than any religious or secular fellowship could be, and make it a harbinger of the kind of community that is possible for mankind.

There are many respects in which what I have been suggesting, so far from being novel, is a return to the early spirit of science with its vista of human liberation in the growth of knowledge. It would be most appropriate on this occasion to think of my proposals as a return to the vistas of Franklin and Rush, of Priestley and Jefferson.

Notes

1. Cf. chapter 14, note 5.
2. For the story of Project Camelot, see Irving Louis Horowitz, *The Rise and Fall of Project Camelot: Studies in the Relationship between Social Science and Practical Politics* (Cambridge, Mass.: M.I.T. Press, 1967).
3. New York Times, September 5, 1974, p. 20.

13. Knowledge and Responsibility in the Professions

We often talk of the importance of knowledge or the use of knowledge in practical affairs. But what if the reverse were the case, that knowledge had to be extracted from the realm of practice, as Galileo wished that he could capture all that was implicit in the skill of the workers in the naval yards and arsenals? Perhaps the use of knowledge in any important area of practice is not merely applying knowledge but adding to our understanding of the character of knowledge itself.

Knowledge (alias *theory*) has had a checkered career since the days when Plato imparted to it the aspiration toward a mathematical model. On the one hand its ideal was to grasp the true, the good, and the beautiful, all in a single unity. On the other hand, it was insulated from the workaday world of the sensory and the empirical. One would have expected that the long centuries of the growth of empiricism would have remedied the situation. But instead the dogmas of empiricism have hardened the philosophical isolation of pure knowledge from applied knowledge, of theory from practice, of thinking from doing. Even the twentieth century has had its supreme bifurcations of formal and material, theory and practice, fact and value—indeed, science and ethics—that have so beset our philosophies.

The theme of responsibility in relation to knowledge is impoverished by these dichotomies. The assumption has been that responsibility is tied to action, not to knowledge. We are familiar enough with the usual disavowals of responsibility so long as we are merely pursuing the truth. Sometimes the value-fact dichotomy is invoked: knowledge is of what is the case, while responsibility is built on value or what ought to be the case. Or thinking gives us alternative possibilities of action; responsibility comes from commitment or choice among the possibilities. Or truth generates responsibilities only when captured by the roles of a profession: the doctor, the lawyer, the teacher, the engineer have responsibilities about what to do with the truth because the aims and practices of a profession generate obligations in practitioners, but truth as such stands above the fray. If we make a profession of the scientific enterprise as the pursuit of truth, its only value is to find the truth and its only responsibility is to know.

The professions are today becoming seriously concerned about their responsibilities. The problem is being forced upon them by their "consumers." In business, Naderism is a far cry from "caveat emptor." In medicine and law the concept of malpractice is almost rampant, and insurance rates have bounded to heights that threaten to bankrupt many a practitioner. Flogging education and demanding teacher accountability have become a popular pastime. The engineering profession is worried by the deleterious effects of numerous chemical wastes long released into the environment, for which it is increasingly called to account. Concern with responsibility used to identify the socially minded professional. Now it may equally mark any professional eager to set limits to his responsibility. It may be just as prudential as the nuclear industry seeking federal legislation to set an upper monetary limit to its total liability for a nuclear accident.

Under these conditions of ferment, the professions constitute a particularly rich area for the study of the relation of knowledge and responsibility. I want to consider first the general relation of theory to practice or—what for our purpose amounts to the same thing—the relation of knowledge to action or of "pure" knowledge to its application. Second, I look at a few of our standard professions, especially how changes in responsibilities come about. In an epilogue, turning to the enterprise of pursuing knowledge, I examine some of our attitudes to it in the contemporary world and touch on responsibilities which adhere to the enterprise itself. The formulation of issues in the second and third parts are consequnces of the theoretical analysis in the first.

I

The traditional problem of the relation of theory and practice or knowledge and its application is far from clear. Some speak as if their separateness can be read off simply from the nature of theory and the nature of practice. Others, attending to the empirical story, are more impressed by the complexity of relations.

There is certainly a sense in which theory and practice are manifestly separate. For example, the theory of nuclear energy is one thing and its application to make war or to make harbors and divert rivers is quite another. Again, it is argued that theory is thinking and practice is doing. (Sociologically, this is the brain-brawn dichotomy, with a host of different attitudes of prestige and status involved.) Thinking is in the realm of mind, doing in the realm of bodily activity.

On the other side, it is argued that theory and practice *cannot* really be separated. This is often asserted on behaviorist grounds: a man's beliefs are not one thing and his behavior another; to know a man's beliefs, look at the way he behaves. Functionalist theory in social science seems to take the view that the meaning of a human institution or custom is to be found in the social forms or type of culture it supports in its social operations. So too an operationalism in the general theory of meaning seems to be tying meaning to practical consequences. Thus it is often argued in social philosophy that the meaning of a social theory is to be found in the practices it permits, encourages and justifies. Marxists, for example, will describe western democratic theory as "bourgeois democracy" because its actual operation supports capitalist relations of production.

On the other hand, some thinkers attack the separation of theory and practice arguing that they *should not* be separated. This clearly supposes that a separation is possible. Their point is simply that we would be better off not separating our theory from our practice.

Such controversies are carried on in a complex and often in a heated way, and this suggests that perhaps the initial concepts of theory and practice themselves are not as clear as we ordinarily take them to be. It is likely that if unpacked they will be seen to carry all sorts of theory and surplus-theory from which the differences spring. For example, it is manifest from these few arguments that belief in the separateness of theory and practice at least in some cases springs from a dualistic conception of mind and body, thinking being aligned with one and doing with the other, while the belief in their relation sees thinking itself as a complex type of doing. Perhaps the pattern of theory-practice relations is itself a function of the underlying psychology of thinking and the underlying structure of our modes of production.

It is helpful to look at other ways of categorizing theory-practice relations in other conceptual frameworks than our own. Take, for example, the Aristotelian *theōria, praxis* and *poiēsis*—contemplation, action (or doing), and making something. This is threefold, not twofold, and we would have trouble correlating it with our usual scheme. To understand Aristotle's distinctions we would have to appreciate his whole conceptual framework. Contemplation is thinking or grasping an eternal structure of what cannot be otherwise—it does not include our tentative inductive generalizations. Yet he treats ethics and politics as a theory of praxis and theory of art as a theory of making. Acting is differentiated from making by havings it end (*telos*) complete within itself; it is essentially a teleological concept. Making has the end product distinct from itself. We should have great difficulty if we tried to parcel out our theory and practice into these three concepts. Would writing a book be a contemplation or a making, or both? Is thinking to be regarded as a form of making if we discover that each act of thought is producing brain changes? Is learning a making or a doing? Is a conversation joint action or joint production? And so on.

There is no reason why we should regard our categories of theory and practice and their relations as less complex, less theory laden than Aristotle's. The theory with which our distinctions are laden, however, is different. If it expresses an underlying body-mind dualism, then it is committed to the sharp separation of thinking and doing noted above. But this gets us into trouble. How much thinking goes into doing without which it would not be human action but simply physical movement? And how much doing goes into thinking which not only expends energy but also—in its preeminent scientific form—involves a reference to experimenting? Is a scientific experiment thinking or doing, or should we place the thinking about doing that it entails on the thinking side? Clearly, the indeterminate zones hold not only for translating our distinctions into Aristotelian terms but in our attempts to apply the distinctions themselves.

It will be helpful to develop the parallel (outlined in chapter 11) between the present problems in the relation of application and theory and the process that the relation of experience and theory went through in the history of philosophical reflection on the scientific enterprise. In the ancient philosophies the distinction was sharply drawn between the theoretical-intellectual and experiential-sensory. Since science was taken to consist in necessary truths grasped by the intellect (on the purely mathematical model), experience had at best the ancillary role of stimulant to reflection, or suggestive source for generalization. It did not have a verificatory role or confirmatory role for universals. It was thus outside the scientific (or rather epistemological) domain, in spite of the expressed desire to "save the phenomena." It did, however, have a secure though humbler status in connection with belief, as against knowledge, that covered all particular judgments. Only later, with the development of inductive theory, did experience come into its own. Its rise, however, came as much from the fall of necessary truths and their equation with Humean "relations of ideas" and eventually with empty tautologies as it did from the elevation of belief to science. In the long run, with the development of probability theory and the acceptance of fallibilism as a mark of science, the whole picture changed.

The idea of *application* is a special way of construing theory-practice relations. It is implicitly authoritarian or at least involves a hierarchical ladder. It assumes that there is already the established theory or knowledge that it is to be applied. The theory is self-sufficient, and if it needs application it is only in the sense in which the master needs the servant or the slave—for achievement of *other* purposes. The theory does not need application in the sense in which it needs experiments or empirical evidence. On the other hand, where the theory is insecure, the line between experiment and application grows thin. We get simply ordinary knowledge, roughly empirical information that is not precise enough to consider as experiment but is cumulative enough to reckon as established belief. Where the application is regarded as a test, the other components in the situation—the motives and values—are no longer the identifying features of the situation; they are merely causal factors which help make the test possible, just as other conditions sometimes completely out of our control (e.g., that there will be an eclipse) also help make the test possible. This suggests that application can be looked on as experiment when there is a focus on the fallibilistic character of human knowledge. Of course there are better and worse experiments, and the difference may lie in the variables under control. Thus the ordinary psychoanalytic session is not the equivalent of an experiment, nor perhaps the ordinary classroom. The best cases of coincidence of application and experiment are those in which the application can only be achieved by the kind of care the strictest scientific standards may enforce, as in delicate surgical operations, or in space shots, or in attempts to control the weather. Human affairs are likely to be the scene of the coincidence of application and experiment, since the purposes we have in mind in application are human uses. We would not be likely to think of an experimental success in such cases as an application if all it furthered was the increase of our knowledge—paradoxically enough, a quite practical aim—although this habit may shift as knowledge as a whole comes more and more to further other human aims. This is precisely what is taking place, with the general acceptance of basic research as a practically fruitful human enterprise. The outcome may well be the integration of application within the framework of the scientific enterprise.

If this takes place—and even with far less than this—the meaning of "application" and "practical" in contrast to "theory" and "theoretical" may well become a relative contextual one, rather than a general categorial one. It will in one context draw a line between some knowledge and that knowledge acted on for a certain class of purposes, in another between a more general or more abstract system and its interpretations in a moral or a welfare context, between a general goal and particular means by which it is sought. Again, what is practice or application in one relation may be seen as theory in another. This is clear in the customary hierarchy of fields. For example, if we line up a pure geometry, a physical geometry, a systematic technology, a blue-print for construction, and the actual operations of the bulldozer, each may be seen as the application of what preceded it; but if we insist on a sharp line, perhaps only the last would be practice. The same relations might be seen between biology, psychiatric theory, theory of therapy and psychoanalytic session.

Several recommendations follow from this revised view of theory-practice relations. One would be to abandon any attempt to give a *general* answer for the relation. Instead, many different kinds of relations are possible for different kinds of contexts. The kinds of relations would reflect the meaning of the concepts of theory and practice and the general conceptual network in which they are enmeshed. It would reflect the historical character of theory and practice themselves and even the results of some of the bodies of theoretical knowledge itself—particularly the psychology of thinking and the intellectual models for practice. It would reflect the purposive structure of human life, both in general and in the special conditions of life at each particular historical period. It would reflect the kinds of areas to which knowledge was being applied and their current state.

A second recommendation from the revised view would be to abandon the claim that the relation of theory and practice has always to be very direct, and the claim that it has always to be very indirect, and to recognize that both patterns as well as many intermediate ones may occur in different contexts. The issue is of practical importance in education. The claim for direct relation is familiar enough in the frequent attempts to formulate philosophical (metaphysical or epistemological) types and assign to each some particular mode of educational practice. For example, it is said that a Platonic metaphysics will mean an authoritarian classroom, or a Lockean epistemology will mean a conditioning classroom. The indirect relation of theory and practice has been perhaps the more pervasive assumption; it is said that theory can never be translated directly, the best it does is to make one sensitive, so that he comes to practice—moral, legal, educational, social—with a greater intuitive refinement. After all, many of the great moral philosophers insist that character is prior to theory, that a man has to develop good habits in order to be a good moralist when his intellect develops. Philosophy especially quickens the spirit, but it is the quickened spirit that acts, not philosophical theory that guides. And the same indirect relation is seen between not only philosophy and political theory, but between political theory and political action; the gap comes between the political scientist and the politician just as it comes between the metaphysician and the political scientist.

That this indirect type of relation holds in some fields under certain conditions of the field is no doubt the case: the relation of theory and practice is routed through the growth of the individual and the acquisition of skill. This is the kind of case in which an apprenticeship system is used; Michael Oakeshott has glorified it for politics as the only mode, branding

Locke's *Second Treatise of Government* and Rousseau's *Social Contract* as cribs for a middle class not born into rule.[1] And Ryle's often used distinction between *knowing that* and *knowing how* seems to rest on a similar separation of the theoretical and the experience acquired by apprenticeship.[2] But the apprenticeship model is only one possibility among others; the appropriate one depends on the exact state of the field and the extent of basic theory and theory of application. Parts of medicine may still be rule-of-thumb, but other parts may be theoretically advanced enough to be applied almost mechanically. Certainly the proportion of the mixture in space travel is quite different from that in seafaring in older days. Indeed, the very kinds of relations between theory and practice may be different according to what theory provides. A successful mathematical model furnished by a theory has one kind of impact on practice; the relation here is a logical one. Where theory is chart-line, it may guide exploration but not predict special obstacles or disruptive factors. Where theory provides instruments, the analysis of the context of application is aided but not given and has to be carried out with the assistance of the theory in each particular context. And many other distinctions of type and of detail are no doubt available. The relation of theory and practice is thus context-differentiated (in the sense that different contexts determine different kinds of relations), context-bound (in the sense that the type of relation reflects the type of context), and so in general context-relative.

A third recommendation is to attend to the more precise analysis of the factors that determine the kind of relation. We need here to pinpoint three dimensions as summing up the kinds of factors generally indicated above. One is the state of knowledge in a given field at the time that application is sought. A second may be called the situational condition of the field to which knowlege is to be applied. A third is the purposive pattern of the persons concerned as participants or else the established purposes of the profession or domain involved. These three dimensions vary to some degree independently, in the sense that separate readings may be taken. The knowledge appealed to may be highly systematic while the field of its application may be extremely complex, or the conditions of the field may be fairly simple while the knowledge is fragmentary and complicated. The purposive pattern may be one of established aims or of even bitter conflict. Of course the relations of the three may themselves be complicated as when purposes differ because of different beliefs about the state of the field or the different possession of knowledge. No fresh absolutistic trichotomy is here suggested.

Human beings engaged in practice in some way work out a synthesis of readings. It is not a simple computation. It is as complex as the way probabilities and preferences are put together in decision theory. We need a concept to refer to the integrated resultant. The familiar concept of *degree of control* will do for this. It means knowledge applied to the conditions of the field to achieve the purposes involved.

The degree of control, as so defined, both determines the kinds of relations of theory and practice that pertain to the particular situation and is reflected in the kind of responsibilities that the situation gives rise to. Consider, for example, the clear responsibility of the nursery school teacher on finding that a few of the children have measles. Contrast this with the uncertainty, in the early days when psychoanalytic generalizations about the dangers of repression were being tried out, about the responsibility for curbing pupil aggression. In the case of measles there is both knowledge of the cause and cure, a mode of identification of who has the illness in the situation and a common purpose to have the patients get well. In the case of aggressive behavior, the teacher might be uncertain whether non-repression would aggravate it, how the present harmful effects stacked up against future character effects, precisely what in the pupils this aggressive behavior represented, and above all this what kind of character was socially desirable.

II

Let us now examine the interplay of the three dimensions taking illustrations from the princely professions of medicine, law, engineering, and teaching. The general hypothesis is that the responsibilitities are increased as the degree of control grows greater.

With reference to the state of knowledge, engineering is clearly most fortunate. Its basis in the physical sciences allows it to share the character of those disciplines. Where technological decisions are precise, it is able to determine its responsibilities with a high degree of reliability. Its uncertainties come from the impact of conflicting purposes, to be noted shortly. In various fields of medicine and on various types of problems, the range in knowledge goes from the completely adequate, the fairly adequate (sometimes sufficient to select several paths of action but not enough to choose decisively between them) to the situation where it is better to take a stab at doing something rather than do nothing, and finally situations in which there is not enough known to determine any effective mode of action. That is why one field may be near mechanically operated diagnostic techniques while another operates with rule of thumb. Law is in the somewhat similar position. A lawyer's advice can range from the legally correct form of a contract or a will (though not without an occasional surprise) to predictions in which the court will prove as arbitrary as organisms in their response to treatment. And a legislator's bill is always fraught with unintended consequences. Teaching is perhaps least well off. Knowledge of how learning takes place is controversial both among teachers and in the psychological base itself. Routine learning through drill still struggles with attempts to make the mind take fire. Reliance on the intrinsic attractiveness of subject-matter competes with success motivations or a devious route through pupil identification with the teacher.

With reference to situational conditions of the field, there are some philosophical complications. The picture of a field is not given in direct observations, but is a way of construing what is going on in some synoptic fashion. The history of medicine shows this perhaps most clearly. As we saw in considering models of intervention and control (ch. 11), at one time medical thought looked on the organism as a self-enclosed teleological process in which the natural tendencies interrupted in illness sought to restore their inherent balance; the doctor's task was thus to remove obstacles to nature so that it could carry out its cure. At a later time, under the influence of the Cartesian separation of matter and mind, the medical model became mechanistic: the body is a machine made up of parts and the doctor's work is to repair and even if possible to replace defective parts. Contemporary medicine is in some fields still using this conception, but it is increasingly pressed to replace it with one that will integrate body-environment transactions allowing of constant feedback. Much of the difficulty in psychiatric theory comes from the conflict of different views of this type. Engineering, since the development of Newtonian science, has been largely mechanistic, though the kind of mechansim has changed with the development of the theory of electricity. Perhaps space science is forcing an integrative view upon it. The legal profession has also had some transformations of its fundamental view of its field of work. Traditionally it was supposed to be concerned with the conflicts of people. Set in one or another political system which selected the kinds of conflict to which there was to be legal attention, law was responsible for developing methods of trial and decision, in line with the aims of the law. Increasingly, with the growth of the legislative side of the political life and the greater articulation of social aims, the law has been given the more general field of providing some forms of social control and social order. Teaching has had a constant field, for the phenomenon of a new generation to be "encultured" is a perennial one; but the amount of enculturation to be carried out through schooling has grown with the complexities of life and the social needs for knowledge, skills and specific attitudes. Nevertheless the character of the field has been envisaged through different models of the nature of its student material: natural egoism and anti-social tendencies to be curbed and disciplined, purely plastic raw materials to be fashioned into desirable social patterns, inherently benign tendencies to be found appropriate expression, and so on. Accordingly, it has felt its responsibilities in different ways—mechanistic manipulation, agricultural seed-scattering, interpersonal dialogue, and social interaction, etc.

With reference to the purposive pattern, medicine has the clearest aim, since the norm of health has immediate interpretation with fairly clear marks. This is probably why, in the Hippocratic oath, it was able early to turn attention to its ethical standards. A general norm does not, however, get rid of problems within its general aim, when field conditions have sharply changed. Witness, for example, the controversy at the present time over whether ideals of mental health are simply vehicles for dominant cultural and social values; problems of abortion, organ transplants, psychosurgery; relations to the aged and the dying; and so on. Many of these have become problems of moral responsibility because of the growth of knowledge and means of control, even though some of them are argued indirectly through analysis of concepts, such as when a fetus is a person or when a human being is actually dead. But the core of such argument is really a moral decision about what the doctor may or may not do. On the whole, in spite of controversy, the medical profession has its inherent basic goal, so that doctors who helped the Nazis in organizing efficient death camps were almost universally seen as perverting medical arts. Even on the crucial question of scope for medical work—that medical services in many countries do not reach the masses of the people—the medical ideal is clear, though the record of the profession less unblemished.

Engineering (cf. ch. 12) lacks an internal goal, other than effectiveness and efficiency. It thus becomes the servant of other aims, furnished by the nation or the prevalent political and economic patterns. Some branches of technology get tied to agreed-on broad social needs, such as civil engineering in construction of bridges and roads. But a great part of engineering has even to compromise its standards of effectiveness to meet business demands for profit, as when the engineer designs shoddy products to achieve a built-in obsolescence. Even more, standards of health are sometimes compromised—for example in questions of safety in mining, and in preparing for possible dangers in nuclear plants and disposal of nuclear wastes.

The legal profession has an integral idea—that of justice. But the conflicts over the meaning of justice and its applications to different areas are so intense that the pattern of purposes is extremely uneven. In some fields, such as criminal law, the institutions of enforcement and punishment are at present under severe criticism for their inhuman effects. Fields of civil law are the scene of conflicts of interest in as diverse areas as between landlords and tenants, industrial polluter and suffering public, taxpayer and civil servant, not to speak of the more familiar conflicts of different kinds of property. Technological change plays havoc with traditional patterns of rights and the new institutions that technology makes possible have broad and powerful effects; witness what the development of xerox and other copying machines have done to copyright laws, and what transnational corporations do to the actual distribution of property rights and powers of control. In public law there has been an emergence of new forms of civil rights and the growth of concepts of privacy (in response to the new technology of surveillance) and individual protection; but the phenomena revealed in the Watergate investigations have shown how far-reaching can be the instruments of political tyranny even in a democracy. On an overall view it would seem that the law at present shares with engineering its lack of a determinate internal purposive pattern; it is rather the scene of a social battleground of classes and interests.

Teaching, tied as it is today to large social institutions, falls somewhat between engineering and law on the one side and medicine on the other. While the concept of learning has perhaps greater definiteness today than that of justice, what is to be learned is itself a function of a great variety of aims. It is unnecessary here to describe to educators what the contemporary social revolution—in which new institutions are coming to be, not merely demanded—is doing to the aims of education, even apart from the wholly new instrumentalities available.

Let me yield for a moment to the old pattern of teaching in which the synthesis of a semester's work was expressed in a set of grades. The following table will summarize the state of knowledge, conditions of application, and pattern of purpose for the four professions we have talked about:

	State of knowledge	*Conditions of application*	*Patterns of purpose*
Engineering	A	A	D
Medicine	B	B minus	B plus
Law	C	C plus	D
Teaching	D	C	C

Such a table will enable us to pinpoint what needs improvement, just as we would tell the parent in the parent-teacher conferences where constructive effort is required. Thus do the readings on our three dimensions furnish us directions of responsibility for the professions—subject, of course, to the vagaries and impressionistic character of the grading process.

Grading, however, is not enough. It has to be accompanied by a diagnosis of the situation which will tell us why the major responsibilities of reconstruction point in a given direction. Why is the central responsibility of engineering today to reorient its pattern of purposes for human ends; why that of medicine to apply knowledge it already has for a more comprehensive health care, as well as progress in basic research; why that of law to break through the narrow defense of established interests to become a powerful tool for social reconstruction; why of teaching to engage in a probably more far-flung reconstruction of its aims and character than almost any other contemporary institution? I offer these suggested paths of responsibility without here attempting to carry out the detailed argument needed to defend them. For example, I can conceive of an engineer saying that his profession would do more for mankind by finding a source of boundless clean energy, such as solar energy, than by all the cultivation of social-mindedness in his socially evasive profession. I would only suggest that the fact this obvious need has not been the object of central effort itself shows where the prime responsibility of effort today must lie.

The diagnosis of our contemporary situation will carry us to fundamentals, not simply because it is a philosophic good to get to fundamentals but because the root of our problems is itself at that deeper level. Let us recognize it frankly: our age is one of widespread social revolution. It is like the seventeenth century with the emergence of physical science and a new world outlook. It is like what happened on the European continent when the French Revolution swept away old institutions in a promise of new freedom. It is like the early twentieth century with its gigantic strides in industry and its promise of growing material progress and democracy. There is the pessimism inherent in the breakdown of past ways and the optimism of the hope of new opportunities. But the social revolution is penetrating and no realm of human life seems able to stand aside. Thus the responsibilities of personal and practical life stem from nothing less than a view of the whole, the form of human life that is sought on the globe, including both the quality of life and its basic principles of social organization. The reckoning of what is now possible and what is now desirable is set in the whole complex of knowledge and self-knowledge as it exists and as it is rapidly increasing.

This basic lesson holds today for all the professions. That is why they can no longer guide themselves by their past internal practices, why a determination of their responsibilities involves even more than self-criticism, and why their understanding of themselves involves a relation to the total state of mankind and its needs. And that is why reconstruction in profession after profession carries it into fresh paths of total reshaping rather than partial readjustment.

This conclusion is of course not itself a determination of responsibilities but an invitation to a way of thinking in the investigation of responsibility in its relation to the three dimensions summed up previously as degree of control. The responsibility for determining responsibilities remains itself a major responsibility for the professions.

Let me add two comments. The three dimensions could all be viewed as forms of knowledge—knowledge of the laws of nature or the degree of systematic order we can find, knowledge of the conditions of existence in terms of which the first kind of knowledge can be applied, and self-knowledge about our aims and hopes and efforts. They need not be viewed in so unified a way, but perhaps to do so will bring them closer together. If they are so viewed, then the concept of knowledge is enhanced and its weight in determining responsibility becomes almost as heavy as when Socrates said that virtue is knowledge.

Secondly, given the interaction of the dimensions, our purposes also are not given, but forged in the processes of life and knowledge. Hence the determination of our responsibilities is only in part the consequence of already established purposes; it is also the active creation of our aims and so of our institutions and forms of life. It might almost appear as if life is itself the application of knowledge, and responsibility the self-consciousness of our human directions.

III

The place of knowledge in the determination of responsibility is recognized, but there are hesitations about knowledge itself. The older social revolutions ended in profound disappointments. The knowledge that seemed to guide them turned out often to be the ideology of a particular class or some special interest. Can we do better today with our greater knowledge and our more powerful technology? Or, as many think, are these new gains further sources of enslavement and indecision—adding responsibilities we are incapable of bearing? Is the lesson that of the ancient Ecclesiastes, that he who increaseth knowledge increaseth sorrow? Our problem is raised to a new height: it is no longer the question of how knowledge increases and sharpens responsibilities but what responsibilities we have for the advancement of knowledge itself.

Several different types of scrutiny need to be distinguished, each resting on a different fear. One rests on a *fear of over-complicating life*. A second rests on a *fear of what increased knowledge will do to man*. A third involves a *fear of ideology*, a mistrust of what poses as knowledge. A fourth while quite trusting the growth of knowledge embodies a *fear of what we will find*.

The fear of over-complicating life may be illustrated from the effects of a single invention. Take, for example, the new methods for dialysis. Before we knew how to remedy kidney disfunction we had simply to become resigned to the fact that diseased kidneys were found among some of us and that death was inevitable. Now we are able at great effort and considerable social cost to save a few, but it is now we, not nature or change, who condemn the others. We have to bear the psychological costs of setting down principles for deciding who shall live and who shall die. Are we really able to carry out such tasks? They are more like tasks for gods than tasks for men. Increasing knowledge solves some problems, but it generates more than it solves. Even if in the long run we develop cheap, simple, and plentiful methods for dialysis, the general effect of knowledge is like that of the intermediate stage, not the happy final outcome. To intensify knowledge is to complicate life to such a point as to render permanent the Age of Anxiety.

This paradigm calls attention to one type of consequence which calls for caution. But there are others in which the consequences of knowledge may be general rescue and the simplification of problems—for example, the discovery of the role of vitamins in human physical well-being. Again, there is no reason why an effort to advance knowledge and control in a given area should not try to think through in advance the sort of problems it will generate—just as bio-medical ethics is trying to think through the question of the consequences of a further breakthrough in genetic control. It is true that growing knowledge brings responsibility of decision, and this may engender anxiety. But it is well to remember, as Margaret Mead once pointed out, that in primitive life the predecessor of the Age of Anxiety was the Age of Terror. It may be quite possible for mankind to assume the burden of responsible decision in many areas, recognizing that all decisions involve losses as well as gains, and tempering optimism by the degree of resignation required.

The fear of what increased knowledge will do to men takes two forms. One attacks science for making man arrogant, the other for turning him into a pure cognitive machine without feeling. Both are much exaggerated. It is true that at an earlier stage the vista of human knowledge engendered undue pride; but the lessons of scientific progress, of the unknowns that lie beyond the newest discoveries and the unintended consequences of inventions and discoveries, work on the contrary to breed a due humility. It is significant that the fantasies of omnipotence come rather with the less advanced than the more advanced sciences; compare the humility of an Einstein with the pride of a Skinner. On the whole, the attack on science as arrogant came from those who feared that their own outlook might be affected by the results of the sciences. The fear of science as pure intellect without emotion is likewise exaggerated. It stems from a separation of the scientific and the humanistic which is the temporary character of a particular cultural stage of scientific development in which the contrast of the qualitative and the quantitative loomed large. That stage is now passing, and the possibilities of science as a humanistic imaginative enterprise are being increasingly realized. The impact of science in the arts is also not without effect in overcoming the two-culture syndrome.

The fear of ideology is of course a serious one. Especially in the studies of man many apparently scientific discoveries turn out to be permeated with ideology. The battle against distortion of knowledge in terms of interests and prejudices is an old and persistent one, and it may even center at times on strongly entrenched views in the current body of knowledge. How many dogmas about the differences between men and women, or about racial differences, have now been shown to rest on evidence that itself embodied the discriminatory habits congealed in traditional institutions. Current controversy about the nature of intelligence tests is a good illustration. The charge is now widely made that so far from measuring an inherent or developed intelligence, the tests constitute an indirect system of certification in a meritocracy in which opportunities are opened and rewards are given to those who serve the values of the Establishment, the tests selecting those strata who are "in" and ignoring abilities and attitudes that are "out." Ideology need not therefore be the overt advancement of beliefs that serve specific purposes (though it sometimes may be) but something much more subtle and much more permeating of the presuppositions in the fabric of knowledge.

Nevertheless, if we look at the history of the growth of knowledge, we can see that the struggle against ideology is a constant battle on many fronts. It is not something to be resolved by a simple touchstone of truth, but the discovery bit by bit of the marks of intellectual adequacy in field after field as we uncover past mistakes and apply the lessons of this discovery to further work. The fear of ideology is a ground for perpetual caution. It is no more ground for the wholesale inhibition of knowledge than the fear of tyranny is for the abandonment of liberty.

The fear of what we will find may well be one of the most profound of human phenomena. It was a classic belief that mankind wishes to be deceived. Sometimes it is cast as a sociological "truth" that man lives by myth. But the current use of the concept in social theory is somewhat ambiguous. Myths turn out to be aesthetic representations of truth quite stateable in more transparent scientific terms, or they turn out to be values sometimes universal and sometimes provincial; in the latter case, often lacking a general justification they are thought of as myths because they seem arbitrary and yet often deep commitments. There is, however, a very real issue in the general fear. A.E. Housman's "I, a stranger and afraid, in a world I never made" signals the basic insecurity of man and the fact that the world may at bottom not be congenial to his hopes and aims. The hope that the more we got to know it the more we could control in the light of our values may be of temporary duration, just as Bertrand Russell once pointed out about the growth of science that you would expect its early phases to move rapidly because it could discover so many of the simple (and perhaps over-simplified) things there were to learn; it did not promise the same when we got to the more complicated. Yet how would this situation be remedied by slackening the desire to know and the pursuit of knowledge? So far as we can see from the history of myth and of religion, the insecurities are equally expressed in poetic and in religious form. If eventually we will come to know our limits more vividly, there is time then for resignation *at that line of development*, not by surrendering the territory that may be won in between. Or better still, there is the old virtue of proper humility or proper pride to be applied throughout to the enterprise of knowledge as a whole.

These fears then—and this probably holds for others as well—are good grounds for caution, but do not in any way diminish mankind's responsibility to advance, consolidate, and perpetuate knowledge as the collective achievement of mankind, and to put it to the service of man's needs.

Perhaps I have been considering the easier half of the task of responsibility for the advance of knowledge. It is easy enough to advocate the pursuit of truth when truth has already so great a hold as a human aspiration. The more formidable job would be to show in detail how the pursuit is bound by specific responsibilities when it thought itself so untrammelled. This is harder because it cannot be done in very general terms. As we saw in dealing with the relation of theory and practice, the kind of relations express the actual state of development of knowledge and its operations in the light of the state of the sociohistorical field and the state of development of human purposes. The responsibilities of the scientist and of science as a profession cannot therefore be derived solely from the general aims of science nor from the intensity of human devotion to it. One has to scan the specific operation—from the selection of problems, the modes of carrying out research in different fields, the impact of the fact that research is going on, the likelihood of results, to the impact of results, the probable uses to be made of results, and so on. At each point there may be dangers and hopes, immediate bad effects may have to be balanced against future advantages, and so on. It is far from a simple task, and the dangers of short-sighted decision are great. But it is an unavoidable task, especially as science has moved into being a basic component in the character and operations of contemporary life.

Notes

1. Michael Oakeshott, "Rationalism in Politics" in *Rationalism in Politics and Other Essays* (New York: Basic Books, 1962).
2. Gilbert Ryle, *The Concept of Mind* (London: Hutchinson's University Library, 1949). In Ryle's treatment, theory itself—knowing that—turns out to be a kind of knowing how, namely knowing how to teach.

14. The Scientist and His Findings: Some Problems in Scientific Responsibility

Current controversy over race in its relation to intelligence runs the gamut from questioning whether there should be any research in the area to proposing specific legislation on the assumption that definite results have already been achieved. For example, Bodmer and Cavalli-Sforza, at one extreme, conclude that "for the present at least, no good case can be made for such studies on either scientific or practical grounds."[1] Shockley, at the other extreme, not merely concludes with confidence that "perhaps nature has color-coded groups of individuals so that we can pragmatically make statistically reliable and profitable predictions of their adaptability to intellectually rewarding and effective lives,"[2] but goes on to propose a program of offering bonuses for voluntary sterilization.

In the light of this controversy, we shall have to discuss—if we wish to pinpoint even in outline the range of the scientist's possible activity: 1. choice of research topic, 2. modes of doing research, 3. modes of presenting results among scientists, whether of experiment or survey or theory, 4. mode of publicizing results and ways of engaging in controversy about them, and 5. the scientist's participation in the formation of policy, whether social or governmental (down to specific legislation).

Since the controversy is today occurring against a background of assumed common values about freedom of inquiry, any account of responsibilities must first make sure we understand this libertarian framework. Briefly, it centers—in the context of scientific work—on two values: one is truth and the other is freedom of thought, inquiry and expression. Some see liberty as an absolute right, others regard it as the appropriate structure for a society determined to achieve the material and cultural progress essential to a growing well-being of its people. The difference may not be relevant to our inquiry, chiefly because both tend to assume that freedom of inquiry is the best means for ensuring the discovery of truth. Concerning truth itself there are also two views: some regard its pursuit as an absolute value; others see it as a very high value to be pursued but not at any price.[3]

A simple appeal to libertarian rights may not itself be enough to locate the points of controversy. We have to recognize that the theory of liberty in our tradition is not simply the proclamation of a single principle of freedom but has involved an often precarious balancing of three elements: 1. a maximum of individual liberty, 2. a framework of rational discussion in resolving disagreement, and 3. a specification of limits directed to self- and social-protection.[4]

Of these, the first obviously needs no discussion. The second element is perhaps too strongly formulated, for it may suggest rationality in the discussion itself rather than the necessary conditions for rational discussion. The necessary condition for rationality is the willingness to listen and to be heard in turn; it is a framework of civility which is something more than keeping quiet and something less than rationality itself. The third element is the most difficult in our tradition, because limits proposed often tend to become seats of dogmatic authority through which the established order holds back change. Generally, freedom of thought and expression does not guarantee complete freedom of action. Concepts of sedition, license, clear and present danger, pornography, etc., have at various times been offered in different fields to pinpoint borderlines of action, to suggest limits, and to legitimize paternalism. Traditional experience in these concepts has not been very happy. "Clear and present danger" was voracious enough in the Smith Act cases to trespass on the first amendment, and it has taken a long while for us to recover. We have never gone so far as Harold Laski once suggested (criticizing the decision in World War I cases)—that dissenters should even be allowed to hand to soldiers actually embarking for the war front pamphlets which urged them to refuse to serve. At least the British soldier, he said, was so indoctrinated that if the pamphlet could persuade him, its arguments were probably correct! In any case, whatever the complexities in setting limits, the rights of self-protection and social protection are clearly as firmly established as the rights of freedom of expression; both have their place in basic rights. The real issue arises when they are in conflict; one must then decide whether dangers are real or people are conjuring up dragons.

We turn now to the question of the scientist's responsibility with respect to each of the five topics in the light of the libertarian framework. In some cases the treatment can be brief, in some fairly definitive, in others at best suggestive of the complexity of the problems.

Choice of Research Topic

As a *prima facie* matter, there seems little reason for not allowing full scope to the liberty in a scientist's pursuit of any intellectual interests that he may have. Society has enough positive incentives to attract a sufficient number of scientists to what may be felt as socially urgent. Moreover, the scientific maverick has on many occasions been a source of fruitful intellectual mutations, even beyond keeping orthodoxies on the alert. Still, it is conceivable that specific scientific responsibilities to avoid dangers of great material and social harm may give rise to an obligation not to enter on specific researches at a given time. The first temptation in the liberal tradition is to rest on the value of truth and say that responsibilities will emerge only in the later steps of publicizing and applying one's findings. Whether it is possible to control these subsequent steps is itself an empirical-scientific question concerning the character of communications and the state of the existing society. Therefore, the general responsibility of the scientist to become conscious of and to consider the consequences of research in an area can be recognized at this point. The types of considerations that should enter into his judgment may be gathered later, but in the libertarian tradition the decision is the individual scientist's own.

A second general responsibility in choice of a research topic is fairly obvious. Although a scientist may generally not be concerned with why he finds a research problem interesting or attractive, he should try to determine, insofar as possible, the extent to which his motivation may involve him in possible ideological conflict. This determination is clearly important in an issue like race and intelligence, in which both racist and utopian attitudes may be found. There is no reason why a scientist should not investigate a field because he hopes the results will be of such-and-such a sort. (We may pass independent judgment on his motives.) However, he should where possible be aware of his hopes so that he will thoroughly test his findings.

Modes of Doing Research

It does not follow immediately that any scientist who has thought through any interest and projected research should be absolutely free to pursue it. Any control over research, whether society's or one's own, is guided by the major countervailing factor in the libertarian framework of the need for protection. For example, a particular biological project might require producing a culture that could, without control, endanger many lives; decisions in research would accordingly depend on safeguards available. The research might involve experimenting on human subjects; the moral problems here have recently become prominent in medicine, psychiatry, and even experimental psychology.[5]

To recognize that freedom is not absolute is not enough. We need a complex ethical inquiry into responsibilities, types of legitimate checks, location of burden of proof, areas of legal coercion or administrative obstacles, and dangers of inhibiting discovery through imposing present scientific orthodoxies dogmatically. Whether there are general answers or only methods for approaching particular situations is not itself clear. Thus we seem to have reached an agreement on ruling out experimentation for biological warfare, and it seems unnecessary to discuss the mad scientist in fiction who wants to perfect a cobalt bomb to end life on this planet. Norbert Wiener was only one outstanding example of a scientist who turned away from research that would contribute to war, but his was a personal moral decision. The social moral problems here are what kinds of research should and should not be restrained, i.e., research to be legislated against, research to be forbidden at public expense, research to be discouraged, and, of course, the kinds of research to be positively stimulated.

Several points about countervailing considerations to absolute freedom of research are worth noting. First, these considerations refer to the impact of the research itself, not to the applications of results, which raise subsequent questions—for example, early fears that computer research, when applied, would produce unemployment. Second, the considerations involve both the present state of knowledge and present patterns of value. With respect to our knowledge, we cannot assume infallibility. With respect to values, coercive restriction on investigations can be considered only when serious social dangers to life and health may be involved. We cannot impose on research a special morality or the values of special class or religion. Third, the extent of the need for controls reflects the state of development of science itself and its role in society. There is a difference, for example, between the times when science was the work of isolated individuals whose investigations seldom affected the character of social life, and contemporary times when science, including basic research, is a major dynamic factor in the functioning of our societies and may generally alter the character and quality of life. Fourth, because of the broad nature and role of scientific research today, indirect controls will come from the need to allocate scarce resources. Finally, there is the moral responsibility of those who decide on controls to avoid dogmatic imposition of special orthodoxies of knowledge and value.

Presentation of Research Results

It might at first seem that the presentation of results among scientists (not publishing to the general public) could be taken as a matter of course: one completes research and lets other scientists know about it. (I am not referring here to problems of top-secret classification which inhibits scientific communication in some areas; that is serious enough, but not relevant here.) The belief that one can simply tell, even in technical language, what one has discovered, is misleadingly oversimplified. There are many different kinds of traps to be avoided. Some are external questions, such as the timing of publication. Attention has recently been directed to a tendency to rush into publication, because, in the current contraction of grant funds for research, a competitive advantage accrues to those who in effect advertise through publication promising directions of inquiry. Such publication is unethical when the results are not secure enough as yet to support the promise. [6] More serious are the internal questions of the adequacy of data and the relation of interpretation to data. In the psychological and social sciences, the phenomenon of conflicting schools of thought and theory often leads scientists to interpret research without regard for possible alternatives, and consequently without stating how tentative the enunciated results may really be. It is therefore, a basic responsibility of the scientist in reporting his results to distinguish between data and interpretation, to exercise the greatest methodological care about data and about modes of manipulation. [7] I am not referring to the difference between a good and a bad scientist, but to the degree of sophistication among good scientists themselves.

The extended controversy on the question of race and intelligence since 1969 when Jensen published his article in *Harvard Educational Review* [8] has demonstrated the tremendous diversity of issues—logical, conceptual, evidential, cultural, politico-social—that are involved in the assessment of his findings. For example: 1. Logical issues, such as whether one can go from individual heredity as a source of individual differences in intelligence to heredity as a source of group differences. Heritability applies to populations, not traits or characteristics alone. 2. Conceptual issues such as how far the interaction of genetic and environmental variables at all stages in a person's development makes the mathematical model drawn from simpler traits in the study of population genetics inapplicable in the case of intelligence. In general, the feasibility and utility of distinguishing between heredity and environment. 3. Methodological issues about the meaning of intelligence and the validity of its measures, the controversial aspects of the belief in a constant general factor, the debates about what the tests test, about the way in which tests are constructed and items revised, and so on. [9] Similar issues about the concept of race itself and the bases of race classification. [10] 4. Evidential issues about the reliability of focal data—e.g., the controversy about the study of twins. (Certainly something has gone scientifically astray when the same material is treated on the one hand as a firm factual basis for definite conclusions about heredity in the I.Q. and on the other as involving sloppy neglect of interfering factors.) [11] 5. Cultural issues about the role of cultural and subcultural factors in intelligence tests and testing in cross-racial studies.[12]   6. Sociological issues about the extent to which the tests reflect middle class selection of capacities to fit into and continue the patterned needs and values of the Establishment.[13] 7. Politico-social issues such as the history of ideological use of theories of racial superiority and inferiority in economic exploitation, colonial struggles, immigration policies, etc. 8. Issues of synthesizing to get a general conclusion out of a diversity of evidence of differing weight—how far one can establish the general strength or tentativeness of the theoretical position as a whole, including the problem of weights and emphases.[14]

A proper responsibility in this controversy is of course relative to the broad state of knowledge of the issues involved. Suppose a scientist surveying arguments on race and intelligence completely ignored the anthropological work on the ethnocentric elements in testing. He would definitely be failing in his scientific responsibility, because this work in anthropology is longstanding and could be assumed to constitute a standard caution. On the other hand, he might ignore currently emerging sociological critiques, which, if sound, might show where he erred, but might not necessarily show that he was irresponsible as a scientist. Suppose he failed to consider published experimental studies of the dependence of I.Q. results on the race of the people administering the tests or the mood of people taking the tests. His failure might be culpable neglect of evidence, if he treated the "fixity" of the I.Q. as an established item of knowledge no longer to be questioned. Suppose he attributed an ideological motivation to opposing arguments in order to avoid confronting them. He could be considered irresponsible unless there were other (e.g., empirical) sound grounds for thrusting the arguments aside. Finally, he might level his charges of ideology against those disagreeing with him and ignore possible ideological commitments on "his side." Such an argument would be naive.

Clearly, there are various degrees and shades of responsibility and irresponsibility in the declaration of scientific findings. Their determination is a matter of concern for the specialists in the field and for those in neighboring fields upon which the findings may draw or in which presuppositions of the research may lie. As a nonspecialist, I cannot presume to judge whether in the Jensen controversy there has been a high degree of scientific irresponsibility.[15] However, it does seem to me that areas of science, especially in human affairs, can no longer coast on an attitude of "I say what I believe because I believe it." If for no other reasons, the institutional uses of psychology and social science will press responsibility on them; for example, findings in psychiatry will be tempered by studies in law, and research in psychology will be reevaluated by research in education.

Publicizing Results

Clearly the scientist has some responsibility for the way his ideas and research results are published. At the very least, he ought to correct popular misinterpretations and misuses of his research, even if he himself does not write for the mass media. Especially today, when scientists have a widespread influence on the public, it becomes important to consider what responsibilities they have for the fate of their findings in public or semipublic media.[16] Whether he or others are writing, the research scientist should make sure that opinions are qualified by degrees of evidence, alternative conceptions are noted as well as the extent to which results are being applied from a domain the scientist has explored to one which he has not. For example, doubts may be raised whether responsibilities are being met in many popular articles by ethologists on aggressivity, by sexologists on the proper relations of the sexes or on the nature of women and the women's liberation movement, and by psychologists on parent-child relations.

In December 1973—to take a complex illustration—the American Psychiatric Association altered its position of nearly a century and declared that homosexuality is not a mental disease.[17] The import of the change, which technically simply removed homosexuality from a list of mental diseases and appeared to be a revision in nomenclature, was not wholly clear. What the public learned in newspaper reports, television programs, interviews, and discussions among psychiatrists was something like the following:

1. The change is not just yielding to the pressure of homosexual organizations.
2. Because many homosexuals are satisfied with their conditions, there is no point in branding them as mentally ill or forcing them to think of themselves as freaks.
3. There are not enough specialists in homosexual treatment to cover the whole field.
4. Only those homosexuals whose sexual orientation distresses them are suffering from a disturbance.

As far as wanting to know whether the APA's previous position had been incorrect and was now being corrected, the public knew as little as when a press secretary announced from the White House that a previous statement was now "inoperative." The American Psychiatric Association at the same time urged the removal of all legislation penalizing sexual acts performed by consenting adults in private and full civil rights and protections for homosexual citizens.

What would a responsible relationship between scientists and the public have required on the theoretical state of the question raised by the APA's revised position? The Association might have issued something like the following statements:

1. There has long been controversy on the basic theoretical issues underlying homosexuality, both on the biological and pyschological aspects. One view treats human beings as having an initial general capacity for sexual response, which becomes varyingly directed in individual development; the other as having an initial capacity for specific heterosexual mating, which becomes diverted (or distorted or crippled) in development and seeks substitute expression. On the psychological side, the first view regards the absence of heterosexual interest like any unactualized capacity, the second as a continuing drive incapacitated by psychologically generated fears. The scientific evidence between these hypotheses is not decisive, so that different analogies can be used and examples found for which each appears to hold.
2. The concepts of health and disease, of normality and abnormality in these domains have large components of value judgment, so that some scientists have even urged that the previous classification was nothing more than an imposition of cultural values through a dubious conception of normal health. Although these extreme formulations are unacceptable to most psychiatrists, the possibility that part of the traditional theory of homosexuality in psychiatry has these value components is a proper hypothesis for evaluation. In addition, there have been historical changes in the medical model, which would have different effects on the concept of mental health.
3. Psychiatry has been unfortunately beset by different schools, each of which has asserted dogmatic truth rather than thought of itself as finding part of the truth.
4. The social background of repression in sexuality has been weakened by social movements of liberation. It is therefore appropriate for psychiatry to be wary of diagnostic branding which impedes an individual's decision about his way of life.

Would the public understand these statements and would an ultimate resolution be seen as a consequence of theoretical and practical trends, with a recognition of the inherent probabilism in most scientific theory and the relative aspects of nomenclature and classification? The public would at least be treated with respect, although its automatic subservience to the psychiatric establishment might be weakened.

In a similar way, any responsible statement publicizing the recent controversy on the relation of race and intelligence must somehow make clear all the kinds of issues involved, together with the paucity of definite evidence, the tentativeness of the conclusions, and the present disagreement among experts. It should separate and define issues rather than treat the whole question as a single package. It might well enter into the history of the problem, showing on the one hand its ideological entanglements and on the other how scientific developments like those in population genetics opened new lines for consideration. In addition to indicating the difficulties as well as the hopes in the investigation, any statement should emphasize the importance of the presuppositions concerning the meaning of "intelligence" and "race" that underlie formulations in the controversy. Most important, perhaps, from the point of view of the history of science, a statement should consider the possibility that formulations will themselves undergo change as science advances and should raise doubts as to whether present formulations may not be an inadequate basis for present and future investigations. Finally, there is no public place for the dogmatism that jumps to a reconstruction of policy in practical fields as if it were a deductive conclusion from an established thesis.

If such cautions are observed, public controversy about a scientific thesis has some chance of maintaining a scientific character. Such cautions were not observed in the public controversy over intelligence and race; it soon became embroiled in charges of ideology and even of conspiracy to repress scientific findings. The charges culminated in a formal statement, "Comment," signed by fifty scientists and published in the *American Psychologist* of July 1972. The rapid development among professionals of this controversy, carried into a public forum, makes essential the analysis of scientific responsibilities in controversy. As to the controversy itself, it has ceased being a scientific exercise and has become a social battle fought with social instruments.

It is worth briefly examining the statement by the fifty scientists on behalf of the study of hereditary influences in human abilities. How far as a specimen of public controversy did it promote the scientific goal of getting strictly scientific solutions to scientific problems as well as scientific understanding of ideological problems? It has, first, a strangely mixed sense of whom it is addressing. If it addresses academics, then why does it appeal to the American Civil Liberties Union? If it is for the general public, then why is there no clarification of the central, crucial distinction between the study of heredity in individual differences and that in groups or races? As an indication of the lack of clarity of the statement, one signer later wrote that the resolution dealt solely with behavior and heredity and took no position with respect to race differences.[18] Further, the statement is prefaced with a background account including the persecution of scientists from Galileo to Mendelian biologists. There is reference to the attack on Einstein in Nazi Germany, yet no mention of Nazi Germany's use of biology for racial persecution and genocide. This is a glaring omission in a statement which supposedly attacks ideological blocking of science. It is sad that this statement by eminent scientists against ideological interference in science was interpreted in its turn as another ideological weapon, but it is scarcely surprising.

Two general shortcomings in the statement prevent it from helping to clarify the confused intellectual-social situation of the controversy and are responsible for its tone of outraged innocence. One concerns the history of science, the other, the character of academic life today.

First, the statement does not distinguish sufficiently between an external attack on science and an internal scientific controversy. It says in effect that in history science has always had to face external intolerance from Galileo's time to our own, and consequently men of good will should rally to the defense whenever science is attacked. This overlooks the marked difference between the time when science was an undeveloped enterprise itself, and the present day when science is powerful and influential. External attacks have not, of course, come to an end, for new sciences arise and often tread on different social interests. However, the model for ethical analysis in Galileo's trial is no longer the major relevant model for analyzing internal controversy like the role of heredity in intelligence. To develop an appropriate model for the latter is precisely equivalent to delineating the responsibilities of scientists in controversy and in social action.

The second shortcoming in the statement is an insufficient refinement in understanding academic orthodoxies. It is important to distinguish between social motivation intruding into the judgment of scientific beliefs and orthodoxies in philosophical or scientific beliefs themselves. The different and often conflicting schools of thought legitimately represent different conceptual frameworks that can be assessed by their relative success or failure in advancing a field. They become less legitimate when they are enshrined as orthodoxies in academic departments and dominate appointments and publication in controlled journals. Yet such domination does not add up to conspiracy to suppress opposing views; it is usually too overt to be conspiratorial and expresses itself in judgments of worth and lack of worth! Such a process has often been found in the conflicts of behaviorism and psychoanalysis in American psychology, of analytic movements and phenomenology in Anglo-American philosophy, of mathematical and institutional economics. If "hereditarians" feel shut out from academic expression, the door is surely less firmly barred for them than for psychoanalytic theorists. Imagine publication of a statement by fifty psychoanalysts about some behaviorist psychologists as persecutors, or a statement by fifty Marxian philosophers on academic discrimination and misrepresentation of their perspective.[19]

A scientific approach to different schools of thought is largely a task of understanding the differences and analyzing their diverse components in order to make comparison profitable and the interplay of differences useful. Instead of breaking off communication there should be analytic dialogue. Scientists should assume responsibility, in the public presentation of scientific findings and in controversy about them, for keeping the spotlight on the scientific character of scientific decision and, when ideology is involved, for turning a scientific light on ideology itself. Scientists should understand what is involved in ideological struggles and not be overwhelmed by them. A scientific statement should not be shrill against the shrill or a hoot against the hooters. The scientist has an obligation to understand why the hooting occurs, not to take the stance of injured innocence.

It is, of course, generally agreed by scientists and intellectuals that, whatever side they have taken on a controversy, the framework of civility is to be maintained; viewpoints are to be reasonably discussed, not shouted at. SPSSI and its officers are just as determined as the fifty scientists in their condemnation of those who disrupt meetings or interfere with the orderly presentation of argument.

Such a general attitude rests on at least three bases. First, libertarians recognize the central role of civility. Second, they fear that any violent interference with freedom of expression will grow from shouts to demands for economic reprisal to threats of physical assault, to terrorism. Third, they generally assume that noncivility springs only from irrationalism and therefore undermines the foundation of a scientific outlook.

What reason leads some people today not merely to engage in conduct disruptive of the framework of civility, but to claim that the disruption is rational? Their claims must be understood, not merely ignored or dismissed. Indeed, the advocates of disruption would claim not to be abandoning the libertarian outlook but to be invoking its component of self and social protection. They also assume that the application of coercion can be controlled and will not result in terror. For our present discussion it is not relevant whether the coercion be private action or economic sanction or governmental penalty following on legislative prohibition; which is to be used would have to be argued separately.

War is the only clear example in our tradition when civility is unhesitatingly abandoned. We do not give freedom of expression to an enemy; we jam his broadcasts without examining the truth of what is being said. We use deliberate and controlled violence to frustrate expression. In peace time attempts to impose legal limit on free expression have been directed, for example, against communists, as in the McCarthy period. The theoretical argument for this kind of coercion has been that those who will not allow liberties when they are in power cannot expect them when they are not. (The same argument was used against Catholicism in the seventeenth century.) Generally our tradition has rejected such limitations by trying to link repressive measures to action rather than expressions. An interesting unsuccessful attempt to legislate against expressions of belief was a bill introduced shortly after World War II to ban from the mails materials that would incite to racial hatred. The argument over the bill split the then broad alliance of liberals and the politically left. Although some labor organizations endorsed the bill, the American Civil Liberties Union opposed it. The secondary arguments against the bill were persuasive: the legislation would give too much power to postal authorities; it could be misapplied (for example, anthropology text books containing analyses of racial equality could be banned in those states where segregationists were prone to violence); if Congress was unready to support bills for removing discrimination in employment at that time, it should not be trusted to legislate on matters of freedom of thought. The most moving argument for the bill came from veterans who had recently seen Hitler's concentration camps and gas chambers. They argued that the sacrifices of the war had been made to overcome racial violence, and it was folly to allow comparable views to be promoted through the United States mail. To the veterans, the bill was not a question of freedom of speech or inquiry; it was a question of fighting a war.

Recently, private action to limit expression was seen in the revolt on the campuses against government spokesmen defending the Vietnam war. The action was effective, for it became almost impossible for a governmental representative to appear on a college campus with any assurance that he could present his views. Those who engaged in or encouraged limiting discussion argued that debate within the framework of civility would be ineffective against a pattern of deliberate lies, undisclosed invasions, and designed deceits on the part of the government in power. They believed that their conduct shortened the war. It helped change the climate of opinion, influenced Lyndon Johnson's decision not to run again for the presidency, and helped force Richard Nixon to slow down and eventually to end the war. They saw their moral choice as between abandoning civility to save Vietnamese and American lives and continuing civility at the cost of continued immoral war. The crux of an evaluation of their position would thus seem to be not whether civility was morally violated, but whether their analysis of the situation as a whole was correct: had the policy of the administration really brought the country to the point of war on the home front?

Arguments to justify the abandonment of civility in the controversy on race and intelligence seem similar to those used in the Vietnam war debate. Their advocates recall the depths of frustration that almost destroyed American cities a few years ago and the continuing battles for integration in jobs, housing, and education. They point to the fate of programs of attempted assistance to the disadvantaged and might even refer to *Life* (1970), quoting Daniel Moynihan, who as a White House advisor, said that "the winds of Jensen were gusting through the capitol at gale force."[20] They would trace the ideological role of beliefs about racial inferiority in repressing blacks and other ethnic groups in American history, trace its penetration into intellectual disputes, pinpoint the racial blind spots in the history of I.Q. testing, give a social explanation of the precise timing of the renewed investigations in an atmosphere of contracting economic opportunity, and show how intellectual complexities were overridden to produce dogmatic conclusions tied immediately to reactionary social policies. They see the central issue not as freedom of speech or inquiry but as preventing pseudoscientific ammunition from being circulated to the racist enemy. The supposed intellectual discussion is to be regarded as no different from enemy broadcasts during wartime. Although they are breaking with civility, they believe they are acting rationally rather than attacking reason or science.[21]

How can scientists, for whom a framework of civility is scientifically important, deal with such a position? They can at least assume a greater collective responsibility to stop their scientific findings from being used as ready weapons of war on the social-intellectual scene. We have seen what this entails in ways of publicizing scientific views and findings and in analyzing controversial questions. It also entails a deliberate scrutiny of projects and findings that are likely to become involved in social conflicts. Scientists can presumably distinguish, better than the lay public, those projects which are manifestly pseudoscience, those that are speculative and not open to determination in the foreseeable future, and those that are scientifically feasible. Scientific agreement is not perhaps always to be expected, but if a debate about a project or finding is responsibly publicized, there may be the same result as if scientists always were given "equal time" whenever ideological uses of their findings got on the media—particularly if the debate is accompanied with a scientific analysis of the ideological uses themselves. Such efforts will not, of course, defuse social conflicts themselves, but they may help defuse the use of the war model in relation to scientific research.

Participation in the Formation of Policy

The responsibilities of scientists in relation to policy and social action and the impact of all stages of their work on the social scene should not be considered as an exceptional or crisis matter. It is already time that scientists face these responsibilities as a continuing collective task.

A number of arguments are, however, sometimes offered as to why scientists should not be concerned with the impact of their findings on social policy. These include: the value neutrality of science, the specialization of the individual scientist, the ideal of progress with the expectation of compensating consequences, and the unpredictability of consequences.

The argument for the value neutrality of science no longer has the strength it did when science was a small and uninfluential endeavor. Now scientific stands do play a consequential part in social policy. The psychiatric reversal on homosexuality described above was not a merely theoretical scientific matter. The previous classification of homosexuality as an illness had been related to laws and practices which resulted in jailings, ostracism, and possibilities of blackmail. The revised classification was an influence for liberation. As a minimum, the scientist has some responsibility for knowing whether his findings are likely to have a social effect.

The scientist sometimes claims that specialization excuses him from social responsibility. "I am a psychologist dealing with heredity and intelligence," he might say, "I am not a sociologist to reckon on whose passions will be influenced or an educator to balance values of the society in educational programs." [22] If so, let this scientist forego making policy recommendations as if they followed purely from his findings alone, or let him enter into an interdisciplinary project with sociologists and educators. Most important let him remember the dangers of others misusing his results, try to foresee where misuse may happen, and do his best to guard against it.

The ideal of progress is sometimes taken to assure the scientist that any further advances in knowledge will be for the good of mankind, and if consequent social dislocations do occur, people will find compensating modes of action. Therefore, scientists did not need to worry about unemployment when the mechanical cotton-picker or the computer was developed. Even if this long-run view is held, the lives of people in the short-run are important. A scientist may not feel inclined to say that he will not investigate a field, because his findings may have short-run bad consequences. He may, however, feel responsible for warning policy makers of his findings in order that they can plan compensating lines of action. The present concern of biomedical scientists with the ethical consequences of genetic discovery and with problems of genetic engineering is a clear determination not to be caught unawares in the way nuclear physicists were unprepared for the military application of their work.

The unpredictability of consequences as well as the prevalence of unintended consequences is well illustrated in the history of the automobile whose technical and sociological consequences meant vast unforeseen changes in our mode of life; and when we thought we knew them all there came the fresh realization of effects in polluting the atmosphere. Responsibility does not mean that the early inventors are to be held accountable for the total outcome. But it does suggest the need for cooperative scientific responsibility to monitor systematically the applications of science in a continuous, not an intermittent, institutional manner.

There is, of course, a gap between science and social policy that is filled by knowledge other than the findings of the particular scientist and by values and value-judgments in the community. The scientist in making his findings public should recognize this gap and indicate the multipotential effects of his own work on different value-judgments and on different assumptions of other knowledge. He may thereby be able to excuse himself from specific social responsibilities, but if he draws conclusions for policy, he should make clear what other knowledge he is using and what the value stand involved is taken to be.[23] Thus we would expect a believer in the role of heredity in intelligence and its application to race to make points such as the following clear, if he is drawing conclusions for education: 1. How his recommendations for educational policy would have any relation to researches on race in an educational system, the professed ideal of which is the treatment of each individual as an individual person. 2. What the underlying, possibly alternative, ideals of educational apportionment would be. For example, is each child to be educated for a rounded development of his native abilities, or only for what we can predict he will do best, or only for what there is current social or economic need? Or does education involve a conception of scholarship as a capacity of an elite group that is to be given the greatest scope? 3. What life rewards of different educational tracks are to be taken to be, and whether they are occupational success and higher gains or the values of a type of cultivation of spirit.

In short, the consequences of the hereditarian findings (setting aside the issue of correctness) may vary widely for different conceptions of the nature of the educational process, the aims of education, the supposed rewards of education, and the presumed social consequences of tracking. If the scientist who is recommending changes in policy does not assume a responsibility for clarifying the concomitant bases of that policy, he should hardly be surprised that the worst view may be taken of his purposes in making the recommendations. In critical social controversies lack of clarity often borders on obscurantism, which is a traditional property of ideological uses of knowledge.

The same cautions are relevant for a scientist who is not himself translating his findings into policy recommendations, but whose findings are nevertheless a ready intellectual weapon in an existent social conflict.

Finally, the scientist who becomes an advisor or expert for a particular legislative campaign or some immediate alteration of executive policy has assumed this status voluntarily. The ethics of campaigning, much in flux today, are then relevant. An inquiry by moral philosophers into this area would be fruitful, and illuminating comment can be found in the ancient Stoic philosophers on what must be endured if one decides to enter politics actively. But this at any rate is clear in our present inquiry: a scientist serving in the political context cannot claim that his freedom of scientific inquiry into race and intelligence is being jeoparized if he is attacked for sending his children to private rather than public schools, or if he has not resigned from an all-white club.

In the light of all these considerations, what can be concluded about the judgment that under present conditions research into the problem of race and intelligence is inadvisable? It was suggested that this type of question is a legitimate one for the individual scientist to ask himself about his own future action. Some physicists have refused to work in a whole area in which war uses of their discoveries are at present unavoidable. In the area of race and intelligence, there is less clarity and greater complications than for the physicists. There are dangers of intensified racial discrimination. Chomsky compares such work in research to that of a "psychologist in Hitler's Germany who thought he could show that Jews had a genetically determined tendency toward usury (like squirrels bred to collect too many nuts");[24] the very raising of the issue of Jews and usury would lend weight to the persecution of Jews.

There is, however, the fact that the question of race and intelligence has long been asked and might best be answered by intensive scientific investigation and criticism. Unfortunately the research itself may not be feasible, because the social conditions under which we live may preclude doing the central studies which involve, for example, control of nutrition, prenatal environment, hormone level as it affects anxiety in mothers, as well as social attitudes. It is also possible that the question of race and intelligence is so inextricable from ideological conflict that it would be impossible to keep it as a purely scientific one for a given period. A further issue is whether the research has any scientific validity. At the present stage of genetic knowledge and of conceptualization of heredity-environment interaction, the formulation of research may be too gross and too speculative for useful work. Generally, the libertarian tradition would be inclined to stress scientific grounds in foregoing a line of research, but a decision based on social consequences cannot be eliminated for individual scientists who find it a regretful outcome of a rational balancing of values. Probably different scientists will reach different personal conclusions.

I have tried to show that the responsibilities of the scientist in the translation of his findings into public policy are great; that the issues are complex; that the route is indirect and involves other knowledge, values, and ideals; that responsibilities extend back to the selection of research areas and the modes of experimentation. However, the scientist's primary responsibilities are scientific and involve clarification of relations. He is not necessarily to be saddled with agitational or activist duties, for how a person partitions his activities is a separate moral question.

Finally, I have not attempted to pass judgment either on the scientific merit of the findings in the controversy over race or intelligence or on the extent of ideological intrusion into the controversy. And I have not attempted to assign blame or responsibility for the present sad state of the controversy. I am inclined to think that considerable responsibility lies in the whole state of the psychological and social sciences and philosophy today—not only in the traditional isolating beliefs of value neutrality, but in the dogmatic habits of schools of thought and their intransigent conflicts. Differences of view, of method, even of beliefs can be enriching if they are examined in a responsible dialogue and scientific interaction. If instead the attitudes and habits of war or the game model of the football field govern the procedures of intrascientific relations, what can scientists expect in the public forum but that youth should absorb this lesson and extend it with corresponding youthful vigor? A framework of civility should begin at home.

Notes

1. Walter F. Bodmer and Luigi Luca Cavalli-Sforza, "Intelligence and Race," *Scientific American* 223 (1970): 29.
2. William Shockley, "The Apple-of-God's-Eye-Obsession," *The Humanist* XXXII (1972): 16.
3. For a concrete situation in which choice would be forced on us, consider legal cases in which the possibility of knowing the truth is rejected because of the illegal means by which it is ascertained. Bernard Botein and Murray A. Gordon, in their book, *The Trial of the Future* (New York: Simon and Schuster, 1963 ch. 2) discuss the likely future situation in which a jury could be dispensed with and hypnosis of the parties or drug-induced revelations substituted for jury verdicts; the choice would be between a ready means of truth and preserving human dignity.
4. John Stuart Mill, the central apostle of liberty in the liberal tradition, assigns so important a place to the atmosphere of rationality that he declares the principle of liberty to have application only when society has reached a stage in which men settle issues by rational discussion. Mill's *On Liberty*, ch. 2, is the classic defense of freedom of thought and expression.
5. See, for example, the on-going discussions by geneticists to defer lines of our research because of their potential hazards; it concerns gene transplantation accidentally creating, for example, drug-resistant germs or new types of cancer causing viruses. For a general study of current problems and controversy, see June Goodfield, *Playing God, Genetic Engineering and the Manipulation of Life* (New York: Random House, 1977). See also the researches and publications of the Hastings Center of the Institute of Society, Ethics and the Life Sciences. For an example of the growing sensitivity to moral aspects of psychological experimentation, see the discussion of Stanley Milgram's experiments in the review of his *Obedience to Authority* by Steven Marcus (*The New York Times Book Review*, January 13, 1974), pp. 1-3.
6. An episode of this type in the field of cancer research is reported in *The New York Times*, April 18, 1974, p. 20.
7. P.W. Medawar, in a review of H.J. Eysenck's *The Inequality of Man*, concludes: "One of the most important things laboratory research teaches us is how very often we are quite mistaken, but this is a disability of which Eysenck betrays very little awareness." (*New Statesman*, January 11, 1973).

8. Arthur R. Jensen, "How Much Can We Boost I. Q. and Scholastic Achievement?" *Harvard Educational Review* 39 (1969): 1-123.

9. A story told about John Dewey's reaction to intelligence testing may still be salutory today. Attending a conference on the question at Teachers College, Columbia, he was asked by Thorndike to comment. He said intelligence testing reminded him of the way in which people used to weigh hogs in Vermont: a strong rail was balanced carefully on a fence, the hog was placed on one end of the rail and stones on the other end. The hog's weight was then determined by guessing the weight of the stones! For a clear presentation of methodological issues in intelligence testing, see N.J. Block and Gerald Dworkin, "I.Q.: Heritability and Inequality, Part I," in *Philosophy and Public Affairs* 3 (Summer 1974).

10. For a study of the genetic and physical factors involved, cf. essays by W.F. Bodmer, John Hambley, and Steven Rose, in *Race and Intelligence*, ed. Richardson and Spears (Baltimore: Penguin Books, 1972).

11. For a critique of the twins studies, see Leon Kamin, *The Science and Politics of I.Q.* (Potomac, Md.: L. Erlbaum Associates; distributed by Halsted Press, New York, 1974).

12. The cultural critique was raised in the 1930s by the research of Franz Boas and Otto Klineberg. It is recently supplemented by a new look at black cultural forms as specializations of their own rather than impoverished derivatives of white cultural forms; see, for example in the case of language, W. Labov's study of the independent character of black children's speech, in his "Logic of Non-standard English" in Frederick Williams, ed., *Language and Poverty* (Chicago: Markham Pub. Co., 1970).

13. Cf. *The New Assault on Equality: I. Q. and Social Stratification*, eds. Alan Gartner, Colin Greer, and Frank Riessman (New York: Perennial Library, Harper and Row, 1974).

14. Jensen has been criticized as adding many weak strands to reach a strong indicative conclusion. As for emphasis, J. McV. Hunt decries the initial position given (in Jensen's first sentence) to the view that "Compensatory education has been tried and it apparently has failed." It is, says Hunt, a dangerous half-truth: "I find it hard to forgive Professor Jensen for that half-truth placed out of context for dramatic effect at the beginning of his paper." ("Has Compensatory Education Failed? Has It Been Attempted?" reprinted in *Environment, Heredity, and Intelligence, Harvard Educational Review*, 1969, pp. 148, 149).

15. The criticism of inadequate responsibility in the controversy is a wide one. For example, the Eastern Psychological Association voted the following resolution as its annual business meeting at the 44th Annual Meeting: "The EPA wishes to reiterate its long-standing commitment to a policy of strict adherence to scientific principles in research. Because this is especially important for research on social issues, the EPA censures the use of inconclusive evidence concerning race and I.Q." (reported in *American Psychologist*, September 1973, p. 767). Cf. also Sandra Scarr-Salapatek's conclusion on this question in her review of works by Jensen, Eysenck and Herrnstein (*Science*, December 17, 1971, p. 1228): "And to assert, despite the absence of evidence, and in the present social climate, that a particular race is genetically disfavored in intelligence is to scream 'FIRE! . . . I think' in a crowded theatre."

16. Of the articles that initially fired the controversy, Jensen's appeared in the *Harvard Educational Review*, which may be regarded as a semipublic medium; Richard Herrnstein's "I.Q." appeared in the *Atlantic Monthly* (September 1971), pp. 43-64.

17. Reported in *The New York Times*, December 16, 1973, p. 1. See also the discussion in *The New York Times*, editorial section, December 23, 1973. p. 5.

18. In correspondence with the chairman of SPSSI's Fact-Finding Commission on the Suppression of the Academic and Scientific Freedom of Hereditarian and Behavioral Researchers and Teachers.

19. The sociology of school conflicts is only beginning to be studied more intensively. It relates to larger theories interpreting the history of science—especially on the question of continuity and discontinuity in development. Some of these issues emerge in Thomas Kuhn's well-known study, *The Structure of Scientific Revolutions* (Chicago: University of Chicago Press, 1962).

20. Quoted in Norman Daniels, "The Smart White Man's Burden" in *Harper's* (October, 1973): 25.

21. A somewhat different approach is taken by Bayard Rustin, who condemns the harassment of Shockley as giving undue publicity to an unimportant issue: "The economic and social environment of black people was altered quite dramatically during the 1960s and black people, as a result, have achieved unprecedented progress. For blacks, the challenge is to continue to dedicate ourselves to winning important, mass gains, not to creating martyrs out of proponents of irrelevancy." (*The New York Teacher*, January 20, 1974, p. 20).

22. In an article on "Reducing the Heredity-Environment Uncertainty" (reprinted in *Environment, Heredity, and Intelligence, Harvard Educational Review*, 1969), Jensen replies to several critics of his original article. In a section on Social and Educational Policy he says: "I am not a social or educational philosopher and I am sure that neither I nor anyone else at present has thought through all the policy implications of my article," (p. 239). This kind of warning might well have had a central place in the original article, but it would also require amplification about what is involved in thinking through policy implications.

23. Medawar points out in the review of Eysenck cited in note 7, above, that usually when we think of inherited differences in intelligence we take refuge in the area of equality of opportunity. (Presumably this is on meritocratic assumptions.) But, he says, we could equally call for greater opportunities when compassion and a sense of justice step in, as we do in the case of children with genetic defects who have greater medical needs.

24. Noam Chomsky, "The Fallacy of Richard Herrnstein's I.Q.," reprinted in *The New Assault on Equality*, supra note 13, p. 98. Chomsky assumes that only a racist society would raise the question of correlating intelligence with race. However, we note that it is the same society that has raised the question of the ideological character of such an inquiry.

15. Preferential Consideration and Justice

The underlying problem is how to increase participation of women and racial minorities in education and business in ways compatible with our conception of justice. This particular direction of inquiry is made necessary not only by difficulties and controversies in the application of affirmative action—that is, specific programs involving governmental regulation and use of sanctions—but by a growing number of legal cases of which the DeFunis case was almost paradigmatic. This paper is devoted to some general theoretical aspects of the problem, chiefly relating preferential consideration to our concepts of justice. It falls into six parts: I. The meaning and scope of preferential consideration; II. Conceptions of justice; III. How we operate with principles of justice; IV. The objectives of affirmative action; V. Issues about Quotas; VI. An Epilogue on basic directions.

I

I use the term "preferential consideration" in preference to either "compensatory justice" or "reverse discrimination." "Compensatory justice" already contains a full answer: it formulates the situation as one of preference to compensate for some, usually past, injustice. It certainly fits some types of cases, such as giving compensation to Indian groups for past forcible seizure of their lands. But it narrows the pathways of justification where one might want to argue preferential consideration for refugees or victims of some natural disaster or even preference on grounds of some sound national interest. "Reverse discrimination" already labels action as discriminatory and is thus pejorative. It seems to speak almost as if whites or men, having discriminated hitherto against blacks or women, should now be at the other end of the stick. In any case it passes too lightly over the question of deciding where in the presence of limited resources, helping A for a good and strong reason and so shutting out B is to be regarded as discriminatory and where as merely an unfortunate consequence of the limited resources. "Preferential consideration," on the other hand, is a term that simply describes a phenomenon. The ethical question whether a given preferential consideration is just, is thus separated off and transferred to the grounds for giving a go-ahead signal to a preference and for the direction of that preference.

The phenomenon is common enough. We have veterans' preference in some forms of employment, whether this is to be construed as an expression of gratitude, a recognition of alternative forms of experience credit, or compensation for time out for a social purpose of overwhelming importance. In college admission, athletic skill has often been a ground of preference, not considered unjust in choosing between two students of equal academic record. Familiar, too, is the selection of a student with high athletic ability and normal academic record over one with only a high academic record. (Musical ability rarely counts in the profile.) The draft in wartime raised many questions of preferential consideration both in acceptance and in exemption. Finer tuning of criteria might have used older men for some jobs in which experience would be more helpful than youth; and it could have exempted youths of higher ability who had not had a chance to go to college as well as those who happened to be already in college. The point is simply that to find and formulate the grounds for preferential consideration in any area carries us into a broader and deeper inquiry. We need to explore the criteria used in the light of underlying aims, usually institutional or social. We have to understand the structure of the situation, the complex conditions under which the selection is being made. We have to predict to some extent the consequences of action along the lines of those criteria, with respect to those aims and under those conditions. Anything less than such a full exploration in deciding the justice of the modes of preferential consideration embodied in affirmative action programs is likely to distort the problem and the issues.

Is there anything novel in the fact of preferential consideration as it occurs in affirmative action programs? In general or on a macroscopic view, almost any policy decision or any permitted social drift means a loss to some and a gain to others. To allow car travel in a large city is to deny the streets to pedestrians and people who do not own cars. (Would it be unjust now to limit driving to certain hours or make a mall out of certain streets for periods, just as some squares in older European cities serve as markets in the morning and parking places in the afternoon?) The claim for prescriptive or traditional rights need not be decisive; for example, we are only beginning to realize the rights of nonsmokers, even though these may cut down the rights of smokers. Comparable decisions will now have to be made in many areas of newly discovered pollution and health hazards. (Should the need for a safe water supply in Duluth be preferred to the rights of industry to pour its waste into Lake Superior? If it should be, should compensation go to industry, or to the workers who become unemployed?) To wage a war is to bring death to some part of the civilian population nowadays. (Would it be unjust to have free insurance at public expense for civilians killed or hurt in bombing, as we have had for soldiers?) As property law takes one form or another, so are the chances of some enhanced or diminished as against others. (Should a medical school have to pay royalties if it xeroxes articles on recent medical discoveries to circulate to doctors to keep them up-to-date?) In general, those who have—whether property, money, jobs—are constantly being assessed for many purposes through taxation and through inflation. (It is perhaps clearer in the former than in the latter where the money goes.) In short, we live in an atmosphere of constant redistribution with preferential effects, often unintended, and the built-in patterns of redistribution are quite literally patterns of preference. They may become explicit and when we are conscious of them we may make them the subject for definite social policy. What then are to be the criteria for considering one redistributive pattern just, another unjust?

II

The theory of justice is itself highly controversial today. There are two chief paths. One sets its formulae of justice in reference to the kind of life that action according to the formulae will make possible. Older utilitarian theories spoke to the formulae that under the existing conditions of life yielded "the greatest happiness of the greatest number." More recently, theories in this vein talk of the quality of life made possible by the arrangements taken to be just. The second path looks for an independent or autonomous structure of justice inherent in human life and personal relations, antecedent to any pursuit of the good and setting limits on the direction the latter may take. It is usually cast in terms of a theory of individual human rights. Though compatible with a variety of underlying philosophies it is often associated in our tradition with a contractualism that takes obligation to emanate from the individual's will and mutual obligations from the meeting or consensus of wills.

The two paths reflect an old conflict in the history of moral philosophy—the struggle for primacy between the good and the right. To give primacy to the good means that moral laws, principles, virtues, policies and proposed rights must all pass the bar of evaluation in the light of their consequences for the good. In the complex conditions of modern life it means room for uncertainties, use of probabilities, greater relativity of answers under differing conditions. To give primacy to the right puts in the forefront stern principles and absolute rights, a greater demand for explicit conformity, narrower scope for excuses and a moral rigidity rather than a moral flexibility. If the primacy of the good sometimes invites opportunism, it may also invite a more sophisticated moral sense and a sensitivity to changing conditions. If the primacy of right sometimes invites dogmatism and a legalism that narrows equity, it may also invite principled sincerity that makes great sacrifices in the name of morality and an unbending resistance to oppression. In the nineteenth century, the contrast would have been formulated as Bentham vs. Kant. At the moment, in technical moral philosophy, it happens to be the bout of utilitarianism vs. a Rawlsian contractarianism.

In these controversies there is too much shadowboxing, too many straw men, and too sharp a contrast which cannot, I think, bear up under analysis. On the whole, the first seems to me to be a more realistic philosophy in its understanding of the panoramic procession of concepts of justice over the ages. And it is not incapable of the stiffening of moral fibre that our contemporary world badly needs. But I shall not pursue it here, nor argue from its acceptance.

The fact is that whichever path is taken, the massive pressure of contemporary world problems and the basic demands of the peoples of the globe—both their needs and aspirations—will force their way through either framework. They will simply find different paths of entry. Absolute property rights will not help much where taxation is justified, where collective bargaining is accepted, and where prices may be freely moved about or manipulated. Tax decisions, wage structures and price structures do the redistributing. Similarly, when ideals and basic goods are translated into institutions, the latter take on an inertial life of their own. Still, the theoretical issue may be important in determining the readiness to change redistributive patterns in pursuit of definite goals. The utilitarian will ask whether property laws or laws governing conditions of employment or public expenditure should be changed to yield greater well-being. The individual rights position will filter the question of change through that of determining what rights are to be allowed or are relevant, or if the list is closed, what are to be the interpretations in applying the ones that are already recognized. For example, does the right to property include a right to any property, as the capitalist world has claimed with an eye to property for production, or is it limited to personal property? (In the latter case, society could grant or withdraw rights to property for production as it thought best at a given period.) This was a serious issue in the debates that preceded the adoption of the United Nations Declaration of Human Rights.

The practical consequence at issue was clearly that where productive property constituted a human right it could not be lightly socialized. In any case, the Declaration widely expanded the number and types of rights, thereby broadening the area of social justice. At least on the theoretical level, individual rights theory has often today been as aggressive as utilitarian theory in seeking to expand human well-being.

I propose therefore that instead of thinking in terms of opposing theories, we think of two models—briefly, the individual rights model and the collective welfare model—and ask in which areas of social life it is desirable to use the one model, in which the other. To take extremes: in the domain of criminal justice certainly no man should be punished for what he has not himself done or willed. Here long experience of humanity has established the individual rights model. But in war a man is drafted whether he wills it or not, and that is suffering if not punishment. Obviously, without here raising the question of the merits of defensive wars, we would feel we ought to use the collective model, with some reservations about conscientious objectors. Now the domain in which affirmative action enters—employment and education—is an intermediate one. What kind of argument can we employ to decide which model should appropriate it and to what extent?

Lest you think I am begging the question on a higher level by using a kind of utilitarian ground for deciding where to use which model, let me note that in the current controversy over utilitarianism and justice the same thing is happening on both sides. There is a growing convergence on the view that the greatest happiness of the greatest number is not enough for there is always the minority—the rest of the people. William James with his usual sensitivity, early offered a counter instance to this utilitarian formula: If some god offered us perpetual happiness provided only one lost soul were condemned to suffer endless torment, we should be outraged. The bizarre speculations of one generation of philosophers usually become the practical problems of a subsequent generation. Is it right that eighty percent of the population enjoy a high standard of living and twenty percent bear the burdens or be thrust out of that society's functioning? The point has been reached where an outraged minority will be driven in despair to rock the whole social foundation and so compel an attention that morally should have come long ago. The liberation movements of recent years have shown that the well-being of the majority is not enough. Interestingly, in questions of intellectual and political liberty we recognized long ago that the right of the minority, even (in J.S. Mill's formulation) of one man against all mankind, should be protected. In the area of economic opportunity and social participation, minorites need also to be protected, even against "the greatest happiness of the greatest number."

In the individualist model the same conlusion can be quite directly built into the theory of justice. This is clear in John Rawls' currently much discussed conception. Without entering into presently raging controversies about this theory and its proclamation of the independent priority of the right, we may note that the "difference principle" which is its central moral claim ties the formula of justice to the view that no change in distribution is justified which does not improve the lot of the least advantaged.

We may stop speaking of minorities at this point, a formulation induced in part by the utilitarian reference to the greatest happiness of the greatest number, as well as conditions in the United States. Aristotle pointed out, in defining democracy and oligarchy, that number is not of the essence. Suppose the oligarchs were a majority, it would still be at bottom a question of the rich and the poor. In the United States blacks are a minority, but women are not. In South Africa blacks are a majority. It is therefore a question of the disadvantaged, not the few.

III

Let us now turn to actual formulae of justice. They will specify the ground for apportionment: burdens to be distributed according to property X, gains or goods according to property Y. Here are some candidates for the criteria: ability, merit, work done, need, class position, chance acquisition, wealth, poverty, social contribution, aggressiveness, lack of aggressiveness, tendency of the formula to encourage socially desirable character, tendency to promote survival or a higher standard of living, intelligence, strength, moral worth, and so on. Whatever one's type of theory of justice, there will be a plurality of principles. The utilitarians will ask us to decide which it is desirable to use in what area under what conditions and with what aims. The nonutilitarians will answer all these questions indirectly by selecting and balancing principles, whether by hierarchical order, fixing priorities, determining relevance, distinguishing jurisdictions, balancing according to weight, fashioning compromises, or enduring the tragedies of conflict. Once again, I expect that social necessities and realities will break through either procedure.

Let us explore the problem of selection of principles from the field of our present concern. The criteria in employment and education have usually been merit and ability, not always distinguished. How do we decide whether these are appropriate criteria, and if they are, how do we interpret and apply them? And where can preferential considerations be allowed to enter?

First as to merit or desert, let us pay respect to the old quip: "If we got what we deserved, then God help us all." And as for ability, it is well to recall that absolutist governments have often been tolerable only because of the inefficiency of their agents. If we were asked for a criterion for appointment to government offices in such states we might recommend ineffectiveness as the most desirable property.

Though to invoke such extremes is close to caricature, still it shows that the use of meritocratic criteria is relative to the kind of job and the ends for which that kind of job exists and the conditions of its operation. For example, for each job there is a level of competence that is necessary; only for some jobs need it be the highest competence attainable. For a job of unskilled labor in a mechanical assembly line, working with a machine at a fixed pace, only an ordinary competence is needed. For the surgeon or the mathematician in advanced research, the highest competence is desirable. Why is that? Because the surgeon's business is a risky one for the patients, and in the mathematician's case there is always the feeling that a higher ability increases the possibilities of richer discoveries. Now in the types of jobs that require only a given level of competence, there may be considerable room for other considerations to enter into employment. For example, it may very well be that for certain jobs preferential consideration should be given to mentally retarded persons who can just manage to do them, precisely because these may be the only kind of jobs they would be capable of doing. Similar considerations arise for physically handicapped. If a private employer adopted such a policy he might very likely be praised for his humanitarian attitude. If the government adopted such a policy for jobs under its control, would this be denying equal protection of the laws to the brilliant, particularly if we were in the midst of a recession? And what if the government, finding itself spending large sums for assistance to the mentally retarded, passed a law reserving a certain proportion of jobs in interstate commerce for such persons? It might add the further argument that instead of spending money for them, the taxpayers would be getting taxes in return. No doubt many further considerations would enter into the decision about preference, but it is hard to see it as ruled out on principle in advance.

Let us turn back to the doctor or the lawyer or the teacher. If each of these had the kind of pure essence that angels were thought to possess, then meritocratic criteria could be readily applied. But human aims in the professions are complex and mixed, and the criteria for the best practitioner and the most promising student—for employment and for admission to schools—are more problematic than the use of a one-word name for the profession would suggest. Take, for example, the relation of teaching and research in higher education. Some countries will separate the research institute from the teaching university. Others keep them together hoping that research will fructify teaching. They are then faced with the questions of choosing in appointment between the candidate excellent in research and poor in teaching and the one excellent in teaching but only moderate in research. Similarly, medical selection might have to choose between average medical competence but greater appreciation of psychological problems and interpersonal relations and highly specialized skill without the other abilities. In such situations, one begins to ask about the conditions of practice. If the surgeon will only be called in to operate, then of course there is no question about the criteria.

Two linguistic formulations are possible about preferential consideration in such complex cases. If the grounds are to be found within the description of aims, then it need not be considered preferential consideration at all; rather the problem becomes weighting the components in a pattern of complex aims. If the grounds are not contained within the aims, then the problem is one of outside, weighty considerations that may or may not prove overriding. The questions are now whether the considerations are legitimate and what weight they have. Even which formulation is to be used may be a difficult issue in cases where the aims of the profession are undergoing change or the conception of a capable member of the profession is being transformed by increased knowledge or major social reconstruction. Many of the current problems about the need for role models in the case of women and blacks—or for that matter, men in elementary and nursery school—are of this sort.[1] In general, as the educational level of society goes up, the proportion of jobs to be assigned on merit alone may be considerably reduced, since the needed level of competence may be widespread for many types of jobs. In certain areas it may well be that the effective principle should be decision by lot in some areas, rotation in others.

The general conditions that most affect the principles of justice, adjusting degrees of relevance among them, are the extent of scarcity and abundance. David Hume, who argued for a purely instrumental value to justice, pointed this out by taking the exteme cases. Given complete abundance, there is no need of distributive rules, given complete scarcity they lose all effectiveness. He paralleled this with a consideration of human attitudes toward one another: given complete affiliative concern for fellow humans there is no need of justice; given complete rapaciousness and aggressiveness, there is no room for it. The instrumental value of justice becomes clear in the intermediate zones. But there it is precisely where the principles become most controversial if there is no fine tuning to public well-being. In our present concern—jobs in business and places in education—the state of the economy and the extent of educational opportunity have obvious relevance. In educational opportunity, open admissions is a recent institution that makes the problem of affirmative action less relevant in that area, provided services within the educational institutions are adequate. A DeFunis case has meaning only when openings are limited. Similarly, given full employment, the problem of affirmative action would not arise, except insofar as there were discriminatory practices in wages, types of position, promotion and conditions of work. As for good will, I do not think its strength and prevalence in our society should be underestimated. It was especially evident during the rise of the liberation movements. But it is not the attitude fostered by our competitive economy and our competitive life, and it has to function constantly against the tide. It is particularly prone to become isolated and minimal in effect when scarcity advances.

Two cautions are required in considering the selection of principles of justice and their application. The first is that the formula for distribution need not be the same for burdens and gains. To say "from each according to his abilities" need not entail "to each according to his abilities." Each formula, its place, its scope, and so on, has to be justified separately. In ancient Athens, when the city needed ships for the navy, the rich were assigned the care of individual vessels because they were rich; similarly, they were assigned the production of plays for the festivals. In return they got honor, especially if their play won or their ship did well. In contemporary times the armaments are part of the defense budget, and the cultural life is encouraged by the National Endowment for the Humanities. Both come from tax funds, and the principles of justice and injustice here lie in the formulae of the tax budget and the derivation of internal revenue.

The second caution is that the complex character of principles of justice and of preferential consideration has to be matched with a clear view of the whole of the resulting social processes, not by looking simply at one favored sector of what is going on. Let me suggest this briefly by simplification. Suppose the body of possible workers in a field is 1000, for whom there are 900 jobs, and the 900 hold on keeping the 100 at bay. By refusing any accommodation, they are in effect subsidizing them through the tax fund for relief, unemployment insurance, etc. Therefore, the 900 work more than they have to, for if the other 100 shared the work they would have to work less. Of course in a planless situation they may gain more than they would have in a wider distribution of work but certainly not without some cost. Some unions today, in fields that are closed and well-organized, have in fact been meeting the present recession by such procedures as sharing the work. There may, of course, be many different kinds of formulae which would yield equitable solutions for different fields and their special conditions.

What happens when a society becomes conscious of a great injustice to a part of its population, when it is realized that women's potentialities are thwarted by all sorts of built-in discriminations, that blacks suffer through unemployment, lack of education, segregation, and so on? The moral problem and the legal problem and the practical problem step out on the scene and decision and action come on the social agenda. But what is to be done becomes the center of controversy. Naturally, those who are entrenched feel that their position will be endangered if a vast expansion of resources is not available; profits will go down and standards of living will be reduced if the injustices are remedied. Counter-principles of justice tend to be invoked in defense, principles recognized as intuitively acceptable or socially useful—for example, seniority rights, prescriptive rights, need to keep initiative and motivation at a high level for production, and so on. There may even be dire warnings of unavoidable strikes or backlash as more ominous than the acts of disruption and despair from the disadvantaged. Or more constructively, there may be alternative proposals for remedying the situation, or specific procedures for seeking remedies. The one absolutely basic criterion in evaluating all the proposals from the point of view of justice is that they cannot be such as to result in the sheer continuation of the injustice to the outs. This is the *minimal* requirement. The *maximal* requirement is the actual achievement of an order that will bring those who are excluded into full participation within the society and its goods. The task of affirmative action, as I see it, is to guarantee the minimum and move toward the maximum, as fast and effectively as possible.

IV

Affirmative action can span a range of objectives from the minimum to the maximum just indicated, and it can employ a range of instruments. It makes considerable difference where it sets its objective and what instruments it employs. Maximally, it could impose directly some formulae of distribution, as in the use of quotas. Less than maximally, it could set goals and longer range strategies for diminishing injustices. Minimally, it could engage in mediation, keeping the problem alive and spurring action. In method, maximally it could use strong sanctions for enforcement, less than maximally it could call to account where progress is too slow or is thwarted, minimally it could engage in educational campaigns (or in the economic parallel what has been called "jaw-boning"). Such a variety of objectives and methods is familiar enough in other sectors of our life—for example, the critical economical problem of keeping prices and wages from advancing in an inflationary spiral. What objective is set and what methods used depend on how critical the situation is and what is the analysis of the forces at work. In the case of discrimination against blacks the situation was recognized as most critical after the violent upsurge in the cities a few years ago.

In the light of the heated controversies there have been over affirmative action it would be useful to look at parallels in our national and international life in which the same sort of thing was done and acceptable. I choose a national one in which in effect we used a super-maximal objective and compulsion as the means, and an international one in which less than maximal objectives and almost minimal means were used with perhaps retrospective regret.

The national example is the development of compulsory public schooling. The principle of a just distribution of basic education was quite simple—equal educational opportunity for everyone. We used the super-maximal solution of public education, education nationalized as it were, not something less such as supporting private schools or giving every child a grant to attend private schools. As to method, we use the truant officer to ensure attendance; and at the present moment (1975) our courts, as in New Jersey, are using judicial power to compel a revision of mode of financing for the schools to ensure a greater equality. To see that all this is a type of affirmative action we have only to look back to the situation that existed before the public schools arose—one in which only a few had educational opportunities and enjoyed all the advantages that education brought.

For the international parallel, look to what happened to many of the former colonies after World War II. The old imperial power was often the one left with a mandate. The mandatory power, according to the Trusteeship Council of the UN, had the obligation to bring the former colony up to the point of freedom, and then it would be set free. But periodically the mandatory power had to report on the progress that was being made and show that it was rapid enough to give freedom by a set date. As objective and method, this fell rather far below the maximum. We can, of course, look back on this procedure and see where it worked and why, and where it produced the kind of situation that was found later in the (then) Belgian Congo, though there may also be other reasons why such catastrophes developed. It does not require much hindsight to recognize that if stronger measurers had been possible—even moderate ones such as requiring a specific budget for education in the former colonies and checking periodically that it was properly distributed—some great evils might have been minimized. More realistically, it is hard to avoid the conclusion that the weak type of affirmative action there employed reflected simply the readiness of the mandatory powers to cloak their continued exploitation of the colonies under new titles. And a similar conclusion seems to be inevitable if present pressures to hamstring affirmative action programs against discrimination are successful.

What level of objectives affirmative action should set today and what methods it should employ are thus policies dependent on analysis of conditions, feasible means, urgency, and so forth. Since conditions of groups may vary, the policy need not be uniform with respect to all groups. For example, if we had a policy of affirmative action in employment for physically handicapped, it would probably have to be less strong today than a few years ago, since sensitivity to discrimination has become greater and we are quicker to see it in groups whose suffering has been hitherto insufficiently noticed. If we had a policy of affirmative action in employment for persons who have gone through a period of psychiatric institutionalization, it would have to be quite strong to counter the prejudice that still prevails. Again, to remove discrimination against some ethnic groups has simply required legal weapons, in other cases a strong executive action. In the case of blacks and women, the habits of discrimination are built into our cultures so deeply that only a rude shock may wake us up. The advantages that the established power groups have as a consequence of the discrimination are regarded as rights not to be touched. In practice, delaying tactics, obfuscating tactics, even shock tactics, are employed constantly. Whether the problem is one of intentional obstruction or deep-seated bad habits is not our present concern. The important point is that something strong has to be done to prevent the greater evils of nothing being done, or of injustice being overcome so slowly as to condemn several generations of the disadvantaged to despair and suffering.

The desirable level of objective for affirmative action can therefore be reached by thinking in terms of a *breakthrough* against the Maginot line of discrimination. Of course it will have to judge both strategy and tactics wisely, but to tie its hands as to means would be like saying that the Maginot line should be broken without tanks, or without planes, or without surprise in timing and attacks. There is always the grim reminder that if it loses it will have to be replaced by a strategy on a stronger level, just as in the opposite direction strategies can become milder if conditions become better—if peace is secured in the military analogy or full employment in our present problem.

One consequence of using the breakthrough notion might be setting a limited time perspective for the experiment. One could conjecture that a quarter of a century may be needed to change our national ways in habits of discrimination—just as ten years may be required for developing new energy sources. On the other hand, it is possible that some permanent regulation would be required, as in the case of compulsory public education. So we shall have to wait for the results of our present experiment in affirmative action to see where to go from there.

Another consequence of the breakthrough notion as a defining objective for affirmative action would be that we make no attempt to specify a hoped-for final state, to determine a formula for a permanently just distribution. For one thing, we cannot quite foresee all the changes in our institutional structures that a breakthrough and associated economic and social changes would bring. This raises, however, the heated problem of quotas.

V

Are quotas in affirmative action a matter of injustice, folly, or unavoidable necessity? There are all sorts of quotas—loose quotas, tight quotas, unjust quotas, reasonable quotas. The term had largely become associated with unjust quotas of exclusion, as a means of keeping out racial and ethnic groups and women from desirable participation and control. Jews have particularly suffered in our history from quotas that aimed to exclude.

There are two questions about quotas which are currently discussed as matters of principle, as involving principles of justice. They fall, equitably, on opposing sides. On the one side it is argued that any attempts at quotas to remove discrimination would have to be stated in terms of race or sex and would constitute unjust and unequal treatment, defective both in morals and in law. In short, it is assumed that preferential consideration of A on grounds of race or sex is *ipso facto* reverse discrimination *against* B on grounds of race or sex. On the other side, it is argued as a question of *right* that if women are fifty percent of the population they should have fifty percent of the jobs. I think that the use of both these arguments in a somewhat a priori fashion obstructs an effective understanding of the pros and cons of quotas.

The use of a race or sex label is certainly invidious when its purpose is to exploit, oppress, or exclude the designated group from participation and social benefits. It is not invidious when its purpose is to liberate, bring into participation, open the area of benefits. The group has to be designated by the race or sex label because the oppression exists or existed against the group as so labelled. In the recent negotiations to allow more Jews to emigrate from the Soviet Union, no one raised the question whether this was discriminating against other groups that might want to emigrate from the Soviet Union. There are many areas in which a loose quota has been readily considered just and reasonable. If Hawaii had a law that every board of education should have at least one Chinese member, one Japanese member, one native Hawaiian member, the ground for it would be quite apparent from the history of the different groups to win a place against discrimination. Perhaps one might object on the ground that it did not also have a Samoan member. This need not be a permanent policy, it would depend on the state of the problem of discrimination. In New York politics, it was long a practice to have a political ticket with one Catholic, one Protestant, one Jew, and later on, one Black and one Italian. This was a "practical" not a "principled" matter; no one in the political clubs would have accused the district leaders of imposing quotas. (In the long run, one would hope this would give way to the "best person" principle; but perhaps probability of being elected is one of the criteria of "best.") I do not recall the quota charge having been raised when the courts insisted in effect, that blacks be on jury lists; the reason was too transparent in the light of the grim behavior of all-white juries. Why then is the quota charge raised when affirmative action calls for women on police forces? Or for blacks in law schools?

The argument for numerical representation as an inherent right has no clear basis. In general there is no inherent relation between being a woman or black and the criteria for job occupancy. If there are criteria in particular jobs, then maybe women or blacks would be entitled to most of the jobs. More likely numerical representation is being invoked as the only way to be sure there is no discrimination. But affirmative action does not need the lack of proportionality to investigate. There should be no bar to its investigating on complaint or even at random, as the Internal Revenue office does. (Its reason is the same: to check for error or because it is operating in a domain where the tendency to avoid obligations is strong and widespread.) What it investigates for is progress in overcoming discrimination and habits of discrimination. Even if all jobs were filled by randomizing devices there would be no guarantee of proportions in each enterprise, though it would surely guarantee no discrimination.

The fear—often expressed in caricature—that every group that can fashion an identity will therefore demand representation, is not well-grounded. Not every shoe of esoteric style pinches, and if it does it might be settled by a little stretching. Complaints have to be well-grounded and serious and a good complaint department will treat according to the case; it will not stop business because it has had enough complaints. For example, should left-handed people demand a quota? It all depends on what is going on in the society. If left-handed people are being compelled to turn right-handed to get jobs, and if this produces a distortion of personality with far-ranging consequences, and if the dominant right-handed community consistently discriminates in education, in the production of toys and tools that favor right-handedness, and if there cannot be ready solutions—then perhaps it is time for left-handed people to demand representation on boards of education at least. Is there any doubt about the complaints of the handicapped in employment? Recent history and the proliferation of liberation movements have encouraged complaints about ever-fresh categories of discrimination.

A humanistic society can only welcome this and look at each on its merits. Representation demands and loose quota demands are the cry of the oppressed and the loud sigh of the neglected. It is well to recall at the time of the bicentennial that "no taxation without representation" was the first demand of this type.

There is a great deal to be said against quotas as an instrument of affirmative action. They are drastic, possibly dangerous, and likely to be extremely disruptive. But they could be effective in breaking through. Recently, I heard a law professor defend them as the milder alternative. The most obvious, and most revolutionary, and most disruptive thing to do when we discover a basic injustice in the distribution of jobs, he said, would be to fire everybody and start afresh. Quotas at least would be only partially disruptive. The crucial point, however, was his further assumption that nothing less than quotas has worked or was likely to work under present conditions. Now whether this is a good general argument or not, it clearly has some domains of application. One of the most serious at present, in the recession and/or depression we are undergoing is that seniority rules will mean that the disadvantaged as most recently hired will be first fired. Hence the demand has arisen for two lists of seniority, with firing from both but so as to maintain some kind of proportion not less than presently existing. It is hard to see this as unreasonable, but it should prompt a search for a more innovative way of settling the issue. Again, there are domains in which even the more extreme procedure of starting afresh might be quite applicable. I recall a case—though not in sufficient detail—of a longshoremen's union that had hiring halls but was dominated by its officers who kept favorites in established jobs. When the union was democratized, it instituted a rotation system: everyone who finished a job went to the bottom of the waiting list, and had his turn. There was no seniority. In such a situation, a disadvantaged group had only to be among the members to be fully equal; good and bad times were equally shared. Of course this type of solution is possible only under certain conditions such as: a closed or limited number, with careful control of entry much as in the old guilds; a strong union with a thoroughly democratic outlook; union control over job assignments; probably also effective unemployment insurance; and some degree of constancy in jobs. There could be social consequences too, such as the union rather than the job center becoming the functioning community. And probably the type of job would have to be one that any representative member of the group could do competently.

In many contexts, then, we can get behind the question of quotas to a very complex set of different mechanisms; there the concrete question will always be which mechanism to use and what results it would secure. There is no gainsaying the widespread objection that is today found to numerical quotas as a general policy. The grounds for this common judgment range from the historical passions roused by past use of discriminatory quotas to the recognition that in practice they would yield an over-rigidity and engender collateral injustices. But note that such reasoning is not based on any intrinsic injustice in the conception of loose quotas as an instrument of breakthrough. There is instead the assumption (or the hope) that other ways will be found to remedy the injustices for which quotas have been invoked.

Let us finally look at the charge so current today that affirmative action is just a devious way of imposing quotas, that no tenable distinction can be drawn. The Department of Health, Education and Welfare generally draws a distinction, and so did the Democratic mini-convention when it settled on the idea of affirmative action to keep peace between the old and the new forces. On the other hand, some labor leaders, especially in education, have constantly expressed suspicion of affirmative action. Such instances reflect clearly the position of the "haves" who are entrenched. But equally, the distinction may reflect the dilemma of interests caught between the "haves" and the "have-nots" which are grasping at straws to prevent a break. We must therefore attempt dispassionately to analyze the possible differences. By "quotas" is meant in this context the use of numerical proportions, and by "affirmative action" programs to break up discrimination in a general sense, abstracting from the means adopted. The issue is whether there can be a viable affirmative action which is not tied with quotas.

There are a number of differences suggested by theoretical distinctions that have a long history in moral philosophy. They may be indicated briefly:

1. Quotas rest on formulae or recipes, affirmative action on more general methods, or principles operating as ways of analyzing a situation rather than as fixed directives for action.
2. Insofar as anything is measured in affirmative action it is a degree of success in the effort to solve a problem; in quotas the numerical relations at the end are measured. In affirmative action results may be looked at to get an inquiry started.
3. Excuses have a greater scope in affirmative action than in quotas.
4. In affirmative action there is greater particularization and differentiation of means than in quotas. There is recognition of stages, processes attempted, close attunement to types of problems, and greater credit for progress and attempts at progress.
5. In quotas, exact results are a mark of success; in affirmative action they may even be a ground for suspicion.
6. Affirmative action stimulates or engenders purposive efforts and inventiveness in solving problems. Quotas are more narrowly confined to a path.

Clearly then there is no logical difficulty in the conception of affirmative action without quotas.

The action of the Democratic mini-convention (1974) was in many respects a paradigm of the whole situation. Though determined by pressures and threats of withdrawal, it was held together by the realization that unless some solution were achieved there would be greater losses for all in the next election. It acknowledged the aim of wide participation and called on everyone to take action to secure it. It opposed mandatory quotas, but allowed challenges to delegates where affirmative action performance was inadequate. It ruled out proportion in results as the sole basis of the challenge. And there was the further guarantee that where an exact equality appeared between men and women it should not be seen as a violation of the previous conditions. (There was a battle over a proposed clause about burden of proof based on the composition of results, but this was dropped, so that there is no mandatory burden of proof.) In our society as a whole, all have as much to lose, as vast a stake in removing discrimination and healing our society as the Democrats had in their hope of election success. The detail of the pattern could fit equally well for investigations about discriminatory practices in employment and education. Any use of proportions could only be as starting points for inquiry. The main force of the program is to make sure that something is being done, that effort is directed to removing injustices, that a plan of action is on its way. That it requires sanctions in order to make the breakthrough is an unfortunate fact of our culture. It would, of course, be much better if we internalized the sense of injustice and the pressure to remedy it. But as a people we have been very blind in these moral matters, in spite of our humanitarianism. The story is now very likely to be repeated on the international scene if we do not quickly appreciate in action as well as in thought and feeling that we are using a tremendous part of the world's energy and resources for our own consumption while a great part of the rest of the world is in a deprived state. The questions of justice in that context are even more complex than the ones we have been examining.

Leites and Wolfenstein, in a book some years ago comparing the movies in different countries,[2] found that in the United States we tend to externalize our conscience. Our Furies, they point out, are not within us, but lie in the police car and the siren. The solution of problem—say the familiar sex triangle of the older movies—is not internal debate and decision but a car accident in which one of the parties is removed from the scene. The problems to which affirmative action is addressed could of course be reduced to a minimum by external means—by full employment, by universal education and wide opportunities for professional education, by the development of assistance in counselling and in providing resources for all types of disadvantage. But our society has not worked out such solutions in an institutional way. It has moved slowly toward social security, universal medical care, guaranteed minimum income, and so forth. It needs the greatest imaginative scope to break up questions, to devise a variety of solutions for different types of professions and vocations, different types of labor and learning and teaching. It needs less wielding of general principles that on analysis turn out to be not formulae of justice but bludgeons that will leave things simply as they are. While we mature in all these ways, we need a conscience that will not let the injustices lie no matter how much it hurts us. Affirmative action is such an externalized conscience. That we need it is the mark of our immaturity. Eventually it will give way — we can only hope — to reconstructed institutions and a sensitive humanity that does not even think in the terms that have constituted the grounds for discrimination.

VI

Will justice allow us to end on a note that is forward looking, but leaves its date of maturity ever renewable, constantly postponable? Conscience, it is true, takes time to develop, but continual emergencies provide everready excuses. Just as the energy crisis makes inroads on the attempts to end obvious and glaring pollutions, so the economic crisis, with its unemployment, makes inroads on ending obvious and glaring discriminations. This is where justice — the most demanding of moral cries — refuses postponement, and that is the beat to which affirmative action marches. If there have to be sacrifices to meet crises then the theory of equitable sacrifices must include readjustments to diminish discrimination.

Reason and experience add their voice to this demand. If anything is clear in our economic development, it is that the numbers in agriculture and manufacturing and industry generally have gone down almost without end — when I last looked at it the agricultural was down below five percent and the industrial moving below twenty-five percent, perhaps by now much lower as automation moves apace. The one growing area capable of almost indefinite expansion, are the service industries. If we look at the picture as a whole, our choice is between supporting a large part of our population on welfare and in unemployment, or developing some rational plan for the total use of our population in useful work. This means the acceptance of the goal that was raised in the 1940s of full employment whether or not we export sufficiently to "cover costs." We do not engage in such reckonings when we are at war; we make whatever social arrangements are required to do the job. So too, full employment is the only alternative to the bitter struggle over scarce jobs in which not only justice and the voice of justice is lost, but all sense of dignity.

To maintain dignity means participating in meaningful work, not make-work or bread-and-circuses. It is time to exercise a collective imagination to develop the kinds of jobs that the society needs. Some are obvious: instead of talking of an over-supply of teachers with present large classes in both the lower and higher schools, we could use an almost endless supply of teachers in the proper pursuit of education. The care of the aged is another domain (in which we could take a few lessons from Sweden.) Keeping our streets clean all the time and our beaches and rivers clean could involve endless labors as well as strong civic passions. And the advancement of knowledge and the development of a rich cultural life for the people as a whole could add a wide sector of satisfying labor. An so on, and on.

Unless we go in for these kinds of projects and policies, we shall be prey to the irrationalities we find about us. What more self-defeating than to demand an arsenal be kept on just because people are employed in it, rather than that there be a reconversion of labor and facilities? The same argument applies to our automobile recession. If there is no demand for cars or large cars—except Cadillacs—and no real need for an over-expanded production of cars in a long-range plan for transportation and for avoiding pollution, what we need is planned reconversion of the industry just as when it was reconverted from war uses. Our present pattern of goals is like hoopers objecting to importation of coffee because it cut down beer consumption and so diminished the demand for hoops on beer barrels. And our strategies of solution are little better than—to take a recent example—one state fighting in the courts to prevent another from seeding the clouds and so robbing it of the rain that the prevailing winds would normally bring into its territory. Both might have worked on a cooperative plan for water conservation and development.

It may very well be that the depth of discrimination in our present life will be the point of pressure at which a sober rationality will be brought home to us in our total national life. Historical causality, as the philosophers are fond of pointing out, follows strange paths. It is time to resolve by self-conscious plan many of the problems that have been left in the past to the blind play of forces. If the situation that has produced and supported affirmative action does have such an historical impetus, justice will have brought more than its own reward.

Notes

1. For a fuller discussion of such issues, see Elizabeth Flower, " 'Affirmative Action and the Principle of Equality' Reconsidered," *Studies in Philosophy and Education* (forthcoming).

2. Martha Wolfenstein and Nathan Leites, *Movies: A Psychological Study* (Glencoe, Ill.: Free Press, 1950).

16. Notes on Terrorism

Because terrorism has become such a serious practical problem, a great deal of energy is today devoted to guarding against it. At every airport we are now aware of this. Journals carry articles on how negotiations were handled in specific cases of hijacking of planes or capture of hostages, or how special police or army units are trained for dealing with terrorists. The psychology of the terrorist has been discussed, the role of publicity in setting the stage for his operations, how the well-behaved hostage should conduct himself, and numerous other matters.

Obviously the terrorism envisaged is an evil and concern with it is largely directed to preventing and eliminating it. But surprisingly there does not seem to be sufficient agreement on its interpretation for general legislation against terrorism, for it would not be wholly clear what was being banned. Even in the case of less than that—the hijacking of planes—there are political problems in securing international agreement for a universal ban. Could there possibly be any reservations about terrorism?

Perhaps among the contemporary inconveniences of terrorism we shall have to include theoretical burrowing into the concept of terrorism itself. The theoretical understanding is likely to have most practical bearings, indirectly if not directly.

Problems in the Quest for a Definition

We have plenty of clear-cut illustrations of the phenomenon: gangs terrorizing a trade into paying tribute, assassination of officials and leaders, Nazi tactics in an occupied town, the wiping out of towns in war to intimidate the civilian population, hijacking planes and holding hostages, letter-bombs, even retaliations to terrorism. Because it strikes us as a distinctive phenomenon we begin to focus on it and think it deserves an essence of its own. This is prompted also by the natural tendency to straighten a bent stick by bending it in the opposite direction; for, as Wellman pointed out,[1] terrorism has been treated largely as a footnote to the treatment of violence.

Certainly analysis of the concept of terrorism is prior to formulating or criticizing general conclusions about its use. Wellman, conjecturing what his hypothetical *New and Improved Dictionary for Philosophical Analysts* would say, defines terrorism as "the use or attempted use of terror as a means of coercion." This fits central clear cases of conspiratorial organization to terrorize a population by surprise violent actions. The fit is less neat for individual threats and actions or for situations in which general terror results without intentional "terrorism." As an example of the former, Wellman even includes threatening to fail students who hand in late papers. An example of the latter would be a community in a state of terror because of the frequency of muggings, so that scarcely anyone dares to go out after dark. This is terror without terrorism. The individual mugger may not be interested in inspiring terror to achieve his end; cool prudence in the victim may be enough to make him freeze. Perhaps we could go on to distinguish carefully between a state of terror (which the appearance of a wild animal might inspire without its being a case of "use," though no doubt some evolutionary properties of inspiring terror might have accounted for the success of such animals), the use of terror indifferently among a number of ways of securing one's end (the highwayman, like the mugger, does not care which feeling makes the victim hand over money or what the victim's affective structure may be), and the deliberate or intentional use of terror as the sole or chief means. But perhaps the strength of the concept is wasted if we associate it with an isolated act rather than reserve it for a more systematic structure. And apart from all this we must recognize that even the success of one definition in opening up interesting questions does not mean that an alternative definition may not open up other serious questions.

To speak of wasting the strength of the concept expresses, perhaps, a different approach to the concept of terrorism and the problem of definition. Terrorism is an emerging concept rather than one endowed with an established essence, and what will emerge will be a quasi-technical usage on which will hang a variety of legal and moral consequences. Meanwhile, in the rich complex of phenomena there are different directions we can go in essence-building and as we fix on one or another significant feature in a striking case we may be carried at high speed past other possible turning points and find ourselves on a one-way highway. The legal and moral consequences may bring regret that the itinerary was not more cautiously experimental. For example, a definition might take as its point of departure that in terrorist phenomena all restraints are abandoned, any means being employed; when this is coupled with the recognition that morality involves some internal restraints, it follows that from the outset terrorism is immoral by its definition. Or by contrast, the point of departure may be the prevalence of violence in the phenomena of terrorism; this yields the familiar treatment of terrorism as a species of the genus violence, hence the debate over the types of conditions that would justify violence and even those which might justify terrorism. (Nielsen's analysis moves in this direction.) Or again, the point of departure may be the psychological state of terror; this is Wellman's starting point, which he couples with the recognition that terror is induced in order to compel unwilling subjects to behave in a desired direction. Since terror may be induced in nonviolent ways—for example, by blackmail with a well-manipulated threat of exposure or in threats of loss of employment and ruin of careers as in McCarthyite charges of communism in the 1940s and 1950s—Wellman is led to exclude the reference to violence from the definition of terrorism while recognizing that violence stands out because it is a very effective means of inducing the terror which then has the coercive effect. Of course defining terrorism as a species of violence may take the saving move of extending the concept of violence itself to include institutional violence and psychological violence as well as physical force. (Of this, more later.) Again, a wholly different point of departure may be a focus on some classification of the ends for which the coercion is being exercised. In this vein, just as there are theoretical attempts to remove the taint of violence from legal use of sanctions involving force and limit it to the illegal, so inducing terror in the "criminal classes" to ensure legal observance (e.g., by high payment for informers, latitude for the police to use physical coercion, arming police with unusual weapons, quick convictions and stern sentences including physical and capital punishment) would not be taken to be terrorism.

Such varied handles for conceptualizing the phenomena suggest that our theoretical concern in definition is not with fitting a well-patterned set of current uses of the term, but with fashioning a concept that will bring the clearest understanding of the phenomena and give us the best handle for coming to grips with the social and human problems involved. To get the investigation going comprehensively is more important at the present stage than a (possibly premature) formal definition; the latter can wait until we have more knowledge as an outcome of the investigation—if we turn out to need it then. Meanwhile, it will not matter much whether we think of terrorism as a species of the genus violence but with striking properties, or as an independent genus whose most important species is differentiated by the use of violence (which as an empirical matter is probably the most effective means of inducing terror). Whichever definitional policy is ultimately decided on, there will have to be a host of contextual or nonformal qualifications. For example, minimally there has to be a certain degree of magnitude in various features: a certain degree of alarm, a certain temporal extent, a certain degree of organization or systematic structure in the process, and so on. The important thing at this point is to gather significant dimensions.

Several dimensions are fairly obvious. One is the *range of potential sufferers*—is it determinate, such as the wealthy or officials who have behaved brutally or a designated race or ethnic group, or is it any random or chance collection? Certainly it is the high degree of randomness which is sometimes central in intensifying the atmosphere of terror. (A specific revengeful murder will not cause alarm, while random attacks will, even though less violent.) A second dimension is *motivation*—it is particularly significant whether the terrorism aims at monetary gain or is directed to advancing some cause. A third dimension is the *institutional setting*—does the terrorism come in peace or war, in prisons and police action, or is it prevalent even in education and family life? A fourth is the *cultural pattern* and the place of violence in it. A fifth is the specific *historical setting*—for example, whether terrorism occurs in the historical moment of a rising class or a declining class, in the context of revolution or a reaction against it, in a period of self-indulgent prosperity or of deep economic depression. Doubtless there are other significant dimensions.

Such dimensions could be of use in different ways to understand the phenomena of terrorism and to get a handle on coping with them. Certainly explorations along all these lines would be necessary in seeking a formal definition. For classification of types, perhaps the dimension of motivation (with some attention to institutional setting) would be the best preliminary basis; thus we shall distinguish later *terrorism in predatory crime, terrorism in war,* and *moralistic terrorism* (that is, terrorism in the service of a supposedly moral cause). But to see whether these involve different lessons and different principles for their understanding and handling or whether they are part of a common problem, we have to go beyond psychological motivation to the question of probably deep cultural roots. Again, in the frequent controversies as to whether violence in general and terrorism in particular are ever justified, we find arguments based on the particular historical situation and what it calls for. The remainder of this paper consists of reflections on some of these questions, beginning with controversies about justification, going on to cultural aspects, and then focusing on the different types of terrorism indicated.

The Question of Justification

Since terrorist phenomena involve so much that is counter to human morality, a question of justification, if it is raised at all rather than a question of explanation, is formulated as a problem for the terrorist. We expect him or her—not the sufferer or the public—to say what are the reasons for such desperate action, what the ends and means involved. In fact, terror is rarely seen as an end, more often proposed as a means, and sometimes affirmed as a punishment. Each of these merits exploration. At some points the exploration can draw on the more familiar study of problems of justification for violence in general, since parallel issues are involved.

To terrorize people is rarely conceived as an end. Even a culture that makes a virtue of truculence and admires the bully probably has strength as the cultural goal and bullying as reliable evidence. Similarly, the joys of sadism, that perennial source of terrorizing others, have been sufficiently explored psychologically to be revealed as clinical symptoms rather than as an intrinsic value that happens to be socially dangerous. In the case of violence generally, particularly war as a seedbed of heroism, the claims for intrinsic value have been not infrequent in the past. But even in the *Iliad*, glory in violence was a fairly transparent avenue to status. And whatever the exaltation of war and the warlike in olden days, by the time of the world wars in our century pre-fascist and then fascist doctrines were clearly glorifying violence as an ideological weapon in the battle with a rationalistic liberalism. We may thus dismiss justification of violence as an end, not by stipulation but as the verdict of evidence about means and ends in human life.

The claim that terrorism is on occasion a justifiable means brings it into confrontation with a pacifist approach which rules it out on antecedent moral grounds and also with historical judgments about its efficacy. Since this is a separate problem, our comments here will be restricted to a few summary remarks.

The pacifist view applies to all violence and so *a fortiori* to terrorism: violence is not to be countenanced morally under any conditions. This has not, on the whole, won general acceptance as an absolute for action, though it is found in a "rather die than" attitude in some. But even those who reject it as such a categorical absolute can retain it as a universal principle in moral reckoning, a kind of aspect rule or break-only-with-regret rule, to the effect that wherever violence is used it clearly diminishes greatly the value total of the situation. Nothing prevents such a rule from weighing universally, even though under certain conditions it is outweighed by other considerations of value. Such a universal formulation, however, is too wide a net to catch and hold back even the terrorist who freely admits the evil of what he is doing but claims that vastly greater evils are tolerated by his forebearing.

The general pacifist position in an instrumental mode—that nonviolence is a much more powerful means than violence—does seem to hold only for limited conditions. Consider, for example, the familiar case in the Indian nonviolent movement for independence, where large masses of people sat down on the railroad tracks to stop the trains. What happened would depend on the character of the train engineer. The British engineer would tend to stop the train, but we may assume that a convinced Nazi would plow straight into the crowd. (There is the story of a Japanese engineer who slowed the train down to a crawl and then jumped off, leaving the demonstrators to make their own decision.) In the early phases of the Civil Rights movement in the United States, nonviolence expressed the highly moral tone among the participant youth; but this proved a passing phase in the move toward Black Nationalism. It is not therefore simply a question whether absolute pacifism is a rationally defensible moral position, but also whether nonviolence as a method is likely to be successful. It surely deserves the same kind of evaluation as method that one gives to violence—especially as its intrinsic aspects of mutual respect are so much superior to the intrinsic evils of coercion in its opposite. Now while it is empirically true that nonviolence is not a dependable instrument, it is equally the case that we have no really successful analyses of why it fails where it does and why it succeeds where it does. The question of its theoretical weight is therefore still unresolved, as well as the assumptions underlying some of its forms. For example, there is the hypothesis found in Tolstoy and Gandhi that nonviolence releases a basic human drive. Whether it is thought of in religious or metaphysical terms as a love force or in quasi-biological form as in the anarchist faith in sympathy and mutual aid, it still remains one of the major theories about the human constitution. Our scientific psychologies have been too busy exploring guilt and aggression to take time to explore sympathy and mutuality, and only their occasional resurgence from the social depths reminds us that our theoretical picture is far from complete.

The claim that there are conditions under which the use of terror is justified is still left open by the confrontation of violence with pacifism which we have considered, for it is possible that terrorism may be the only viable means in some limited and very unusual situations. In those cases it would have to face on its own the historical arguments referred to above, which were directed against violence generally but might seem to hold *a fortiori* for terrorism. Chief among these has been the view that resort to violence is a sign of weakness in social struggles. This need not, perhaps, apply to war in which it has often been the excessive strength of a nation that prompted it to war against those who refused its demands, and in the course of the war to employ terrorist tactics where popular resistance was strong. (Not to be able to win without terror here might, of course, be construed as *relative* weakness.) In the case of revolutions, confidence of strength has also played a part, though occasionally an uprising expresses suffering and despair. Nevertheless, the view that resort to violence is a sign of weakness has a certain plausibility. If you can win an election or have the army solidly behind you or (in the history of the labor movement) have an unchallenged and firm picket line, why use violence? The history of sabotage in labor struggles would seem to bear this out; often it appears as a rear-guard action after a strike has been lost, expressing frustration or else desperately attempting to keep the action going. Perhaps the qualification with respect to relative weakness should be extended here. It is not just weakness that is involved in the resort to violence, but sometimes the kind of relative weakness that is found in a precarious balance of strength. For example, a revolutionary terror may come from a governing group that fears a counterrevolution, or a rightist terror from a governing group that is losing its grip. Terror in either case is directed toward preventing the consolidation of the opposition.

The view is sometimes urged that terrorism is not as effective as is often thought, that it is wasteful, that alternative methods for achieving a good end can usually be found, and so on. But there still remains the possibility that although it expresses weakness, it may sometimes be the only available resort of those who are laboring under injustice, and the only alternative to permanent resignation. Here the hope is that the instrument of terror will keep the issue alive, and even serve to integrate the oppressed. But suppose it would, does that constitute a justification? And is it the same kind of justification that would be offered for a declaration of war?

There remains the element in terrorism that suggests the treatment is deserved. Terrorism then poses as an unusual form of punishment. Even where the terror is indiscriminate in its victims, it chooses them out of a certain class, as when public places in England have been bombed because the British do not withdraw from northern Ireland. If this is meant to be a mode of punishment, it clearly assumes collective responsibility; the morality of such an assumption would have to be considered. Or else it is punishment in the sense of deterrence—making one public suffer as it is asserted the other public has suffered—so that it will be stimulated to urge the withdrawal. In any case, it sets a limit to the indiscriminate character of the terrorism, in specifying a reference class.

In spite of such claims, the general moral objection remains that in terrorism people are being treated no longer as persons but as means. I do not know how far this will carry the argument; it is quite possible for the reflective terrorist to affirm his respect for the victims in the same sense as one side in a war respects the soldiers on the other side (though more often this has been explicit in the relations of opposing officers). Nietzsche, it will be recalled, wanted all crime treated as rebellion since it would respect the criminal, whereas a reformatory theory seemed to him to be manipulating the person. Presumably, an imaginative terrorist might declare a day of respectful mourning for the victims of his efforts, using it also to blame the reference group for these consequences of injustice.

In any case, terrorism is a great evil. From the terrorist's standpoint it is a necessary evil. His task of attempted justification is harder than the justification of occasional violence: terroristic violence is not violence that is resistance to violence, it is not protective violence or ordinary deterrent violence, or violence in the service of an established institution or of a revolutionary movement as such. It bears the stamp of a violence that is beyond the line of duty and beyond ordinary demands. It need not be uncontrolled, and can be duly measured; it need not be sadistic and probably loses its character if it is such. It has a kind of ultimate character. It deals in death, often in random surprise, in utter disrespect for the person, in what seems like a calculated irrationality. When the terrorist undertakes to argue about it, his theme is that his end is good, there is no lighter means available, and that the means should be endured because even with its evils the net result of advancing the end is better. He is likely to be dogmatic in his justification, he will not stress the difficulties involved in judging, nor make subtle differences of degree between violence against person and violence against property. Nor will he usually consider the possibility that the means have so high a cost to people that the end should rather be given up or left to a later period when more moral means become possible. Nor once embarked on terrorism is he likely to consider the moral need for less rather than greater violence.

The approach through justification, from the point of view of the terrorist, has not carried us far enough into understanding what is going on. Perhaps the language of means and ends and that of cost-benefit analysis, and even that of respect for persons, misleads us in the situation under discussion. It talks as if terrorism were a mistake in calculation or else perhaps a correct reckoning where it succeeds (something like deciding whether to use nuclear fission or coal for generating power to carry on the normal processes and achieve the normal ends of society), or else neglect of a proper interpersonal attitude. Now all these categories of analysis are perfectly proper but they are too general. The understanding of the specific phenomena that is needed for their evaluation may be more specifically cultural and social; and the broad philosophical attitude required for developing a normative policy, that is, asking what is to be done about terrorism, may come from looking at terrorism not only from the point of view of the terrorist's application for justificatory credentials, but also from the standpoint of the sufferer or patient.

Cultural Basis of Terrorism

Terrorism gains wide currency in a society in which violence is an acceptable means. (Neither is an expression of human nature in the raw.) Philosophical controversies over what violence is legitimate and what not, or even over what is and what is not violence, are indicative of a growing self-consciousness in the culture, though often in an oblique way; they are rarely simply verbal disputes to be resolved by linguistic analysis. This may be illustrated by taking two controversies—one old in political science, one new in sociology.

The first of these attempts to distinguish between force and violence. Force is declared a neutral term, whether the force of the policeman or of the burglar. Violence is the use of force that is not legitimate; it is used by the burglar, not the policeman. Such an analysis, however, merely passes the buck to the concept of legitimacy. If it has moral connotations, that is one thing. But if it is purely legal, then it raises all the questions of the Nuremberg trials about mass killings in accordance with law; similarly, about repressive violence by a government taking power through a coup, as in Chile; but also even about an established democratic government doing what the Americans did with napalm to intimidate civilian populations in Vietnam. Surely the latter would have been violence (and terror) even if there had been prior and explicit congressional authorization for the use of this weapon. The attempt to limit "violence" to what those outside of an Establishment do, and to deny it of the behavior of Establishment figures and supporters, represents the efforts of the Establishment to take refuge in a moral garb, usually when it is under moral attack. How much of that garb covers real social needs, how much simply the nakedness of an order in trouble, is precisely the issue that no linguistic legerdemain can dispose of.

The second controversy concerns the sociological use of the term "institutional violence." This is at the other end of the spectrum, for it extends the term "violence" to cover areas in which there may not even be overt force involved. The extension is warranted only if we exhibit continuities (often discovering them with the aid of theoretical advances) with the overt uses of force. No doubt this involves considerable analytic ordering in the nest of related concepts. The sense of "institutional violence" is both theoretical and moral. It is concerned with what actually goes on in prisons, in schools, in the very habits of male-female relations as well as parent-child relations. Take, for example, the time when spanking was an approved (legal and moral) procedure in dealing with children. It took a Plains Indian wholly outside our culture to recognize it as degrading violence—though perhaps our children sensed it too. It took the growth of psychological sophistication among adults to see it as their own aggression, not as a neutral instrument for the child's well-being in preserving a moral fibre. It took a wrench to see it from the child's point of view, much as photographers have sometimes tried to compose pictures from the child's perspective that would show the threatening look of a gigantic adult. Similarly, it has taken a whole liberation movement to uncover the character of male violence in relation to the female. If the concept becomes stretched when it is attached to the violent word as well as the violent deed, perhaps it nevertheless calls our attention to the common psychological essence. We must not be in a hurry to dismiss notions of psychological violence. In the current concern with what goes on in prisons, isolating a prisoner is none the less doing him violence even without beatings. Whatever be the conceptual outcome in the refinement of "violence" it should not be such as to obscure the theoretical findings nor sidetrack the moral critique.

Such illustrations show that our analysis of terrorism cannot stop with the first three dimensions of the list given earlier—the range of potential sufferers, the type of underlying motivation, and the institutional setting. To understand and evaluate for practical action we have to go on to the cultural pattern and the specific historical setting. With respect to the first, it has been noted often that in the United States we have a violent culture, compared even to other branches of the Anglo-American tradition. Growing consciousness about the incidence of violence not merely in penal institutions but in familial institutions (studies of child beating, for example) point in this direction, as do attitudes toward guns and their control. It is not a simple judgment, and we cannot here go into the many aspects that have to be explored both concerning practices and concerning attitudes. Its relevance to terrorism is simply this: to condemn terrorism taken as an isolated phenomenon may be simply to condemn a particularly aggravated symptom without understanding the human illness that underlies, produces and supports it. This does not mean that the symptom should not be treated, but that the disease should also be diagnosed and understood.

The cultural dimension is not enough. We have to look to the specific historical setting. This is often indicated in the recognition that the Vietnam war had a particularly brutalizing effect; discussion of the My Lai episode made this very clear. In the other direction, television brought the horrors of the war into the home, though as a spectacle and interspersed with advertisements providing prescriptions for the body beautiful. But the Vietnam war was not itself an isolated historical incident. Its own analysis would carry us into the present state of world conflicts between the haves and have-nots, the transition from colonialism to the problems of the developing countries, the relations of strength and belligerence in the advanced industrial powers. To take a recent analogy, we are no more likely to understand terrorism without such a depth analysis than we are to understand the recent revelations of bribery of government officials abroad by large American and multi-national corporations without a full analysis of business motivations and business practices in the whole range of their operations. To keep to a purely general moralistic critique of terrorism would be like offering a critique of the bribing as simple individual dishonesty.

It does not follow from this that the vision of long-range understanding must stand in the way of short-range policies. They have to go hand in hand. We thus turn to the types of terrorism distinguished above, treating briefly predatory crime and terrorism in war, then the terrorism associated with social causes.

Terrorism in Predatory Crime and War

The finer distinctions that have to be made in types of terrorism can best be approached initially from the standpoint of the patient or sufferer, that is, not the immediate victim—unless he survives and his experience serves to intimidate him—but the group of potential victims whose intimidation by the act and the way it is done constitutes the agent's purpose. Now one thing is clear: the potential victims do not regard the contemplated suffering as merely chance misfortune, like discovering that one has cancer or has caught a disease in an epidemic (where questions of blame for carelessness are not at issue), or for that matter being shot at by a madman firing from a tower at anyone coming into range. People may be alarmed in such cases, but they are not terrorized. They can try to meet the dangers only by greater caution, systematic diagnosis of causes and probabilities, research for preventive measures.

In predatory crime, any terrorism that occurs is carried out for gain. How will a shopkeeper look upon demands by a gang for payment for "protection"? Conceivably it could be seen as informal taxation, or as a part of the cost of production to be passed on to consumers. But more likely it will be seen as straight crime. Against such crime, the familiar policy is protection, deterrence, punishment. In extreme cases where these are insufficient, people will be driven into a kind of garrison life. Only in very extreme cases will the relation of potential victims and criminals begin to resemble war. Usually it is something less: there are defensive acts that the potential victim cannot take—for example, to shoot to kill rather than wound, to gather evidence by unauthorized eavesdropping—and if the criminal is captured, the victim must allow him to be brought to fair trial.

Terrorism in war engenders different attitudes in the potential victims. It tends to be more indiscriminate, and with fewer restraints. War is an all-out affair, it is fighting to the death to win. (I am referring, of course, to the traditional conception; contemporary war is getting much more complicated, and third parties often try to get it ended as quickly as possible before it spreads and gets out of hand.)

Even in war there are differences in the kind of situation from the sufferer's point of view. There may in some cases, for example, be options. If the underground in an occupied country threatens assassination for brutality by occupying officials, the latter have the choice of not being brutal. Worst of all is the indeterminate war terror—wiping out whole villages to intimidate the civilian population, taking random hostages and shooting them, and so on. Whereas in the case of crime to obey the terrorist demand is often a viable option, in the war model it often amounts to surrender. The innocent victims, if they manage to survive, are drawn into the war on their side. Their suffering is now no different from that of the soldier who is shot at and hit; they have been conscripted by a different process.

It is important not to treat war as crime nor crime as war. There are occasions where the concentration and organization of crime may approximate to war and there are kinds of actions in war that may, as in the Nuremberg trials, become identified as crimes and punished as such. But the categories remain distinct, as do the programs of action for overcoming each.

From the society's point of view, can there be moral demands laid on the terrorist whether in crime or in war, which limit the scope of his action? There is a type of moral rule which takes the form "Never do this, but if you do, observe such-and-such cautions." The sanctions involve, of course, adjusting punishment to the degree of conformity to the regulations. For violence in many of its forms this is quite possible, where crime and war are regarded as professions and invited to work out the rules according to which their ends can be achieved with a minimum of suffering. (The highwayman is a gentleman, the jewel thief never carries a gun, the murderer spares his victim's feelings; the enemy never tortures the prisoner for information when he refuses to give more than name, rank, and serial number.) The history of rules of war shows the difficulties of such moral attempts to minimize suffering. We expect less is possible in the case of terrorism, but even here there is a difference between exploding a bomb in a crowded bar and exploding one in an empty building, or even a last minute phone call to clear the building. The general difficulty with such attempted regulation of war and crime and terrorism, such attempts to domesticate them, is that too often the rules will begin to interfere with achieving the end that the terrorist has in mind, for which he is engaging in the activity. Even in "normal" war, the pressure to use forceful methods to extract information from prisoners when it is believed they can tell about an imminent and dangerous large-scale operation is tremendous, though sometimes inhibited by fear of like treatment of one's own captured soldiers. The moral issues are of course the familiar ones of ends justifying means in critical situations, and cannot be discussed in abstraction since the answer to the general question whether the end justifies the means is obviously that some do and some do not, and in some kinds of cases for some kinds of means it never will. But which are which is precisely the question for the particular situation. There is no point in pursuing this issue here. To attempt a moral regulation of terrorism that distinguishes the respectable from the disrespectable or the legitimate from the illegitimate in war warrants the description of "shuffling deck chairs for greater comfort on the Titanic." One cannot be sure that short-range goals will not be achieved, and certainly lives may be saved, but as a central direction of effort there is a danger that it will divert attention from tackling the underlying causes of the situation that begets terrorism. Within a clear perception of the whole it may be possible to isolate factors in terrorism and put obstacles in its path—for example, since much of it rests on publicity, to limit its advertisements in the media. If, however, terrorism is an expression of weakness or relative weakness, as suggested above, the lessening of its effectivenesses may rather lead to its escalation.

Moralistic Terrorism

The most difficult type of terrorism to handle is that involving a cause, which we spoke of earlier as moralistic terrorism. In the old days, political prisoners were distinguished from and treated differently from ordinary predatory criminals. There was a time when even a foolish cause that was felt and followed as an ideal was taken to be ennobling. But the Nazi experience showed what depraved causes can invite and get people to sacrifice themselves. Even the potential victims of moralistic terror have therefore to ask themselves what is the underlying cause or ideal in the name of which they are likely to be victimized. And they have not merely to identify it but also to evaluate it if they are to form a sound policy for dealing with it.

Of course the various dimensions of terroristic action will make a difference in the public attitude. When an ROTC building was burned in the growing protest against the Vietnam war, it was specific and directed against part of the process of war-making. When revolutionary groups in Argentina or Uruguay kidnapped executives of large corporations they regarded as exploiting their people and asked for large ransoms to be distributed through some ameliorative social channels, the terror was still pointedly specific, though people now became its counters. Suppose they had kidnapped any American and made their demands against the American government? When a white man was shot at in the so-called Zebra cases in San Francisco, the terror (given that analysis of the incidents) was directed indiscriminately against a given broad class. The same was seen in the attack on Israeli athletes at Munich or the shootings at European airports by Palestinian terrorists, or the capture of Dutch trains by South Moluccans.

It is possible, of course, to drop all differentiation and simply feel that we are back to the Hobbesian state of nature, the war of all almost against all. While this reflects the general insecurity accurately at times, it is too general to provide guidance for what is to be done; nor is the comparison to war helpful in spite of all the warring that may go on in many of these cases. Even though the black terrorist may regard it as a war against all whites, or the Palestinian terrorists as a war against all Israelis and all whose activity may help the Israelis, it does not follow that the whites or the Israelis should regard it as a war. This need not inhibit protective action or even punitive action—that is a separate problem. But important differences in the long-range picture will flow from the attitude that is invoked.

Moralistic terror has then to be analyzed with attentiveness to the ideals or causes embodied in it. At the very least it is significant when a terrorist is ready to give up his own life to serve his cause. There are the familiar questions: Is he intellectually muddled about means? Is he mad? Is he expressing a frustration that has led to desperation? Is he just evil? I doubt that one can stop with the individual. As we have seen in identifying the cultural and historical dimensions, one has to look into the whole social background in which individuals come to take such a turn. Certainly there is enough familiarity by this time with the conditions of exploitation and oppression that give rise to racist and nationalist ideologies in the oppressed as well as the oppressors. And sometimes too it is not a question of oppressors and oppressed but the tragedy that lies in the conflict of opposing claims, where each is reasonable if taken alone. To use the analogy of war for the terrorism that grew from such a seedbed would be to be callous to the problems that generated them. It would thrust aside the responsibilities for allowing them to fester over generations of neglect and systematic turning aside, not to speak of the selfish gains that may have come from paying no attention to them. Indeed, for many types of moralistic terrorism, it would not be far amiss to feel them as an historical punishment for the callousness of people to the problems of people. The problems and attitudes of Job in his anguish may be more appropriate than ringing declarations of war. Of course some of the individuals who now suffer from the terrorism may not have shared in the callousness; they may even have spent years in less than successful struggles against the specific injustices. But many issues in the modern world assume wholesale proportions and may not be amenable to retail treatment.

If the act of moralistic terror thus becomes understandable, what can the potential victims do about it? Are they left with no resort but repression, care, protection, sometimes even acts of war and reprisal if not the full application of the war model? It would be too easy to say, "Solve the underlying problems." But at least harder work to open the channels is possible. There are parallel issues in human history that might point a direction. For example, when John Locke enunciated the right to revolution he regarded it as an appeal to heaven when all appeals on earth had failed. But there is a sense in which this right was in part domesticated with the development of democracy so that the room for appeal to the electorate and the public was immeasurably broadened. And in our own time the increased recognition of individual conscience and the phenomena of civil disobedience gave greater social scope to individually deviant commitments of deep intensity. Again, in the history of the labor movement it was a long road from criminal syndicalism to collective bargaining. What is there to be learned about the ways in which threats of war were defused in such developments? Is there anything to parallel these stories for the areas that underlie today's major moralistic terrorisms?

A number of preliminary suggestions come to mind. First, there is the negative warning not to be distracted from the basic problems that generated the terroristic action. In this regard, the kind of liberalism that spends its best energies on opposing busing and affirmative action and has nothing to offer for advancing the liberation of blacks and women, or the conservatism that looks for strong men as dictators in Latin America, is heading for the war model and enshrining callousness.

Second, even the means of repression of terrorism or of reprisal as deterrence have room for sensitivity. What, for example, are the components in Israeli attitudes to reprisal? Is the reprisal directed to destroying camps from which the terrorists emerge? Does it wipe out houses rather than people? Is it pressure on the government of a neighboring state to take a firmer line against terrorism? Is it needed as a channel for expressing popular Israeli feelings? The longer-run settlement has to solve the problem of the Palestinian Arabs, but the short-run character of action is of vital significance.

And finally, there is the seriousness of attitude in seeking solutions to the problems that generated and support terrorism. This may be a long-range matter, but the short-range contribution lies in the authenticity of present effort, the movement not "with all deliberate speed" but with effective immediate acceleration, the breaking through age-old ideological barriers to see the problems, and the movement toward cooperation in formulating solutions and projecting plans of action. It is true that Rome was not built in a day, but times have changed and construction workers know how fast a skyscraper can go up. Worse still, we all know that nowadays Rome can be destroyed in a day.

Note

1. Reference is to Carl Wellman's paper, "On Terrorism Itself," presented at the Conference.

17. Approaches to Environmental Ethics

This paper criticizes the pervasive rights-and-duties way of doing environmental ethics and suggests a reconstruction. It falls into three parts. The first considers the relation of current philosophical labors to the environmental movement. The second reformulates the philosophical questions. The third points out some of the advantages of the reformulation.

I

The environmentalist movement has grown in recent years from the early complaints of individuals and the eloquence of Rachel Carson's *The Silent Spring* to a present social, political and economic power. This is largely due to increasing knowledge of the dangers and immediacy of threats to life and well-being. Our attention is usually directed to some partial front, but the array becomes formidable when put together. Consider even a partial listing: dying lakes and dwindling species, poisoned atmosphere, the wasting of the Vietnam soil, the failure of the nuclear industry to deal adequately with waste disposal and the possibilities of disaster, garbage dumping and the creation of dead areas in the ocean, oil spills from tankers and offshore drilling, intensified fishing and its devastation. Crises come closer and affect our lives more directly. It is no longer a matter of hearing about alien peoples in distant lands—even if we can witness their suffering on TV while it takes place. The energy crisis rocks the industrial countries. The dangers of nuclear reactors proliferate among us. France's plutonium fueled breeder reactor—the kind of development that President Carter has tried to stop in the United States—is going up within fifty miles of Geneva. The Seveso affair, described in the European press as "a chemical Hiroshima" was a major environmental disaster in the industrial area near Milan. The release of chemicals into streams and lakes has become a serious health issue affecting animals and people in parts of the east and midwest of the United States. Radio and TV now track the fallout of Chinese atmospheric nuclear tests as it moves from the Pacific to the Atlantic coast, hoping there will be no rain to wash it down. Our ideas are profoundly affected by the convergence of these and kindred developments. Instead of indefinite growth and progress, the concept of limits is now in the ascendant; and doomsday models, first launched by the Club of Rome, are the intellectual sputniks of our popular philosophies. Equally threatening is the coming paradox that the petrochemicals of our technology which have formed the basis of so much of our progress are proving in so many cases to be a source of danger to our health.

What part has our technical philosophy been playing in this human drama? While the foe works with nuclear reactors and computers and submarines and petrochemical derivatives, we philosophers battle valiantly, still usually clad in Kantian armor with the twin-edged, sharply honed blade of rights and duties. Yet we are not without notable victories. Working back from the fact that human beings have rights we have established (at least to our own satisfaction) that animals have rights. Suprisingly, an occasional judicial decision has accepted the view that a tree has rights, with some devoted person to act as trustee. And of course embryos have rights, and future generations that do not yet exist have rights, so that resources have to be conserved as a duty on our part to correspond to their rights. In early English law it used to be said "No writ, no right." The philosophers have reversed it: they first create rights and conjure up more than writs—whole armies of constructed persons to fight their battles. Jove scarcely did better with Minerva, springing from his brow fully clad in armor. The extreme case I have come across is a defense of the right of rocks not be quarried for a building or crushed into a pavement or shaped into a statute.[1] This is on the assumption that rocks, like everything else, want to stay as they are. We could no doubt find a certain subtlety in this kind of argument. If the basis of a human being's right is his individual will, and a will expresses a need or desire or striving, then as we go down the scale from human consciousness to animal life and below, we go to appetite and effort and eventually to pressure and resistance and inertial shape. But then we end up simply with the view that everything is what it is and not another thing.

Now how does this philosophical activity relate to the environmentalist movement? Does it have any influence or serve any clarificatory role? I am not concerned primarily with the question of causal influence in the historical process. Russell noted that stupidity has a great effect on human affairs so there is no reason why its opposite, intelligence, should not have equal effect. Kant thought human history was pushed along by man's unsocial sociability, that is by Hobbesian passions, but the philosopher could read the moral directions demanded of the process and maybe help things along a bit. Hegel developed the notion of the cunning of reason that utilizes men's selfish passions and local aims to build its historical patterns. These concepts fit the environmentalist situation fairly well; for example, the treatment of garbage and the manufacture of pollution devices themselves become a profitable industry as well as bidding for government funding. Thus there arises a receptive and powerful interest for environmentalist rhetoric and philosophical argumentation. Sometimes the practical alliance is explicit: for example, in New England an argument against polluting a lake has been successful when allied with the evidence of loss of income derived from the tourist industry, whereas in the past it met simply with the reply that local industry was more important than desires of summering visitors to go swimming. No doubt the relations of intellectual factors to other historical factors are multiple and variable according to context— sometimes influential, sometimes powerless—rather than one of uniform pattern. None of this relieves the philosopher of the obligation to make his argument reasonable rather than ideological, and to clarify the situation according to his best available philosophical canons.

Now as to clarification rather than causal influence, there should be little mystery. The moral philosopher in laying down rights and duties with regard to the enviroment is attempting to guide human conduct. If men have a duty not to pollute waters, then they should not pollute; if future generations have a right to nonrenewable resources of the planet, then the present generation has the obligation to preserve them in due proportion or to invent substitutes. The rights-and-duties framework is the usual way of doing moral philosophy, at least since mid-century; the formulation of environmental ethics in its terms seems therefore quite natural. Still, in embarking on a new field there ought to have been some hesitation about what categories or modes to employ. Perhaps some other option would hold in this case. For example, cost accounting categories are used in many if not most areas, but love is not subject to them. Contractual categories are employed in business, not in friendly relations, and when they are sometimes suggested nowadays in the husband-wife relations about daily tasks it is only because this is felt to have long been an area of sexist exploitation. It would have been fitting for environmentalist ethics to take a fresh look, since it was a very fresh set of problems, at how to organize and formulate its inquiries.

Why should one feel uneasy about the rights-and-duties framework in this context? Well, for one thing, there is something bizarre about having to defend the environment against pollution or destruction by the rights of trees and the rights of animals. Of course it is wrong to pollute a lake, but the "wrong" here is the mark of a negative moral decision, not the assignment of a right to a lake. Perhaps an historical glance at the framework that the rights-and-duties analysis pushed aside and the character of its special appeal will help us get the question into perspective.

Before World War II our dominant moral paradigm was a public welfare utilitarianism. We have seen (ch. 9) how in the depression years of the 1930s our social security systems were built up in those terms. Given the desperate needs of the times it is hard to see how it could have been otherwise, since it was clearly impossible to think in the older terms of each individual solving his problems by his own efforts. We have noted also (ch. 8) the shifting character of paradigms of ethical justification and how after World War II the demands for this kind of security throughout the world and for the end of colonialism in Asia and Africa found expression not in a similar utilitarianism but in a human rights outlook. This is the French revolutionary philosophy brought up to date. We cannot enter here into the reasons, whether historical or psychological, or intrinsic to the conception of injustice and oppression. But in any case, from the UN Universal Declaration of Human Rights and its associated declarations on through the liberation movements and the waves of antidiscrimination movements, the lists of rights kept growing, and rights became the typical form of moral assertion. In philosophical discussion utilitarianism was condemned for its laxness, for the opportunism it allowed, for its lack of rigorous moralism. The wide reception given Rawls' *A Theory of Justice* [2] came in part at least from the fact that its refurbished contractualism gave a nonutilitarian base for a rights theory. Nozick's *Anarchy, State and Utopia* perhaps expresses the current mood of practice more clearly: he begins with the fact of rights and does not worry about how they are to be grounded. [3] But his treatment is significant in another way too: as has been widely recognized, his book attempts to recapture the rights concept for the more traditional property rights and rescue it from all the upstart rights. There are always, however, safety-valves and one never knows how far they will open up. For example, Nozick quite casually allows a principle of rectification for past injustices: [4] one wonders how he would apply it to the present suits of the Indian tribes for restitution of lands in Maine and elsewhere. Problems of the proof of injustice along these lines might very well turn into a new route for establishing property rights. Pandora's box of historical robberies once open could keep us endlessly busy. And what could be more evidently in need of rectification than desiccated soil, polluted water, and the like? Can we not envisage a legal-moral instrument (if we play this game) whereby the very fact of environmental injury creates a quasi-property right of restitution? Happy as it is to contemplate, it shows us what the game has become.

Where do we stand with the growth of rights? Rights are now very widespread, if every creature has a right and every rock has a right and nonexistent beings have rights. Old universal rights are being supplemented with new very detailed rights—for example even the UN Declaration includes periodic holidays with pay. One right battles with another—perhaps one of these days all the agricultural land lying under the concrete roads throughout the land will have a right to be restored, only to find the same soil suing through a trustee for restoration of virgin forest. With such proliferation, the existence of a right will become the basis for little more than a weak prima facie duty. The serious question will be how to estimate comparative strength of rights, priorities of rights, relative importance of rights, and so on; we will be back to reckonings and trade-offs and all the rest, in no better a position than what Kant called contemptuously "the serpentine windings of utilitarianism" when he warned us against them.

I do not think we need go back to utilitarianism, and I shall try to explain why shortly. But I do think the rights "paradigm" as fundamental has run its course. In the 1930s it gave way to utilitarianism because the rights framework had been captured by a conservative outlook and was being used (for example by the Supreme Court) to delay necessary social reform. At the present time its revolutionary value on the global scale has been diluted, both in its use on the political right and its overextension in all directions. I do not mean that we do not have to work out our rights and duties and I do not mean that there is not a profound sense in which we have human rights. We have rights and duties in any field *as a consequence of some more fundamental analysis of the field*. For a while we were in a position in which we did not have to ask what the background was or the underlying theoretical framework. Sometimes it still does not matter. When in the middle of the night a gang breaks into your home and takes you violently away, your human rights are violated—whatever your theory of the nature of rights—whether the gang is sent by the state or by a kidnapping syndicate. That is the kind of image which gives strength to the appeal to human rights, but it does not furnish a theory and it does not mean that any theory goes or that no theory is necessary or that rights stand on their own.

Our problem here is not to furnish a theory of the nature of rights[5] but to suggest an alternative way of dealing with the ethics of the environment.

II

In Chapter 9 a distinction was suggested between *macroethics* and *microethics* to indicate a contrast in types of problems. Questions which do not primarily concern a single individual but which reckon with the kind and quality of life of a community are problems of macroethics. Institution-building, constitution-framing, national defense, preserving resources for future generations—all such tasks at least under present conditions of life are not primarily individual problems. (It does not matter that individuals may in the end cast votes about them.) It was noted that conditions change and that questions which at one time are microethical may become macroethical. For example, defense was largely microethical under frontier conditions. Preservation of natural resources for the future was not even a problem at all then. What makes a question micro- or macro- involves reference to existing conditions as well as to the value patterns of the people concerned.

Two points about the distinction need additional comment, one concerning overlapping, the other concerning reduction. Since micro- and macro- approaches are methodologically different, the question arises whether it is not possible to employ both approaches at the same time or as complementary rather than force a choice between them. In many cases, of course, this is not possible, since a whole field gets organized in one or the other way. In others, however, it rests on empirical considerations. For example, in national defense, even under modern conditions, in a small nation in constant danger of terrorist attack every person might very well arm himself for individual defense in addition to the usual organized system of national defense. Mixed economies in the modern world typically combine macro- and micro- approaches. As to reduction, there is no necessity to press on with reducing one type to the other or to decide in a hurry which is "fundamental." No theory of reductivism or of methodological individualism, no bias of moral individualism, no metaphysical nominalism or anything else of that sort settles for us in advance the essentially empirical question of which way will prove most fruitful for treating which kind of questions. Suppose we decide that environmental problems under today's conditions of life are large-scale and concern the kind and quality of life of society as a whole and so are best treated as problems of macroethics. We would then investigate them under the criteria we would find relevant for such problems. Whether our results would be derivable from the desires and acts of will of individuals would be a quite separate question, an interesting one to be sure but not vital to the separate pursuit of macroethics. (Similarly, one could believe biology eventually reducible to physical chemistry and continue biology in its own terms, letting the future work out what the reduction would be like before refashioning the science.)

The rights framework as it has been used in environmental ethics is essentially microethical. In some contexts, a rights framework may take a macroethical form, particularly when a whole society sets about organizing a system of rights to embody a conception of freedom in a set of institutions for its people. But this is not the way the rights conception has operated in environmental ethics. It has not been an attempt to work out the kinds of rights and duties that will yield a predesignated goal of a given kind of environment for the people of the society. Hence if environmental ethics is conceived as largely macroethical, a different framework is required. I am not suggesting that we jettison the rights framework entirely for environmental ethics. Doubtless there are parts which it may organize well, just as in moral problems with respect to animals questions about pets may be best approached as microethical while those about saving a species are best approached as macroethical.

What kind of macroethical framework shall we try out for environmental ethics? The first temptation is to go to utilitarianism, because the standard contrast in current ethics is between a rights theory, whether intuitionist or contractarian, and teleological utilitarian theory. This would be a serious mistake, for although utilitarianism aims to end up at a social theory of a macro- sort, its Benthamite heritage shows clearly that it derives its macro-character (its principles of legislation and its institutions) from a micro-base (the summation of the pleasures and pains of individuals). We would therefore be saddled with the very derivation of the macroethical from the microethical that hampers the free development of macroethics to work out models of its own in an expermiental way in the light of what is after all an area of relatively new problems. We should not commit ourselves in advance to one direction of future reduction. It is even possible that solutions in microethics today will presuppose solutions in macroethics rather than the reverse; certainly the individual in formulating and solving many of his ethical problems takes for granted the institutional structuring and solution of social problems.

One procedure that comes readily to mind is to use in setting the ethical stage a metaphor or model which already involves a group or community. For example, one might think of the planet as the *home* of mankind and try to develop the consequent environmental rules from the appropriate ways of taking care of one's home. While this might very well yield us a planet to be proud of, it would have to be the country home with sufficient land to bury the garbage, not the urban apartment! I doubt whether this model would prove feasible. The home is part of a wider communty which makes it difficult for it to represent a planetary totality in a situation where so many environmental problems are planetary in scope; we would have to shift at least to a space ship. Again, the values of the home are too readily translated, it being a small community, into individual values, and so the temptation to slip into microethical justification (though not in the rights form) is unavoidable.

A more promising suggestion may be offered by comparison with the way we do in fact deal with questions of national defense. National defense we noted, is a macrogoal. It rests on many different complementary bases, in some cases perhaps alternative bases. Nor is the value of national defense affected by the fact that it is also a source of profit to some in what President Eisenhower felicitously called the "military-industrial complex," though that is a ground for caution lest the treasury be pillaged. Macrogoals as the answers to macroproblems have marked features. For example, there is a readiness to budget for them; we do not say we cannot afford them, any more than we ask whether we can afford it before we fight a fire, or contract for that army which is the lowest bidder. Indeed, we allow cost overruns in defense contracts, where we will not in welfare contracts. Of course in part it means simply that industry is in the saddle. But it means also that we accept defense as a macroproblem but have not yet realized that health and welfare are today macroproblems—we still treat them as private affairs of one-person-at-a-time-assistance. This is in line with our historical development; defense became a national task early in the growth of the nation, while assistance to the impoverished remained private charity even long after a person's economic fate was at the mercy of impersonal economic forces. We have not yet overcome this lag. Most significantly, we make no attempt to justify national defense by translating it into sequences of individual benefits. We do discuss its costs and problems of peace and international relations and working for situations in which defense costs can be reduced. But none of this is in microethical terms.

Our reformulation yields so far only a first approximation. It is the macrogoal of a sound environment, as sound as possible for human beings in a balance with the possibilities of sound social human living in the resources that mankind has and can maintain in the state of advancing knowledge that it is capable of continuing. Questions are to be formulated about the qualities of human life in a community and the system of goals of life. This approach will be assumed for the remainder of the discussion.

III

The central paradigms for environmental ethics are to be sought in communitywide, often worldwide, problems that touch important human needs and values. In our time they have been apparent in matters of population, public health and resources, in questions of energy and food supply, in pollution as a consequence of urbanization and the rapid development of industry, in the utilization of the ocean and problems of water policy, as well as in the treatment of land and animals. In any of these areas in the contemporary world, a working solution that will not mean slipping back into increasing conflict and growing chaos involves several successive steps. We have to project some common goals. We must use scientific knowledge or research to discover ways of attacking problems. We will have to develop institutions and practices or utilize existing ones in fresh ways to deal with what are often new problems. Finally, we will find it necessary to adjust systems of rights, duties, and responsibilities to achieve the goals and support the kind of life aimed at. Such a paradigm obviously fits what we ought to do with respect to energy and the search for new forms, whether through fusion or solar energy. As debates about nuclear and solar energy have clearly shown, the issues go beyond providing energy; they concern how it is provided—the large-scale centralized methods versus the possibilitiy of dispersed delivery—and the problems of disposal of waste and all the changes that the differences make in the quality of life of the community. To take a quite different illustration, the increasing list of industrial substances indicated as carcinogens (at this writing in excess of two thousand) poses a problem for the whole chemical industry. It is debated in terms of saving lives, cleaning up working conditions, costs and possible unemployment, recompensing those who suffer, but also in terms of whether a cost-benefit reckoning is morally permissible. Increasingly, it has become clear that there is a human cost for all the benefits of industry, although there may very well be limits at which the cost becomes too great. The costs are to be judged, however, in terms of the kind and quality of life we lead.

Questions of rights and duties with respect to the environment need not be avoided. They arise when the policies are settled and are being applied to roles and functions of individuals and their relations. If the building of solar homes is to be encouraged then of course some protection is required so that they will not have the sun cut off by a high building next door. A right to sunlight is one (rather extreme) way of doing it; zoning rules is another. A rights framework is a conceptual instrument to be used where it works best. Sometimes its use is unpredictable. For example, the Soviet Constitution includes the right to a job. In the United States, surprisingly for a rights-oriented society, we follow a socially oriented policy of aiming (governmentally) at a reasonably high level of unemployment, and if the Humphrey-Hawkins bill is passed we shall go in for a national job budget. Still the paradox will remain, that a socialist society uses an individualist rights formulation while an individualist society uses a social welfare formulation! On the other hand, there are many cases in our society of the settling of environmental policy by claims of legal rights. Disputes over seeding the clouds as diverting rain from one state to another are a striking illustration.

On our revised framework, rights, responsibilities, and principles of distribution are seen more transparently to stem from practices determined in the light of holistic survey of the total scene and the total operations of society. What, for example, would be the fate of the question "Have I not the right to buy any car I want and drive where I want, provided I violate no law?" Instead of subsuming this under a right of doing what you want with private property, the analysis is more likely to calculate the tax investment in road building, oil subsidies, shipping subsidies, etc., also air pollution and the cost of its reduction, and end up with a figure on how much per mile of your travel is being paid for by taxpayers (and what proportion of them do not have a car). A moral decision may still be that it is worth letting you have a right of freedom to travel where you wish with what car you wish, not because it is a human right but because within certain limits it feeds into a pattern of human relations which encourages a desirable individuality; this individuality, which is a social product, is taken to need encouragement by social practices and its value to make it worth the cost. (But then you, the driver, must not complain when other tax benefits for other equally good reasons go as rights to the person who has no car—say, supplemental travel benefits or welfare benefits.) In a similar mode of analysis it would not be the strip mining industry's duty to restore the environment (or in some cases to work out alternative ways of getting the coal even if more expensive) or the coal company's duty to take adequate precautions against black lung disease. It would be part of their cost of production in the same sense as they bear the cost of machinery and its wear and tear, of interest on their bonded indebtedness. In short, these are not secondary or external impositions by an interfering state or a threatening union. They are part of the whole enterprise of mining in a decent or civilized society that is concerned with the quality of life of its people. (Of course this gives rise to institutions embodying such goals and to duties of persons entrusted with carrying out courses of actions required by them.)

Such brief illustrations show how the problems of environmental ethics, once they are formulated as problems of macroethics, stand out clearly as issues of goal-establishment and consequent institution- and practice-formulation. They involve a whole-life view, quality-of-life assessment, part-whole relations, as often as means-end relations. They do not close issues with the peremptory pseudofinality of an assertion of rights, but allow the ready entry of experience, experiment and advancing knowledge with fuller social scope for learning. And above all they permit this knowledge to be brought to bear on the solution of the moral problems insofar as possible, so that morality in its efforts may take advantage of the best of our scientific resources.

It is perhaps possible that a rights-and-duties framework could emancipate itself from its overindividualist setting to tackle more realistically the kinds of issues that face environmental ethics. There is some tendency today to begin to think of group rights as well as individual rights and with respect to juristic efforts to allow class suits. But this is a slow movement and in any case even if it began to make some system out of the rights and duties it would not yet give us an underlying rationale. For one thing, it rarely looks to the historical conditions that shape its own problems in a given period and often even explain why they arise in the first place; whereas on the other hand a concern with the quality of life under given conditions of life naturally turns to the socio-historical examination of those conditions. Take for example, the widespread concern in discussions of environmental ethics with whether the unborn have rights to nonrenewable resources and whether we have duties to beings who do not but will probably exist. The purely analytic discussion engages in conceptual analysis of rights, of legal usage, of the possibilities of the group of future beings having the rights even though no single person has it, of the kinds of obligation-language that would be hospitable or inhospitable to rights for the unborn, and so on. There is an intensity to the issue with many defending the right and some opposing it so strongly as even to be regarding it as dangerous nonsense. A listener might be left wondering whether analysts were dividing on whether they did or did not care much about their great grandchildren! A political historian might wonder mildly why in the eighteenth century people like Jefferson were worried about the dead hand of the past while in the twentieth century the analytic philosophers were worried about the nonexistent hand of the future.

There is something strange about this debate. It reminds us of the controversy about triage or lifeboat ethics, or again whether we have a moral right to shoot a man trying to force his way into our shelter against atomic fallout (ch. 8). They pose moral issues tangentially related to what is going on but somehow distracting in the passions they rouse. Human concern for future generations is not really the issue. The man of eighty planting a tree he would never see full grown was a striking sight to the ancient Stoic philosopher. Yet it would have been presumptuous to ask him whether future generations had a right to his labor and he a duty to plant the tree. There are contexts in which the use of such language becomes a symptom of something gone astray in the situation. The language of rights is not the only way to express a moral concern for the future or for animals or for the aesthetic quality of the landscape or any of the other matters on which we have seen rights claimed. The constituents of a morality are much more ample than the rights framework allowed. For example, every morality contains as a necessary element the specification of a moral community and it is just as much a part of the morality to indicate its composition as to assign rights and duties. The question whether animals and nature fall within the moral community is one that can be asked quite separately; it needs careful refinement, as does the notion of the moral community itself.

We have to formulate the issue as a problem of macroethics and direct our attention to the large-scale changes in the conditions of life of the societies concerned. We then realize that the problems are real enough. The twentieth century witnessed the sacrifice of whole generations for the future, not merely in the traditional practices of war (on the much larger scale of world wars) but in the new form of intensive industrial development. The Soviet Union, cut off from the developed nations by their political hostility, had to pay the price for a rapid buildup of industry by sacrificing its then present generation; and later developing countries had to make similar sacrifices. After the devastation of World War II there was a repetition of sacrifice for reconstruction, with a postponement of gratification. "Rising expectations" were thwarted for the masses by the growth of population, the general threat of depletion of resources, the ecological problems of pollution. The increasing life span placed the growing burdens of the old upon the young. The technology which brings advantages to a part of the present population carries implied genetic threats to the future. Alongside all these problems there is the uneven division among nations between the haves and have-nots, and the extremely wasteful habits of consumption found in some of the former. The result is projected as a growing conflict of interest between present and future, but it is really a serious moral problem about the character of present life and its future consequences. It can be analyzed as such only by a full examination of what a sound life in a sound environment would be like. To debate the right or absence of right of the future is simply to fight the present battle on a fantasy screen rather than to analyze ways of solving the present conflict.

If in the United States today we are not prepared to evaluate our present waste of energy and its impact on the rest of the world, our overweening pride in our gas-guzzlers and high speed, if we will not reassess our diet and what it does to grain levels in the rest of the world, if we consider the small proportion of our aid to the underdeveloped parts of the world compared to some of the other smaller industrial countries and the political returns we expect of it, what shall we make of our sudden concern for the *rights* of future generations? When the Club of Rome's first report called or seemed to call for zero growth in thirty years without thinking of what it meant for the Third World whose growth had scarcely begun, critics were quick to point out that if we had not shared in good times, would we be ready to share in times of sacrifice? In that situation did the debate about the rights of future generations mean that we would sacrifice for the future but let the starving present generation continue to starve? In short, have we done anything but question our present mode of life? The moral issue is the character and quality of our present life, not whether it is logically possible for nonexistent beings to have rights now, given the probability in terms of present evidence that there will be such beings coming into existence in the future. That question is no doubt well-intentioned in pointing to the rapid draining of resources without forethought for the future. But it fights its battle on the wrong front. It raises the wrong image of sacrifice almost as if we were now living at the margin of necessity and had to share our bread with the future. It is not explicit enough about the present waste and the uneven distribution within our society and our world. In brief, it is the wrong image on the wrong question at the wrong front. Its role can only heighten ideology, not further moral clarity, whatever answer one gives to the question.

Formulating the problems of environmental ethics in terms of macroethics will not necessarily ensure a ready solution, but it will pose them clearly instead of in an oblique and hidden way. Nevertheless, so persistent a mode of formulation as is found in terms of rights and duties cannot be wholly without some explanation, and I should like in conclusion briefly to suggest a likely one. The use of a rights conception seems to come in the earlier phase of an historical struggle when there is a demand for recognition. The peremptory component in the notion closes out argument against it when the weight of the establishment usually has the weight of tradition locking the door. The newcomer asks for reason, but traditional reasons are with the traditional institutions. It asks for place, but the traditional places are occupied. It makes promises of social utility, but they are branded as wild or utopian ideals for which tried and true ways are not to be jeopardized. It can plead the sufferings of its members, but poverty and suffering and even unfairness, we are told, we always have with us, so why just reverse discrimination? *Rights* is an ethical concept of demand, not of request or of argument, so it is not implausible that it is associated with a rising movement from first demand to revolutionary power. It was probably natural therefore for so new a movement as environmentalism to slip into this pattern on its philosophical side, especially as it was the usual way of doing moral philosophy in any case. But it will be a mark of philosophical maturity to grow out of it as the movement takes a responsible place in society.

Notes

1. Roderick Nash, "Do Rocks Have Rights? Thoughts on Environmental Ethics" in *Small Comforts for Hard Times, Humanists on Public Policy*, eds. Michael Mooney and Florian Stuber (New York: Columbia University Press, 1977).

2. John Rawls, *A Theory of Justice* (Cambridge, Mass.: The Belknap Press of Harvard University Press, 1971).

3. Robert Nozick, *Anarchy, State and Utopia* (New York: Basic Books, 1974).

4. Nozick, pp. 152-53. Cf. p. 231.
5. The approach to rights taken in this paper is developed more fully in chapter 4 of Volume I ("Some Reflection on the Concept of Human Rights").

Epilogue: The Fact-Value Dichotomy as a Chapter in Social-Intellectual History

The dichotomy of fact and value has been one of the recurring themes in our explorations. The case against it has been amply made. The underlying program for ethics on which it rested has been analyzed and an alternative proposed. This epilogue is not intended to reopen any part of the argument. The aim is rather to raise questions about the history of the dichotomy: the conditions of its emergence, the changes in its formulation with social changes, its ideological roles, the conditions of its loss of force, and in general what we may regard as having been its social meaning. I speak of raising questions since the social study of philosophical ideas and their careers has not on the whole been a well cultivated field. The history-of-ideas movement developed it largely on the intellectual or ideational plane. On the other hand, the sociology of ideas has been too busy with social and political and religious ideas to tackle abstract philosophical ones. Even important philosophical categories that were the battle-cries of major ideologies have not been adequately treated. For example, where are the social analyses of the war of Nature versus Convention (alias Reason versus Custom) which has a crucial intellectual-social role in the historical transitions of the seventeenth and eighteenth centuries? So important a controversy deserves analysis not by centuries but by decades. The same is doubtless true of the Fact-Value dichotomy in twentieth century philosophy; fully to understand its significance we have to see its finer textured changes in their sociohistorical context.

As we have seen (chs. 1 and 2), the fact-value dichotomy, which is in the twentieth century the most interesting theme in the more complex confrontations of the fact-value problem, is itself a bundle of concepts and theses and takes myriad forms. Fact and value are two sharply separated domains. Science and morals are (respectively) brought under that separation. Fact and science are value-free; science is value-free and objective. Values cannot be derived from facts, and conversely. What ought to be or be done cannot be derived from what is, and conversely. (These last two formulations are often assimilated, though technically they may be quite different.) Science is descriptive or explanatory; values and morals are expressive or prescriptive. Etc. Even when the "same" formulation appears over a span of time, the meaning may change as the scope and character of science change or as the prevalent view of the *separate* value domain changes. Science, for example, at one point deemed strict measurability a necessary condition and ruled out much of the qualitative aspects of life from its domain; at a later period, with the development of statistical research in social science they were readmitted. A separate value domain in Kantian times was inhabited by universal moral laws; at a later period ultimate (logically arbitrary) value commitments of individuals gained admission. Both share the same feature of not being responsible to factual or scientific evidence, and so the same thesis of the fact-value dichotomy.

A full social-intellectual history of the dichotomy would have to track down each fine shade of formulation, every guise and disguise. Here I want mainly to ask many questions, offer some clues, and suggest the terms of some answers. I concentrate on the technical philosophical ideas because that is the area in which practically no work has been done in this way and it is time to stress its need. But I want to show its continuities with the wider problems of the social bases and context of the social sciences. Our topics are: the birth and growth of the modern fact-value dichotomy; the emergence of the special twentieth century form of the dichotomy and its career; death or temporary eclipse?

Birth and Growth of the Modern Fact-Value Dichotomy

We must not assume too hastily that the dichotomy accompanied the rise of physical science because we are familiar with twentieth century claims for the value-neutral character of science (ch. 11). Too often we have the habit of regarding contemporary ways as the adult form, fully evolved, and look upon earlier history as if it were trying to reach our mature position in successive stages. This is simply the residue of a typical misuse of historical evolutionism that was common in the earlier twentieth century. Once we learn to see our own ways as responses to our own problems at our own stage of development, we can apply the same mode of analysis to science at any period. Thus even value-neutrality would have a different meaning in relation to different problems and different battles. Copernicus and Galileo faced religious pressure to regard their theories as fictions for practical purposes rather than representations of the real world. When the early struggle with religion is over and freedom of inquiry is supported by the utility that science has in industry and war, in mining and navigation, then science can safely carry on its internal operations on a policy objectivity and freedom from all values except the pursuit of truth as defined by its strict methods. But this does not mean the value-neutrality of science either in the sense of scientists approaching their labors without value presuppositions or in the sense of scientific results being thought to have no value bearings. Many of the seventeenth and eighteenth century scientists in Britain and America assumed that science would reveal God's plan of the universe and therefore the universe was orderly (and many theologians defended science on such grounds and even encouraged "natural theology," the proofs of God's existence by natural evidence). And some scientists thought that their results refuted cherished doctrines of freedom of the will, and so would have profound effects on social philosophy. This was a long step from the later thesis that divided the scientists into the scientist-as-scientist who was neutral and value-free and the scientist-as-citizen (or as person) who was allowed to have values.

The fact-value dichotomy in its modern form was actually born in the eighteenth century about the time when, science being stabilized as Newtonian physics, the study of man (later called the social sciences) emerged as an extension of science. Of course people have always distinguished the way things are from the way they would like them to be. But there is a profound difference between the desire for a better reality and a separation between two domains of existence and value. Behind this difference lies a metaphysical change that was long in coming. Plato had propounded a theory of ideal Forms of which existing things were just copies, but he had insisted that the Forms were all the reality and that existing things were shadows. Aristotle restored the sober view in which our ordinary world of existing things is the real world and any serious talk of possibilities refers to potentialities of this real existent world, that is, alternative directions in which things might actually go. Only from the time of Leibnitz do we get the new concept of purely possible worlds which are not a function of the existent world; in fact the existent world is thought of as a selection out of purely possible worlds, with existence added. Leibnitz used this notion immediately to think of God, having all possible worlds before him, as choosing this one (because He is good) as the best of all possible worlds, a conception which Voltaire's *Candide* satirized. The new approach provided an intellectual ambience for development of the view that geometry deals with abstract systems and not physical or existential spaces, and so alternative geometries are possible. But from the point of view of our present problem it furthered an absolute dichotomy with two kinds of worlds, a metaphysical habit of mind that could be receptive to the fact-value dichotomy, although it need not have compelled it.

The dichotomy emerged in a time of turbulent social change, both in Britain with a growing industrial revolution and in France with a gathering violent revolution. A new picture of man is presented in the economic and ethical writings of Hume and Adam Smith, in the ethical and legal writings of Bentham, in French encyclopedists and materialists, political thinkers and philosophers of history. Neutral or not, it was socially explosive. The old order did not see the new view of man as neutral, and later analyses saw it as ideological projection of emerging classes. Philosophers generally do not worry about the social backround of the birth of the dichotomy because they think they have a valid birth certificate. It is a striking passage in Hume about the is-ought relation.[1] He expresses surprise that writings in which the term "ought" does not appear, suddenly end up with declarations of what we ought to do, and he asks where we get that idea. He is interested in this, just as he was interested in how we got the idea of law and necessity. On the strength of this passage, contemporary philosophers have assigned Hume the paternity of the is-ought problem. Controversy has raged over the assignment, and it would be unprofitable for us here to pursue it. More important is the whole conception of that age about the human sciences. The study of human activity from physiology to probabilities of action, including economics, ethics and law, was regarded as *the moral sciences*, so that here we find no divorce of fact and value. If the economics studied the workings out (both how it tended to be under different social arrangements and which were most effective for human aims) of certain psychological properties such as self-interest, so too moral writings studied the workings out of a universal human sympathy and how it took shape in various emotions and duties. Thus Adam Smith's lectures, including *The Wealth of Nations* and *The Theory of the Moral Sentiments* are a full coverage of the moral sciences. As has often been noted as a reservation in the value-free tendency, Smith expects his science to help men adjust their institutions so that the beneficent tendencies of the natural economic system shall work more effectively. Claims of objectivity for the science as science therefore did not mean the separation of a realm of value. And all this is quite apart from underlying theological presuppositions we have noted about a natural order which will ensure the working out of valuable uniformities or results.

Similarly, when Bentham carries out an ethical critique of institutions and traditions and ethical theories based on what he regards as scientific knowledge of human nature, that is, the laws of psychology, nothing could be further from his view than a fact-value dichotomy. If there is any dichotomy that is really pertinent to his work it is the hidebound past versus the open future. He is calling on reasonable men (legislators) to apply scientific method to ethics and law and remake their institutions from scratch by a rational calculation of the happiness of collections of atomic individuals. It would seem to reflect the possibilities of a new and changing world, more than anything else in the social situation of the latter eighteenth and early nineteenth centuries. At the same time, his pleasure theory and his felicific calculus identify pleasure and the absence of pain as the only marks of value. Yet value as pleasure becomes something to be extracted from all its relations and the qualitative differences of context and to be treasured and measured by itself and in isolation from its source. Marx and Engels noted this in their *German Ideology*[2] and compared it to the bourgeois process of extracting money as the only value out of all sorts of situations and objects (which were thereby treated only as commodities). They accused Bentham of imparting into ethics the exploitative attitude of the bourgeoisie, with the felicific calculus as a pecuniary logic. Such criticism contains the interesting suggestion that the isolation of value in the fact-value dichotomy reflects phenomena of serious alienation in human life. But even apart from this, the identification of the good in various fields with the one element of pleasure did have a unifying and possibly isolating effect and in this respect Bentham initiated the general (scientific) study of a unified domain of value.

With Kant, the sharp is-ought distinction has emerged without any doubt. Perhaps there is some significance in the fact that the place where it emerges clearly is not the more advanced British and French world but the then less developed and politically fragmented Germanic world. Kant was a scientist as well as a philosopher—the fields in any case were not much distinguished then—but the point is that he had practised science with eminent success, and he was steeped in the natural and socio-political studies as well. Both his major critiques and his shorter works on religion, politics, and philosophy of history have the same character and effect; they digest traditional opposing philosophical standpoints and reformulate the questions. In this way he shapes the directions in which philosophy is to go for the next century and a half. And what is more, he carries out his reformulations in a very self-conscious way in the light of the whole human situation. We should have no trouble finding out why he made the is-ought dichotomy and what he thought to be at stake.

Kant makes the dichotomy to assure separate realms to science on the one hand and to morals and religion on the other, so that there will be no conflict in their claims. In his epistemological inquiry he vindicated scientific knowledge as against the alleged knowledge claims of metaphysics and religion. We do not have any proofs or knowledge about the noumenal or "real" world; scientific knowledge is all the knowledge we have, but it is of the phenomenal world. Traditional metaphysics with its rational or empirical "proofs" was a kind of transcendental illusion. On the other hand, analysis of our moral consciousness reveals a moral *autonomy* which has hitherto been obscured. The usual accounts of conscience present it as a kind of moral cocktail in which one puts a dash of the fear of God, some social sanctions, a mingling of sympathy, and other ingredients. Kant's analysis pinpoints a pure respect for moral law, from which he tracks down a demand for (or postulate of) free will and self-legislation in a moral community of selves. To make sense of this, he finds, we demand or postulate God, freedom of the will, and immortality, but to postulate is not to prove them. (In our science, too, our organizing concepts, such as causality, are imposed upon the phenomenal world.) Hence by this partition of realms for science and morality—the *is* and the *ought*—and their strict separation, science can go its own way in depicting the phenomenal world without interference on moral or religious grounds, and morality requires no scientific justification for its verdicts and in turn becomes the basis for assigning what reality is possible for religion.

It is important to note that in this redistribution of realms religion is the great loser. Not only has science been established on its own, but within the domain of morals and religion there has been a great reversal of authority. Instead of religion being primary and morality consequent upon its dictates, morality has become primary and it sets the terms for what is appropriate in religion. In fact Kant held that no historical religion can go counter to moral imperatives. This is not only the emancipation of morality from religion but its ascension to a position of dominance. This is the beginning of what in contemporary philosophy is called "moral autonomy," although the concept becomes generalized and autonomy is declared from many another authority claimant.

The implications of Kant's original version of the is-ought dichotomy as a treaty of peace between science as one signatory and morality and religion as the other are thus more far-reaching than their mutually ignoring one another. Science is not only relieved of metaphysical responsibility (and so of religious responsibility) but of moral and social (applied moral) responsibility as well. Morality by its autonomy is relieved of religious responsibility and even of social responsibility unless its own dictates assign this responsibility. Morality is its own supreme judge. Having freed itself from religion, it is not likely to hand itself over to the dictates of other institutions. Kant felt perfectly comfortable about moral autonomy because he had so explicit and authoritative a moral theory. The theoretical and practical problems were bound to arise when the ethical theory changed but the autonomy of the moral domain was kept. It could, for example—and did—yield an unbridled individualism without responsibility.

In general, then, the matrix of the modern fact-value problem is the relation of religion and of morality to science and to one another in the western world at the period when science was growing and expanding beyond the physical sciences, and beginning to have profound implications for human life and institutions. A number of points about Kant's own scientific and philosophical work reenforce this understanding. His chief scientific contribution—the so-called Kant-Laplace theory of the origin of the earth—dealt with change, and some of his papers (such as on a topic of great current interest, how much the earth is slowed down by the friction of the tides) also focused on change. It was all quite a shift from the eternal Newtonian scheme of the cosmos. Developmental ideas about geology and the animals were thus clearly in the offing. His papers, too, in politics and history have a well-developed philosophy of history in outline, hindered somewhat by his basic duality of the moral ought and the factual. The ideal is a regulative principle which we morally demand of history, not a determined outcome: a race of fully moral men to be evolved in the historical process. Since the process is in the phenomenal world, its operations are causal, and the forces are the Hobbesian passions of men. But Kant sees science forging ahead, material needs ensuring its greater liberty, and even religious liberty granted by sovereigns because liberty for science is needed to carry out wars and ensure defense, and liberty is not divisible. Kant's outlook is universalistic and he envisages an ultimate federation of states and world peace through the exhaustion of warring states. Fact does the work in lowly causal fashion, and the demands of the ideal may be realized. At least this is the demand of the moral self.

It is not clear that the nineteenth century before Darwin used the dichotomy in its full Kantian strength in building up the social sciences, even when it proclaimed the difference between fact and value. Comte fashioned the language of positive science against metaphysics in constructing the new science of sociology, but, as we noted, Kant had already disposed of metaphysics as transcendental illusion. And Comte did expect the results of his science to be as convincing as any religion, indeed to be the basis of a positivist religion. But with the impact of Darwinism we may expect new repercussions, for it threatened the balance of the Kantian settlement on which the dichotomy rested. Would not the military analogy suggest the building of a Maginot line that was stronger, with defenses more absolute, techniques more intricate, devices more technical, using the latest hardware, ready to shift from one defence to another, to make sure the line is not breached?

The effect of Darwinian theory on the development of the dichotomy was fundamental. It removed all the background assumptions that a teleological outlook had embedded in ideas of nature, the order of nature, the goodness of implanted instincts and passions and sentiments, and the initial equipment of the species. Instead, they were now seen as the outcome of evolutionary processes, of natural selection operating on chance variations, of a causal mechanism yielding an order that is misread when interpreted as teleological. The order we find in the world was, so to speak, factualized, and values were left without underlying justification and made to stand on their own. This was not an inevitable reading. It could instead have been read as the recognition that our fundamental conceptions of the world are always value-embedded, that value and fact are inseparable and that values naturally grow up in the development of the world. That an attempt was made to expel the value components utterly was itself a social fact. Perhaps it meant that the values of that age were being challenged by the consequences of the new scientific results, and this was evident enough in the battles that ensued over Darwinism and evolution generally.

In spite of this fundamental influence toward a separation of fact from value, philosophers and social scientists who invoked evolution did not hesitate to apply their science to build up an ethical or a value theory. Herbert Spencer is a complex case: his ethics came originally from utilitarianism, and his biology was interwoven with it; it aimed to justify his ethics and so there was no separation of science and ethics but a fairly sophisticated interrelation in which (rather unusual for philosophical ethics) the concepts are also expected to evolve as part of his scheme. On the whole, then, in spite of the use of the fact-value distinction in the context of the rise and defense of the social sciences, the dichotomy does not stand much in the way either of the attempts to fashion a scientific ethics or to offer scientific foundations for ethics. Nor in the case of opponents of a scientific ethics does it stand in the way of their offering a different interpretation of reality, in effect a different kind of fact, to which ethics will be geared. The Kantian dichotomy was being eroded on all sides.

What happened can best be seen by comparing the Kantian dichotomy with the post-Kantian idealists. For Kant, metaphysics is eliminated, so the realm of fact is coextensive with the domain of science, and the dichotomy of fact and value is at the same time the dichotomy of science and value. When the post-Kantian idealists reintroduce metaphysics, the realm of fact is broadened once more. A claim that the *ought* transcends the *is* would have to establish that it transcends metaphysical as well as scientific fact. This is not Kant's problem, and certainly not the idealists.' The latter, when they think of the *ought* as transcending the *is*, have in mind the spiritual transcending the natural. Value is thus not something alien to fact, only to scientific-phenomenal fact. It fits snugly into metaphysical fact. I think that on the whole some sort of fact-value fit is characteristic of the nineteenth century in spite of the Kantian proclamation. We can look through its fascinating conglomeration of moral outlooks and value theories—religious, utilitarian, biological-evolutionary, socialistic, anarchistic, Hegelian, neo-Hegelian, Marxian, and so on. Kantian influences are strong; they enter into many movements and for many purposes, and when they operate directly in neo-Kantian movements their tendency is toward fashioning formal structures with absolutistic values. But so far as I am aware, what we do not find is explicit formulation of the transcendence or autonomy of the ought (or of value) in such a way as to take it completely out of or beyond the world as a whole—sensory and metaphysical reality alike. When the language of value as transcendent is prominent it is chiefly directed against the material or sensory world in order to transfer values to the spiritual structure of reality in the fashion of philosophical idealism.

T.H. Green's *Prolegomena to Ethics* late in the nineteenth century illustrates this kind of transcendence well.[3] Green is aware of the power of Darwin's naturalization of man and what it may do to the interpretation of the spiritual life. He argues against it by tracing the unity in all desire, effort, and knowledge. He interprets it as a unity of self; and then claims the transcendence of self as a condition of knowledge and morality. He puts the choice for morality bluntly: either recognize the transcendence of the self in moral consciousness or else grant that morality is nothing but a roundabout form of a sanction of fear. Green is clearly relying on Kant, but equally clearly the unity he finds is Hegelian. It is Rationality working its way as a system into consciousness.

In spite of the difference of schools, there is thus a striking unanimity in the nineteenth century. None in the end isolates morality utterly from everything else. On the contrary—so it is believed—if we find the fundamental path of the world or existence or reality or human nature or history it will show us what morality is like and so what we ought to do, how we ought to live and organize our lives. The various schools appeal to different sources to learn about reality: to theology, to metaphysics, to psychology, to biology, to history. God's will as revealed, the dialectic of Reason in Hegel, the pleasure principle in Utilitarianism, the direction of evolution in Spencer, the growth of productive freedom and the resolution of the class struggle in Marx—all of these unfold the path of what is moral and what is worthwhile to their adherents. None of them say that value or good is what it is no matter what the character of the sensible or the metaphysical or the theological or the historical world may be. And yet this latter view is precisely what seems to be emerging at the end of the nineteenth and the beginning of the twentieth century—a strange reaction against *all* the moral theories that had prevailed. This is a new transcendence, a step ahead of the traditional ones. If sensory experience is the first transcendence which makes science possible, and intellectual experience is the second transcendence which makes possible the rational reflection that gives us metaphysical knowledge, then we must recognize that there is a quite distinct third type of experience by which we encounter values. This transcends both the other kinds which are now grouped together as factual. It is indeed an "encounter of the third kind." It is this new type of encounter that begets the technical and most absolutistic form of the fact-value dichotomy that haunts twentieth century philosophy, but not without fusion with other philosophical movements.

We may speculate that in the latter part of the nineteenth century and into the twentieth analogous tendencies were emerging in different branches of philosphy. In epistemology, for example, a logical realism made a transcendent realm out of possible facts and numbers and other logical entities. The conception of intentional objects of consciousness was developed (as in Brentano) and the beginnings of phenomenology projected the ideal of a neutral description of experience as a real isolated from physical and psychological presuppositions. Even the so-called commonsense philosophy of G.E. Moore in its discussions of perception made a fundamental point of considering whether we observed things or surfaces, as if the surfaces were separate objects of our perceptive acts. (And Moore's *Principia Ethica* later became practically the bible of the fact-value dichotomy in philosophy.) Attention had shifted from the substantial to the experiential and was now moving to the limits of the experiential and the ideal. As it hovered on the edge, the question of interpretation became pressing: was what it now discerned a further dimension of reality or transcending reality? Phenomenology, which had first bracketed the physical and psychological eventually took an idealist turn, that is, went back to a familiar "reality." Moore and the commonsense philosophy, both in epistemology and ethics, was regarded as realist; in ethics Moore spoke of *good* or *intrinsic value* as a nonnatural quality which we intuited on inspection of a thing or situation. But there is sufficient ground in their formulations and certainly in their relation to the fact-value dichotomy for interpreting the ethical movement at least as an attempt at complete transcendence, strange as that view may seem. Perhaps an historical observation is helpful. The whole movement of dualistic epistemology stemming from Descartes sees experience as transcending nature or the material world. As we have seen, beginning in the late nineteenth century we have a parallel movement which puts experience and nature in one camp and sees value as transcending them both. There is no more mystery about the new attempt at transcendence than there was about the old. Perhaps the new is simply a fresh battleground to fight the old battle now that man and his experience have been naturalized in the Darwinian evolutionary theory.

This is not a Spenglerian proposal to use the intellectual categories of an age to understand its transformations. But categories may signal deeper transformations as symptoms somewhat earlier than overt practical changes do, just as in onset of illness subtle symptons may be early. So too the intellectual defences of an institution may change before there is a full consciousness that the institution is slipping. In short, sometimes in social change the intellectual aspects crack first.

The Twentieth Century Form of the Dichotomy and its Career

It is not a simple matter to understand the late nineteenth and early twentieth century movement I have been indicating. First let us give it a name to mark its special character. To call it "value transcendence" will fit some of its proponents, especially those who broke from the idealist tradition in continental European philosophy, but not those who reached a similar position in empirical and analytic philosophy. Sometimes the designation "moral autonomy" is employed. But this admits of different interpretations on different ethical theories, and it is not fair to appropriate it for one interpretation. Perhaps the best would be in terms of the effect of the position in maintaining an isolated realm of value. I shall accordingly use the designation of "value isolationism" for the position. I use "value" rather than "moral" because it fits the more general term in the dichotomy. It is worth noting that the treatises which advocated the isolationist view were often works in ethical theory and sometimes in value theory, that they used as basic concepts either "good" from the tradition of moral philosophy or "intrinsic value" from general value theory, and there was free translation between them. (The isolation concerned intrinsic values, not the application to duties, which always involve some reference to the world, but the latter furnishes largely means.)

As I see it, value isolationism was in the first instance a reaction of an almost total sort on the intellectual side which did not commit one to a specific social usage. It could go in almost any direction—revolutionary or conformist or reactionary. The last decade of the nineteenth century was distinctive enough to have been labelled *fin-de-siècle* as designating a mood, and if we extend it into the first decade of the twentieth century we have a restless time of intellectual and artistic experiment and revolt, a growing fascination with the idea of violence, the oncoming shadow of a world war and meanwhile imperialism and the redivision of the world. (Even Spencer and the middle class feared what was happening.) Socialist movements were beginning to threaten the establishment or were perceived so in the organization of the working class.

In examining the career of value isolationism in the twentieth century I want first to reflect briefly on the uses it may have served for the social sciences, then turn for the most part to speculations on its relations to the value domain. The philosophical doctrine—I shall deal chiefly with the American scene—went through three stages. In the first (a), lasting for about the first third of the century, value-isolationism was one among various philosophical views, drawing largely on British philosophy as we shall see. In roughly the second third of the century it dominated the philosophical scene in what was afterwards labelled as "the revolution in philosophy" or more generally called "analytic philosophy." It divides into two parts, an earlier one which was logical positivism or logical empiricism (b) and a later form of Oxford analysis or ordinary language analysis (c). Now this whole movement embraced value isolationism in its ethical theory. Although logical positivism came from the Vienna Circle, its Anglo-American adaptations as regards ethics made extensive use of the work of G.E. Moore and frequently invoked his conception of the "naturalistic fallacy." In general, the influence of Moore's *Principia Ethica* (1903)[4] is great enough to regard him as the patron saint of twentieth century philosophical value-isolationism. It was the fusion of the Moorean and the positivist views that produced the most absolutistic isolationism and gives rise to the most interesting speculations about social context. The Oxford style of analysis persisted in the isolationism though it had no right to it, and we shall have to ask what were the social conditions under which the isolationism and the fact-value dichotomy finally became obsolete.

Value-Isolationism and the Progress of the Social Sciences in the Twentieth Century

It is not implausible that as the social sciences became more established in the early part of the century, the fields they explored—institutions of family, religion, political life, labor and capital—raised increasingly controversial issues. The more established the sciences became, the more they were a tempting prize in social struggles, whether for state and dominant control or for revolutionary endeavor. A value-neutral stance could serve both to resist control and, for that matter, as an ideology to disguise control. (Only detailed analysis of the specific context could determine which was the case.) Both American and European universities have had ample experience of struggles to resist political control, as well as internal struggles of opposing social outlooks. Repeated unmasking of the claims for objectivity as themselves the ideology of the establishment show the magnitude of the struggles over the social role of social science and social thought in social and political life, and higher-order debates (such as about the end of ideology) have the same character on the metalevel. Recently, the sociology of science has begun to explore comparable problems for the growth of physical and natural science as well. (Chapter 14 above presents a case study with respect to the use of genetics on the nature-nurture issue and its educational import.) Perhaps too the intensified value-neutrality served as a shield for the scientific study of values as social facts, which became widespread in the mid-century, extending after World War II from anthropology to sociology and social and personality psychology, and in general to the interdisciplinary behavioral sciences.

In turning from the science part of the dichotomy to the value part, we might have perhaps speculated profitably about the value realm in general. It is not implausible that a stronger isolation was needed—the Maginot line referred to earlier—because of internal unrest. The familiar changing pressures of industrialization, urbanization, class-conflict, demands for democratization, effects of imperialism and world war, revolutions, were so intense that the old values and the institutions that expressed them could be kept only if they were regarded as fixed or absolute and this was possible only if the value realm were isolated and entrenched on its own. But this may be only part of a complex scene. Our plan calls for treating the clear philosophical shifts as the organizing topics and speculating about the social relations that cluster around them.

(a) *Value-isolationism in the first third of the century*. We start with G.E. Moore's position in ethical theory, which came over from England to America and was one of the conflicting positions in the milieu of moral philosophy. Because of the role it was destined to play, our discussion of this part concerns mostly England (and Germany) rather than the United States. As of the 1920s, for example, there seemed no more reason to bet on Moore than on the idealism of Urban or Hocking, the naturalism of Perry, the instrumentalism of Dewey, with only Moore among these representing value-isolationism. Now the question we are headed toward is whether there are plausible social value grounds to explain the appeal that Moore's theory had at a specific time in American life.

Moore's theory, analyzed in chapter 2 above, need not be reviewed here. We saw how his conception of the *naturalistic fallacy*—that it was a fallacy to equate the central concept of good or intrinsic value with any term of psychology (desire, pleasure, interest) or of metaphysics (God's will, Reason) or of history (trend of evolution or historical development) or of biology (trend of nature, human need), and so on—barred any reference to existence or reality in value theory. It left the judgment of what was intrinsically valuable to the individual's direct intuition of an isolated whole that he envisaged, a whole separated off from questions of means and consequences and judged by itself as if it were looked at as a world by itself and its value simply seen. Of course to decide what to do in matters of action involved attention to causes and what would bring about the greatest value. But the ultimate bases of the value system were set in these isolated judgments for which no evidence—scientific, metaphysical, historical, and so, about the world—was in any way relevant. Value was *autonomous*.

This was not idiosyncratic with Moore in England. Later on in Germany we find Nicolai Hartmann's phenomenological ethics yielding parallel results.[5] He distinguishes domains quite explicitly: an Ideal Ought-to-be which is independent of the world of being, a Positive Ought-to-be which is the ideal adapted to the structure of this world, and an Ought-to-do which is appropriate to our personal situation. Perhaps the separateness of the value domain is conveyed tellingly in Hartmann's criticism of traditional religion for regarding God who is the ultimate Reality as the source of Goodness. This, he says, fails to realize that value transcends being and so from the point of view of axiology (the science of value) the goodness of God would have to be assessed by independent value standards. There is thus a tension between the perspective of religion and that of value which, we can readily see, was implicit in Kant's reversal of the relation between religion and ethics and which in general reflects the historical change in the position of religion in modern society. (A contemporary reader of a textbook in philosophy who now sees a chapter headed *Axiology* alongside of one headed *Ontology* can scarcely realize the blood, sweat and tears that went into the parallel position of these chapter headings.) Finally, Hartmann, like Moore, has no way to bridge the conflict that arises when different persons give different verdicts in their value judgment. They both suspect that the differences arise from looking at different worlds and this should be clarified. But if it is the same object being judged, then Moore simply gives up and says he cannot force you to see what is obviously there, while Hartmann talks of a moral blindness analogous to partial blindness.

Now this type of value theory is a very special one, a technical variant of several specialized traditions in moral theory. Why, in Anglo-American philosophy, should Moore have caught on particularly in the middle third of the century, and been relied on over and over again, but in the very peculiar way that writers kept saying his arguments were wrong but his position was right if we reinterpreted it? The idea of the naturalistic fallacy survived, while Moore's view of definition and of intuition and of nonnaturalistic qualities all fell by the wayside. And the one argument that was left to support the naturalistic fallacy, though interpreted in different ways, was the so-called "open-question argument"—that if good or intrinsic value is equated with any content, one could always ask of that content whether it itself was good, and that question would be meaningful.

In chapter 2 I suggested briefly (and elsewhere in more detail)[6] why this structure of Moore's should have appealed the way it did. Its technical construction was such that the judgment of ultimate intrinsic value was the individual's and that he could not be questioned by anyone else, or if he was, the reconsideration and judgment were his own. Moreover, the open-question argument had as its consequence that he could not be pinned down to any feature of existence or reality—he was left unbound. The value theory as a whole may be seen therefore to have a profoundly individualistic orientation of a libertarian sort. This is in spite of the fact that it was cast by Moore in a logical realism that spoke of intuiting objective value qualities, and in spite of the paradoxical fact that Moore himself derived from his basic theory an extreme moral conformism on the ground that the complexities of proof did not provide sufficient grounds for departing from customary rules of conduct even when one knew they were wrong.

That this type of theory can readily have an individualistic value orientation explicitly is seen in the case of Bertrand Russell, who held to it in the first decade of the twentieth century. (He changed his mind about the realism, he said, on reading Santayana's criticism of it.) Russell followed Moore in his technical account, but his popular essay, "A Free Man's Worship," conveys the spirit most clearly.[7] He attacks those who worship power just because it exists and can influence the course of events. He blames Job for accepting God's treatment of him as good, for all God had exhibited to justify himself had been supreme power. Man's period of life is brief, while matter rolls on its relentless way, but he should spend it in the vision of the ideal that he sees, not in surrender to what power happens to exist. The one point that neither Russell at that stage, nor Moore at any stage, investigates, is what a person is, how his values arise, and the character of the process by which he sees an ideal.

It should not surprise us that a value isolationism could yield an orientation of this sort as well as an absolutist orientation. We noted that double potential earlier and the common stand of both alternatives against appeal to scientific justification, in commenting on the fact that Kant could feel comfortable with the is-ought dichotomy because he had an authoritative moral theory to go with it. It is interesting to observe that not long after Kant's time an anarchistic individualism had emerged, speaking in the name of the individual as sole reality and challenging all claims for God, Humanity, Law, Nation, and Ideal, as simply illusions or attempts to entrap the individual. I refer, of course, to Max Stirner's *The Ego and his Own* in the 1840s.[8] It was not, however, a transcendence theory, but a claim for the individual as the single reality. But it would not have been possible to set off such a solely valuable if value had not first been isolated in principle. The Moore-Russell position of the later period has transcended even the ego. The ideal is claimed to be what it is, irrespective of what the ego is.

Russell's active life, morally critical of many institutions, shows by its contrast with Moore's conformism, that quite different social outlooks are consistent with the general orientation of value isolationism. Much depends on the further premises that supplement the major structure. It can be a revolutionary structure precisely because it is tied to no existent institution. But because it has no account of the judging self, values will have the social character of arbitrary commitment. Some types of individualistic anarchism thus come closest to it in social theory. Among conservative outlooks, perhaps the kind most likely to be found is one that is intuitively held, that feels it needs no justification.

To examine the social origins for individual figures in our story is beyond our present scope. Some picture of Moore's personal values and influence can be found in Keynes' memoir.[9] For Russell, there are copious materials in his autobiography.[10] While in part we are dealing with the work of an isolated academic stratum, occasionally it reflects the special breadth and insight that academic aloofness sometimes achieves. Certainly if Moore was criticizing the prevalent moral philosphies of the later nineteenth century on analytic grounds, Russell was conscious of their social shortcomings: utilitarianism was still tied to an individualism that opposed social legislation through fear of the state, religious morality was largely bound to established institutions, idealist metaphysics justified conservative obedience to the state, evolutionist ethics sanctified predatory struggle in ecnomic life. There was nothing in such a scene to tempt the ethical theorist, well aware that ethics had only recently escaped from religion, to bind it to the alleged verdicts of psychology or theology or metaphysics, or history or biology. It seemed better policy to declare independence forever in an absolute autonomy.

Each country has its own cultural-intellectual atmosphere. In the United States during the first third of the twentieth century, no importation of the sort we have been considering would be needed to accentuate the tradition of individualism. Indeed, all schools of moral and social thought, whether idealist or naturalist, theological or secular, in the early part of the century, fall toward the individualist end of the spectrum, just as the economy is laissez-faire capitalist. But corporate institutions were growing. With the Great Depression of the 1930s and the growth of labor unions and social security mechanisms, the intellectual atmosphere changed, and with it some part of the individualism altered. What had been in the first third of the century an expansive individualism, resting on a conception of the individual as the atom of society (the "rugged individual" of popular philosophy) gave ground. On the one side a social individualism saw the individual as a socio-cultural product and called for a rich individuality as a moral ideal; on the other side the increasing pressure of organizational structures on people yielded what is now widely recognized as the alienated individual. The career of our value-isolationism in the United States appears to be correlated with this shift. During the first third of the century it remains simply one among many philosophical outlooks. But in the complex philosophical situation of the 1930s it converges with the rise of the positivist approach to value to make value isolationism the prevailing philosophical position. Let us turn to this partner, and consider the suggestion that their joint effect came from a libertarian appeal to an alienated individualism.

(b) *Value isolationism and logical positivism*. The background of logical positivism is to be sought in European sources, chiefly Vienna and Berlin, where it had been developed in reaction against the metaphysical atmosphere of the continent. It was a scientific philosophy, exalting science in direct answer to the old Kantian compromise, and demoting religion and incidentally morals. Hence the important part of the philosophy was the epistemological analysis of science and scientific meaning, intellectually exciting because it rested on the new logic and the new science. The expressive theory of religion and ethics followed from the fact that they did not meet the critical standards of science. The theory was gradually developed into a new type in alliance with the study of the varied uses of language. Of course the positivist theory of value was not the "third encounter" type we saw at the turn of the century. But the positivists had taken for granted the description of the field they found around them in which the Kantian is-ought distinction had long been elaborated, hardened, and the domain of value cultivated on its own in different ways. Given the commitment to a separate value domain and the rejection of value statements as technically "nonsense," an expressive, imperative, or emotive interpretation seemed a plausible path.

In spite of the fact that logical positivism came to the United States as a scientific philosophy and formed intellectual alliances with pragmatism (in the International Encyclopedia of Unified Science), it stood in the way of philosophical efforts to relate science and ethics, as for example Dewey was urging. Its value isolationism prompted a mistrust of the claims of science in ethics. By the middle 1930s it had already formulated the view that ethical terms were purely expressive or imperative or in part optative (Ayer, Carnap, Russell),[11] so that judgments of value were neither true nor false and there could be no traditional resolution of disagreements. This was developed by Stevenson in the emotive theory of ethics, first in articles in the late 1930s and then in his influential *Ethics and Language*.¹² Emotivism became the standard and dominant form of value isolationism of the period in the United States, combining Moorean, positivist, and linguistic influences. It reflected the social context most transparently, for it focused on disagreement as the central phenomenon to which moral language is addressed, and concluded that concerning ultimate ends where there is no agreement there is no rationally valid method, only persuasive methods causally operative, or else conflict. (Dewey, by contrast, focused on a community attempt to solve its problems by working out practices and institutions, utilizing scientific knowledge and methods in facing its problems rationally.) Now quite clearly the situation of the 1930s did seem one of basic conflicts in which men were called on to make ultimate choices: total system choices of fascism, nazism, liberalism, communism; economic system choices of socialism or capitalism; political system choices of dictatorship or democracy. These were no longer matters of theoretical debate but of past or imminent shifts in a world of depression and apparent collapse. A belief in the ultimate relativity of ethics seemed much closer to the realities than confidence in a scientific rationality for ethics.

As for the sciences themselves, neither in practice nor in theory were they at that time in condition to undertake an ethical role. Technology had been blamed for increasing unemployment during the depression. Basic scientific research for human well-being was neither appropriately appreciated nor supported; an expansion of scientific research was to come during World War II in connection with war needs. Again, the structure of the professions that usually furnished the popular model of the scientists as moral advisor—medicine and psychiatry—hardly had the democratic character to inspire confidence to an ethic that emphasized autonomy. The social sciences were turning to questions of value, and there were controversies similar to those in philosophy about cultural relativism and its consequences for moral or ethical relativism. The anthropological concept of patterns of culture was relativistic in that every culture had its own values, but it seemed to carry along a general liberalism in that it recognized an inherent dignity in every culture, as against the Nazi doctrine of racial superiority. By contrast, the psychologists tended to look for intra-individual sources of universal moral patterns, and we have seen how moral philosophers built upon their work (ch. 6). But on the whole, the relativist theme remained dominant.

The analytic ascendancy in ethical theory, which was quite clear in the 1940s, came from the juncture of the positivist and the Moorean views in the 1930s and maintained the fact-value dichotomy in its extreme form. My speculations concerning the special basis of this transformation have been suggested earlier and may be amplified as follows. The expansionist character of American individualism changed its quality with the depression and world situation of the thirties. The social individualism which would recognize the individual as a cultivated social product demanded a much sharper transformation in American institutions and a much more cooperative approach in American life than apparently was practicable at that time. Whatever the historical possibilities and alternatives, the next half century showed the power of American capitalism and its use of its old individualistic ideologies. Given the actual growth of corporate forms of organization and the increased limitations of effective individual action by organizational structure, and the many areas of individual ineffectiveness, alienated individualism is a well-recognized form of reaction. An intellectual response, particularly in the intellectual domain, is an intensified voluntarism. This, I suggested, explains the appeal of the value-isolationism in the Moorean structure, once isolationism is separated from its original setting and from Moore's own moral conformism. The positivist partner keeps science in its own very important, but separate place. The result is not simply an up-to-date Kantianism for the science and the morality of the twentieth century. It is a finely tuned structure that restores to the alienated individual the right of judgment on questions that ultimately count in the changing and basically insecure world of the 1930s where countries are being sold out and races persecuted, colonies exploited and invaded, death camps constructed, bizarre doctrines promulgated as intuitive truths, while unavoidable world war is in the offing. It strikes exactly the right note until the war comes. Then it begins to sound a bit hollow. But it goes on in revised form until it peters out.

(c) *Value-isolationism in its latter-day career.* During World War II there was a felt need to justify the struggle against nazism, not simply to see it as defense against attack or one side against another. Interestingly, occasional philosophers and scientists who undertook such tasks (for example, Ralph Barton Perry or Julian Huxley)[13] believed in the relation of science and ethics, not in the dichotomy of fact and value. They could not, as is sometimes popularly done, take an almost proprietary view of ideology and say that "our side" was defending our ideology against hostile foreign ideologies, and an ideology is the ultimate value stance of a people; for that could equally be said by the other side. (This view of ideology was examined and criticized in Vol. I, ch. 9.) They had rather to argue that our side was right and the other side wrong, and they had to appeal to a minimal all-human ethic. To do this involved carrying through an analytic critique of nazism. They had to face the fact that the world was changing during the war and that the old order could not simply be restored, hence that moral reconstruction would be needed and an understanding of morality and its functions on a far broader basis than had hitherto been dreamed of.

In the United States, however, these tasks were not carried far. In the popular forum conservative outlooks quickly came forward. At first it was said that the war had been fought to restore free enterprise throughout the world. But as the Cold War developed it was more often cast as Individualism vs. Totalitarianism. (The latter concept proved stretchable, so that the American Right could throw in socialism and new dealism as well as fascism, nazism, and communism.) On the whole, the ideology of individualism was so well established in American life that after World War II even a neoconservatism using the familiar symbols of old-world conservatism—tradition, the attack on rational intellectualism, religion, morality—did not get very far. As the Cold War deepened, and particularly during the McCarthy period, the exaltation of American individualism became a dogma. Value in effect was identical with individual self-assertion.

On the whole, the vista of change in moral and social philosophy did not find much place in the universities in the United States after World War II, during the 1950s and into the early 1960s. In philosophy it is to be understood largely in professional terms. The new and interesting movement of ordinary language or Oxford analytic philosophy had taken over from positivism. The center of philosophical gravity again was not in ethics, but in epistemology, and it was hoped that some of the same importance would come from the extension of humanistic and linguistic methods as had come from the postivistic analysis of scientific revolutions. Oxford analysis persisted in the fact-value dichotomy because it embodied its Moorean heritage. It really had no right to so sweeping a dichotomy, since it was committed to tracing informal uses in varied contexts in which all sorts of relations are to be found. Persistent informal investigations played their part in the eventual undermining or bypassing of the dichotomy. But in any case moral philosophy devoted itself to metaethics, rejecting substantive or normative ethics as the task of the rhetorician or the preacher. Discussions of the naturalistic fallacy continued to flourish.

The world was in turmoil, colonialism was on the way out and institution after institution was under attack. That moral and social philosophy did not react to all this requires explanation. Perhaps it lies in the intensity of the Cold War of the period and the McCarthyite repression of critical thought. Social philosophy in the United States practically died out, while the logic of the social sciences, a largely technical field, began to attract good students.

In the mid-1960s the situation changed. With substantive problems engaging the young—liberation movements, new conceptions of moral rights, changed conceptions of the good, experiments in modes of life, active protests against the Vietnam War—practical or normative ethics was bound to make some inroads on metaethics. In due time the fact-value dichotomy almost vanished from the philosophical scene. When Rawls' *A Theory of Justice* was published (1971), "naturalistic fallacy" was not even to be found in the index.

Why did the fact-value dichotomy find itself unemployed? Perhaps it was played out as an intellectual dispute—its moves had become standardized and its byways overrefined. But it is much more likely that its whole basis had been destroyed. We saw initially that in its earliest form the fact-value dichotomy had served to insulate science from social reponsibility and isolate value from currents of change. Now a critical point had been reached and passed. The part played by science in society had become so great that scientists could no longer be uninvolved, and values had in reality undergone such massive changes in large sectors of life that isolation was not the appropriate strategy, whatever the social purposes involved.

This did not, of course, happen all at once. On the value side, after World War II, much of the public rhetoric in the United States talked of restoring throughout the world the old values for which the war had been fought, and even spoke of them as the values of "our Christian civilization," forgetting that the war had involved a variety of people of many religions (and none) and cultures. But there was not merely global variety, there was change within America itself. For a time this could be obscured by regarding it as simply sinning against old values, but by the end of the 1960s the massive changes, their frank and sincere espousal, and the fact that value change penetrated into almost all religions, finally forced the recognition of change. In the case of science, Nazi racial doctrines in the early 1930s prompted organized scientific protests and promoted self-consciousness about social responsibilities of scientists. The development of atomic energy and the use of the atomic bomb was decisive. After that, the doctrine of a value-free science changed its social import: it was no longer a defense to protect science but an ideology to avoid responsibility. Its full collapse was bound to come with the further intensification of scientific discovery and technological development. Whole areas of morality were opened up as a consequence in which the scientist now has to cooperate to determine his duties and rights and responsibilities—bio-medical ethics, environmental ethics, ethics of technology, etc. In more general terms, science and technology are no longer an outside activity operating on their own; they are institutional activities that utilize a great part of the resources of a society and on whose work a great part of the actual functioning of today's society depends, not merely extra features but basic production and needs and processes. The distinguishing feature of recent developments is that basic research, not merely technology, has been brought within this scope, as issues about research into nuclear energy and recombinant genetics have shown. There may now be questions about whether scientists should shoulder responsibilities collectively or whether some scientific responsibilities should be supervised by public agencies, but these are special questions. Nobody raises general questions about a value-free science where a public investment of several billions is involved.

Death or Temporary Eclipse?

What is the likely fate of the fact-value dichotomy? Is it waiting in some intellectual limbo until philosophers have time to go back to more general theoretical problems? (Perhaps it will take shelter under some friendly broader dichotomy as it did for a while under that of the cognitive and noncognitive.) Or is it a dichotomy whose strength and coherence lay in its ideological and social uses, so that once these have lost their purposes its intellectual unity is rapidly dissolved? I think the latter is the case. The rich technical exploration of the problems in the arguments about it revealed such complexity that there probably never was any unity apart from the desire to keep science and value apart. The basic philosophical problem that replaces the problems generated by the fact-value dichotomy is the need to integrate knowledge and action in morality and social policy. This involves many tasks in integrating the intellectual and practical, the technical and the valuational; for the fact-value dichotomy ran a fission through the whole conceptualization of ethics and value theory and there is a major reconstruction to be carried out in the methods of ethics and the study of values.

The reconstruction that is required with the lapse of the dichotomy is more than theoretical. The dichotomy, as is well known, permeates our general culture, our institutions and attitudes. For example, our educational system typically separates science and humanities (the "two cultures"). Our newspapers separate news and editorial opinion, ignoring the role of selection in the former. Perhaps the *reductio ad absurdum* of the dichotomy is seen by the high school teachers who find their students' conception of fact to be the TV quiz show answers and of criticism (evaluation) simply to say that they like or dislike something. Hence the problems of a transition in which the dichotomy is to be overcome will be as difficult as those experienced in any movement toward integration. There will continue to be a separateness as of two parties interacting until the process is finished and perhaps only then will it be clear that there is a difference of enterprises—describing or explaining on the one hand and evaluating on the other—and not two parties or two domains. Perhaps the best way of going about integration is to have each of the "parties" extend itself over the whole field so that its role or influence is omnipresent. And in fact that is what seems to have been happening. Let us look briefly at science and value in turn.

Of the present value fields perhaps the arts have been most receptive to scientific offerings. At first science was the helpful servant, providing better tools (printing, writing, painting, communication). Then it was discovered that the whole scientific enterprise was a humanistic intellectual adventure; that made it a brother of the spirit. Soon after it was along the front line in artistic labors, with its techniques of sound production and visual effects. And we tend by this time to overlook the fact that it created the arts of film and photography, and transformed the materials and techniques of architecture. Among the social policy disciplines, education has begun to look to science for teaching aids, and of course it has long leaned (at times too heavily) on psychology and its conflicting schools. Politics has reached eagerly for tools that could help in its internal political struggles—particularly for the statistical techniques of voter canvassing and polls and psychological lessons of campaigning and image making. Law has moved more slowly with respect to science except perhaps in crime detection. Morality, and most of all ethical theory, have kept its influence at bay. We have considered sufficiently in this volume the work that has to be done in reconstructing ethical theory in relation to science. In general it has become clear that science is everywhere relevant throughout the world, in every enterprise. There may be debates about what form of technology is appropriate to where, whether large-scale technology or "intermediate technology." But when this whole process is over, and the scientific aspects of every situation become a recognized part of any deliberation, the integration with respect to science will be complete. One will no longer think of science as a separate esoteric domain but as the continuous extended knowledge that mankind has achieved collectively and is available for use in all inquiry where relevant.

The reconstruction for the value side of the dichotomy is likely to catch up with the science side though the integration has been less continuous. The concept of value has undergone marked extension of meaning, while morality has rather extended its scope. Briefly, value had begun with conscious pleasure or desire (or the arousing quality of the object) and then went on to individual interest or selection. But in the work of the social scientists it came to be used for social or cultural selection or orientation, particularly where other cultures exhibited alternatives (for example toward the past rather than toward the future or toward the simple life rather than toward the progressively complicated one). In the psychology of personality, differences in attitude, in depth formations, in characteristic response, were seen as values or as reflecting values. Broad intellectual categories were taken to be fit objects for value study since the element of selection in classification rests on a purposive base. Indeed the initial organization of our world of things and events out of a matrix of continuous experience was recognized to have a value character. Such changes in the concept of value made it relevant to all symbolization, all experience and all theory construction. In this broad sense, value has long been an integrated aspect of all experience. By contrast, morality seemed confined to narrow quarters. With explosive force, however, moral questions were raised in area after area of social life: the Nuremberg trials, the Vietnam War, Watergate, the various liberation movements, new lists of human rights, industrial responsibility for pollution, spread of the concept of professional malpractice and accountability. (Though many areas once under social control are now turned over to individual decision—for example, in matters of religion or sex—they are not deemed morally arbitrary but left to individual *moral* decision.) With even questions of scientific research raising moral problems (cf. ch. 14), there remains no corner of human life in which moral considerations may not be relevant. Thus in the long run, when the effects of the dichotomy have run their course and the integration is complete, one would expect as the normal part of any enterprise, in undertaking a project and developing its "budget," that there would be a study of the effects of the project on the people involved and their ways of life and a formulation of the distribution principles of the consequent gains and burdens. The morality would be in the budget, not an aftereffect nor an "external cost," and the knowledge which enlightens would be in the morality as well as in the data of the project. Of course one would bring moral standards to the making of the budget and to the selection of projects, but the standards themselves would already express the knowledge at the base of our human outlook. And that knowledge itself would embody human purposes. But what the shape of the integrated ethical theory would be like, requires a full-length study of its structure, and that is another task.

It should be noted that our conclusions about the passing of the fact-value dichotomy and the transition to an integration of knowledge and action stand independent of the specific historical speculations offered of the twentieth-century career of the dichotomy. The conclusions required only the more general view that the dichotomy served primarily a social role which has now been outworn. Our speculations were offered to suggest the need for analyzing detail of theory and its interlocking with social context over shorter periods of time—a task which has not been sufficiently undertaken for philosophical ideas. It was set as an epilogue because it was not a conclusion of the book or a summary of the book, but an independent suggestion which can add to the modes of analysis considered in the two volumes. This kind of analysis of social context in the case of a philosophical idea helps us see what the idea was accomplishing on a fuller scene and so encourages critique, develops self-consciousness, and enables us to consider how desirable objectives could be undertaken in a more direct philosophical way instead of a philosophically devious and ideological manner. It thus enhances the philosophical goal of clarity, but in a fresh and broader way, not in the narrowing manner that sheds moral and normative responsibility by claiming to be purely logical or methodological.

Notes

1. David Hume, *A Treatise of Human Nature*, Book III, part I, sec. 2.
2. Karl Marx and Frederick Engels, *The German Ideology* (London: Lawrence and Wishart, 1965), pp. 448-54.
3. T.H. Green, *Prolegomena to Ethics* (Oxford: Clarendon Press, 1883).
4. G.E. Moore, *Principia Ethica* (Cambridge: Cambridge University Press, 1903).
5. Nicolai Hartmann, *Ethics*, trans. Stanton Colt (New York: Macmillan 1932; original German publication 1926).
6. Abraham Edel, *Method in Ethical Theory* (Indianapolis: Bobbs-Merrill, 1963), ch. 5.
7. Bertrand Russell, "A Free Man's Worship," *Independent Review*, 1903. Reprinted in *Mysticism and Logic* (New York: W.W. Norton, 1929). Russell's ethical theory at the time, acknowledged to be under Moore's influence, was formulated in his *Elements of Ethics*, included in his *Philosophical Essays* (New York: Longmans Green, 1910).
8. Max Stirner, *The Ego and his Own*, trans. Steven T. Byington (New York: Boni and Liveright, 1907; German original 1845).
9. J.M. Keynes, *Two Memoirs* (New York: Augustus M. Kelley; London: Rupert Hart-Davis, 1949).
10. Bertrand Russell, *The Autobiography of Bertrand Russell*, first American edition, 3 volumes (Boston: Little Brown, 1967-69).
11. A.J. Ayer, *Language, Truth and Logic* (London: Victor Gollancz Ltd. 1936); Rudolf Carnap, *Philosophy and Logical Syntax* (London: K. Paul, Trench Trubner and Co., 1935); Bertrand Russell, *Religion and Science* (New York: Holt, 1935).
12. Of the articles, the central one was: C.L. Stevenson, "The Emotive Meaning of Ethical Terms," *Mind* 46, 1937; *Ethics and Language* (New Haven: Yale University Press) was published in 1944.
13. Ralph Barton Perry, *Our Side is Right* (Cambridge, Mass.: Harvard University Press, 1942); T.H. Huxley and Julian Huxley, *Touchstone for Ethics, 1893-1943* (New York: Harper, 1947).

Index

Absolute, 131
Ackerman, Robert, 56n., 189n.
Affirmative action, 182, 298-301; quotas in, 301-05
Allport, Gordon, 83-84, 89n.
American Psychiatric Association, pronouncement on homosexuality, 275 f.
Analytic method, *see* Ethical method
Application, as a way of construing theory-practice relations, 256-58
Aristotle, 66-70, 74n., 78-79, 89n., 142, 160, 199, 223, 255-56, 294, 341
Atlee, Clement, 224
Augustine, Saint, 79, 89n.
Austin, J.L., 56n., 63
Authenticity, 132-34
Autonomy: of morals, 38; Durkheim's view of, 147-48; Piaget's view of, 148; *see* Kant
Ayer, A.J., 355, 363n.

Baker, Russell, 248
Bateson, Gregory, 106, 120-21n.
Benedict, Ruth, 74n., 89n., 155-56n.
Bentham, Jeremy, 37, 41n., 64, 67, 72, 89n., 272, 331, 342-43
Berkeley, George, 7
Berlin, Isaiah, 183
Biddle, Francis, 229-30, 237n.
Bodmer, Walter F., 269, 285-86n.
Bosanquet, Bernard, 109
Bradley, F.H., 138, 155n.
Brandt, R.B., 17, 41n., 156n.
Brentano, Franz, 19
Buber, Martin, 186

Burke, Edmund, 161, 223
Butler, Joseph, 112

Carnap, Rudolf, 355, 363n.
Carritt, E.F., 80, 89n.
Carson, Rachel, 325
Cartesian dualism, 11, 261
Categorial dichotomies, how to evaluate, 46-48; *see also* Ethical method, programs and dogmas
Catholicism, 157, 229, 279
Cavalli-Sforza, L.L., 269, 285n.
Chomsky, Noam, 284, 287n.
Clement of Alexandria, 170n.
Club of Rome, 220, 336
Collective welfare model, 293
Computer decision systems, and evolutionary processes, 115-16
Comte, Auguste, 92, 345
Conscience, 32-33, 144; and the scientific enterprise, chapter 11; its social development, 227-31; externalizing of, 305-06
Contraception, 166
Crime, and terrorism, 319-20

Darwin, Charles, 57, 96, 166, 345-46, 348; Social Darwinism, 93
DeFunis case, 289, 297
Dewey, John, 16n., 25, 34, 39, 53, 56n., 70, 75n., 77, 80, 83, 87, 89n., 96, 106-07, 110, 113, 119n., 142, 150, 155n., 156n., 179, 190-91n., 193n., 200, 203, 214, 216n., 230-31, 240, 286n., 351, 355-56
Drives, and drive mutations, 98
Durkheim, Emile, 123, 147-48, 153, 156n.

Ecological mode of thought, 224
Edel, Abraham, 16n., 42n., 74-75n., 95, 119-20n., 155-56n., 170n., 190-91n., 216n., 363
Edel, May, 74n., 156n., 216n.
Egoism, Moore on, 35
Einstein, Albert, 266, 277
Ellul, Jacques, 199-200, 202-04, 215-17
Empirical knowledge, place in ethics, chapter 3
Ends and means, *see* Means and ends
Engels, Frederick, 89n., 156n., 216n., 343, 363n.
Engineering, 260-64; engineers, social responsibility of, chapter 12
Environmental ethics, chapter 17; growth of, 325-26; role of philosophy in, 326-28
Epictetus, 60, 89n.
Epicurean, 73
Equality, 182
Erikson, Erik, 85, 89n., 103, 136-37, 144, 155n.
Ethical method: analytic method for dealing with moral change, chapter 9; metaethics separated from normative ethics, 178; programs and dogmas contrasted, 176-78; present ahistorical character, 178-79, and proposed revision, 179; present modes piecemeal and isolated, 179-81; network analysis and context-regarding analysis, 181-86; excessively individualistic, 186, and proposed revision, 187-88
Ethical relativism, 356
Ethical theory: use of science in, 33-38; place of empirical knowledge in, chapter 3; commitments in, 58 f.; scientific bases for, 93-97; scientific constraints, 94; scientific assistance, 94-95; scientific foundations, 95-96; scientific methodology, 96-97; requirements for contemporary ethical theory, chapter 8; a broadened conception of morality, 159-63; recognition of empirical and value components, 163-65; humility regarding concepts, 165-68; a holistic view, 168-69; an activist approach, 169-70
Euclid, 39, 57, 64-65

Evaluation, permanent possibility of, 20, 22
Fact-value dichotomy: patterns of relation, chapter 1; variety of formulations, 3-5, 340; entities involved in, 5 ff.; configurational types, 7 ff.; explanatory hypotheses, 10 f.; relation to self-formation, 11 f.; inadequacies of separatist program, 48-53; an integrative program, 53-55; separating science and ethics, 124-25, 226; as justifying scientific neutrality, 242; a dogma of empiricism, 253; social-intellectual history of, Epilogue; rise of, 340-41; alleged Humean paternity, 342; Kantian development of, 343-45; Darwinian impact on, 345-46; post-Kantian idealism and, 346-47; novel form of, 347-48; twentieth century career, 349-60; resultant value-isolationism, 350-60; death or temporary eclipse, 360-63
Fear of knowledge: as overcomplicating life, 265 f.; of effects of its increase, 266; of ideology, 267; of what we will find, 267-68
Francis, Saint, of Assisi, 8
Frankena, W.K., 161, 178, 189-90n.
Freud, Sigmund, 59, 84, 93, 100, 105, 109, 144, 149, 155-56n.
Friedman, Milton, 229, 237n.
Fromm, Erich, 85, 90n., 95, 119n., 133, 155n.

Galilei, Galileo, 241, 253, 277-78, 341
Gandhi, M.K., 314
Glass, Bentley, 234, 237n.
Good Life, the, changing concept of, 174
Gottlieb, Gidon, 165, 171n.
Green, T.H., 109, 347, 363n.

Hallowell, A.I., 85, 90n.
Hare, R.M., 61, 231
Hartley, David, 92
Hartmann, Nicolai, 45, 56n., 61, 74n., 89n., 352, 363n.
Hartshorne, H., 81, 83, 89n., 128, 149, 155-56n.
Hegel, G.W.F., 13, 101, 103, 109, 347

Hobbes, Thomas, 37, 39, 72, 107, 183, 228, 322, 327, 345
Hocking, W.E., 351
Homosexuality, 275 f., 281-82
Housman, A.E., 267
Hume, David, 37, 43-44, 52, 72, 89n., 256, 296, 342, 363n.
Huxley, Julian, 114, 357

Individualism, types of, 354-55; in American thought, 357; individual rights model, 293
Intelligence, and race, 273-74

James, William, 130; James-Lange theory, 56
Jensen, Arthur, 273-74, 280, 286-87n.
Justice: compensatory, 289; distributive, 205, 208, 212 f.; formulae of distribution, 294, 297; Hume on instrumental character of, 296; theory of, 291-92

Kant, Immanuel, 6, 8, 44-46, 60-61, 63, 79, 89n., 140, 142, 146-47, 160, 187, 230, 292, 326, 329, 343-47, 353, 355
Kardiner, Abram, 85, 90n.
Keynes, J.M., 177, 189n., 354, 363n.
Kierkegaard, Soren, 203, 231, 237n.
Kluckhohn, Clyde, 84, 89n., 145, 156n.
Kohlberg, Lawrence, 83, 86, 89n., 135-37, 139-40, 148, 155-56n.
Konvitz, Milton, 229, 237n.
Kropotkin, Peter, 93
Kuhn, Thomas, 75n., 176, 286n.

Ladd, John, 27, 41n., 156n.
Laird, John, 77, 82, 89n.
Laski, Harold, 270
Law, 260-64; requirements for contemporary legal theory, *see* Ethical Theory, requirements for contemporary ethical theory
Legal realism, 159, 160
Leibnitz, G.W.F., 341
Leites, Nathan, 305, 307n.
Lewis, C.I., 39
Liberty, 163, 182-83; traditional theory of, 270; in scientific research, 271-72; in presenting research results, 272-74; in research on heredity, 277 ff.

Lifton, Robert, 86-87, 90n.
Locke, John, 170n., 258-59, 321
Luther, Martin, 131
Lysenko, T.D., 229

MacBeath, A., 138, 155n.
McCarthy, Joseph, 229, 279, 311, 358
McDougall, William, 116
Macroethics, 187-88, 329-32, 337n.
Maine, H.S., 168-69, 171n.
Maritain, Jacques, 168
Marković, Mihailo, 183, 192n.
Marx, Karl, and Marxism, 56, 92, 93, 123, 146, 179, 190n., 200-02, 204, 214, 255, 278, 343, 347, 363n.
Maturity, 31, 134-37.
May, M.A., 81, 83, 89n., 128, 149, 155-56n.
Mead, Margaret, 89n., 106, 120-21n., 155-56n., 266
Means and ends, 23, 34-35; assimilating consequences to means, 24; holding to an end, 24-26; conflict of ends, 26-27
Medicine, 260-64
Merit, 294-95; meritocracy, 175
Microethics, 187, 329-32
Milgram, Stanley, 234, 285n.
Mill, J.S., 92, 214, 285n., 293
Models in ethical theory: casual, 60-63; heuristic, 63-65; structural, 65-68; conclusions about, 69-74; moral blindness metaphor, 61; debt metaphor, 61-62; musical metaphor, 64; Perry's monetary model and Aristotle's craftsmanship model compared, 66-68; Euclidean model, 57, 64-65; medical model, 65; military model, 60; models as theory-surrogates, 70-71; four central models for morality, 124, 141-53; the goal-seeking, 143-46; the juridical, 146-49; the self-development, 149-50; the decisional, 150-51
Moore, G.E., 8, 17, 19-22, 27-28, 35-36, 41n., 42n., 45-47, 56n., 153, 156n., 177, 189n., 348, 350-54, 356-58, 363n.
Morality: and technology, chapter 10; what morality is, 198; and values, compared, 198; moral community,

198, 205, 208; and human relations, 206, 208-09, 211 f., 213 ff.; and the good life, 206, 209 f., 215 f.; moral development, 135-37, 139-40; moral and immoral as criteria, 137-39; present state of, 174-76; moralistic terrorism, 321-24
Moynihan, Daniel, 280
Muller, H.J., 224
Münsterberg, Hugo, 7, 16n.
Murray, Henry A., 84

Natural norm, 112
Naturalistic fallacy, 17, 19-24, 350, 351, 352, 359
Negative income tax, 229
Newton, Isaac, 46, 57, 72, 80, 241, 261
Nielsen, Kai, 311
Nietzsche, Friedrich, 14, 41n., 62, 64, 74n., 79, 89n.
Normal and abnormal, 131-32
Nowell-Smith, P.H., 6, 16n.
Nozick, Robert, 192n., 328, 337-38n.

Oakeshott, Michael, 191n., 258, 268n.
Oppenheim, Felix, 183
Ought: analytic treatment of, 63-64; ought and is, *see* Fact-Value dichotomy
Oxford analysis, 350; Moorean heritage in, 358

Pacifism, 313 ff.
Paley, William, 161, 170n., 184, 192n.
Parker, DeWitt, 7, 16n., 64, 74n.
Pepper, Stephen, 75n., 95-114, 120-21n., 191n.
Perry, R.B., 28, 41n., 66-70, 74n., 76, 102, 120n., 157-59, 170-71n., 191n., 351, 357, 363n.
Philo Judaeus, 62
Piaget, Jean, 123, 147-48, 153, 156n.
Plato, 43-44, 58-59, 62, 69, 77, 89n., 199, 258, 341
Pleasure: functional interpretation of, 34; psychological interpretations of, 37
Plotinus, 62
Possibilities, Aristotelian and Leibnitzian concepts compared, 341-42
Pound, Roscoe, 160, 170n.
Practice: relation to theory, 225 f., 254-60; Aristotle's analysis of the relation, 255-56
Preferential consideration and justice, chapter 15
Price, Richard, 161
Privacy, 163, 175, 182
Professional responsibility, chapter 13
Progress, 219, 282; inevitability of, 244
Prometheus, 197, 202, 203, 220
Promising: Rawls on, 35; Nietzsche on, 36
Public welfare utilitarianism, 328
Pugh, G.E., 114-19, 122n.
Purposive activity, as basic value phenomenon in Pepper's theory, 98-114

Race, and intelligence, *see* intelligence
Rawls, John, 35, 42n., 160, 162, 175, 179, 184-85, 192n., 292, 294, 328, 337n., 359
Reason, as slave of the passions, 52
Religion, how affected by Kantian moral theory, 344
Responsibility: of science, chapter 11; of scientists and engineers, chapter 12; of professions, chapter 13; in science, *see* Science
Reston, James, 224
Reverse discrimination, 289
Revolution: John Locke on, 323; domestication of, 323; French revolution, 161
Right and wrong, 165
Rights-and-duties framework, 327-29; rights of future generations, 335 f.
Rousseau, J.J., 259
Royce, Josiah, 330, 337n.
Russell, Bertrand, 8, 228-29, 237n., 249, 267, 326, 353-55, 363n.
Ryle, Gilbert, 259, 268n.

Santayana, George, 353
Sartre, J.-P., 25, 32, 40-41, 127, 143, 154-55n., 189n., 230-31
Scarcity, effect on moral theory, 207
Schools, in science and philosophy, 278 f.
Schweitzer, Albert, 8
Science: and social conscience, chapter 11; changes in scientific enterprise, 222-26; professional obligations of,

234 ff.; social responsibility of, chapter 12; responsibility for social applications of, 243 ff., 246-51; responsibility in research, 271-72; in presenting results, 272-74; in publicizing results and public controversy, 275-80; in policy formation, 281-84; patterns of use of science in ethics, chapter 2; science and moral judgment, 23-29; scientific study of nature and tasks of morality, 29-33; use in metaethics, 33-41; scientific research and moral judgment, chapter 7; scientific establishment of criteria, 124-41; models of morality, 141-54; impact on interpreting moral terms, 128-29
Searle, John, 189n.
Selective systems, 99, 110-11
Selye, Hans, 5, 16n.
Shockley, William, 269, 285n., 287n.
Sidgwick, Henry, 19
Singer, Edgard A., 249
Skinner, B.F., 266
Smith, Adam, 37, 52, 72, 342
Sociobiology, 114
Socrates, 25, 78, 215, 231
Spencer, Herbert, 92-93, 190n., 200, 216n., 346-48
Spengler, Oswald, 348
Stevenson, C.L., 71, 75n., 355-56, 363n.
Stirner, Max, 353, 363n.
Stoics, 60, 65, 73, 79, 142
Structuring a moral problem, 126

Tawney, R.H., 31, 41n.
Teaching, 260-64
Technology: and morality, chapter 10; what technology is, 199-204; and technique, 199
Terrorism, chapter 16; conceptualization of, 309-12; justification issue, 313-17; cultural basis of, 317-19; in predatory war and crime, 319-21; moralistic, 321-24
Theory, relation to practice, 225 f., 254-60; *see* Practice
Tillich, Paul, 189n.
Tolman, E.C., 95, 102
Tolstoy, Leo, 314

Topitsch, Ernst, 75n.
Triage, 167

Unavoidability, 130
United Nations Universal Declaration of Human Rights, 184, 229, 292-93, 328-29
Universality, 129-31
Urban, W.M., 351

Value: intrinsic, 28; and the goal-seeking model, 145; as surrogate criterion for survival, 115-16; analysis of, 117-19; value-isolationism defined, 349, and its career, 350-60. *See also* Morality
Vatican II, 166
Veblen, Thorstein, 200, 216n.
Violence: in conceptualizing terrorism, 311 ff.; and pacifism, 313 ff.; distinguished from force, 317; institutional, 317 f.
Virtue: psychological underpinning, chapter 5; historical parade of, 78; variety of theoretical analyses of, 78 ff.; twentieth century eclipse, 80 f.; effect of changing psychological presuppositions, 82-86; directions of probable development, 86-88; impact of anthropology on, 81; impact of positivism and behaviorism on, 81-82; personality theory and, 83-86
Voltaire, 341

War, and terrorism, 319-21
Weber, Max, 31, 41n., 123, 156n.
Wellman, Carl, 310-11, 324n.
Wilson, E.O., 114, 116, 120n.
Witkin, H.A., 85, 156n.
Wolfenstein, Martha, 155-56n., 305, 307n.